ROUTLEDGE LIBRARY EDITIONS: GEOLOGY

Volume 4

CATASTROPHIC FLOODING

T0179411

CATASTROPHIC FLOODING
Binghamton Geomorphology Symposium 18

Edited by
L. MAYER AND D. NASH

Routledge
Taylor & Francis Group

LONDON AND NEW YORK

First published in 1987 by Allen & Unwin

This edition first published in 2020
by Routledge
2 Park Square, Milton Park, Abingdon, Oxon OX14 4RN

and by Routledge
52 Vanderbilt Avenue, New York, NY 10017

Routledge is an imprint of the Taylor & Francis Group, an informa business

British Library Cataloguing in Publication Data
A catalogue record for this book is available from the British Library

ISBN: 978-0-367-18559-6 (Set)
ISBN: 978-0-429-19681-2 (Set) (ebk)
ISBN: 978-0-367-89640-9 (Volume 4) (hbk)
ISBN: 978-0-367-89652-2 (Volume 4) (pbk)
ISBN: 978-1-00-302032-5 (Volume 4) (ebk)

Publisher's Note
The publisher has gone to great lengths to ensure the quality of this reprint but
points out that some imperfections in the original copies may be apparent.

Disclaimer
The publisher has made every effort to trace copyright holders and would welcome
correspondence from those they have been unable to trace.

Catastrophic Flooding

Edited by

L. Mayer and D. Nash

Boston
ALLEN & UNWIN
London   Sydney   Wellington

Allen & Unwin, Inc.,
8 Winchester Place, Winchester, Mass. 01890, USA
the US company of
Unwin Hyman Ltd
PO Box 18, Park Lane, Hemel Hempstead, Herts HP2 4TE, UK
40 Museum Street, London WC1A 1LU, UK
37/39 Queen Elizabeth Street, London SE1 2QB, UK

Allen & Unwin (Australia) Ltd,
8 Napier Street, North Sydney, NSW 2060, Australia

Allen & Unwin (New Zealand) Ltd in association with the
Port Nicholson Press Ltd,
60 Cambridge Terrace, Wellington, New Zealand

First published in 1987

British Library Cataloguing in Publication Data

Catastrophic Flooding.——
    (Binghamton Symposium in geomorphology; V. 18).
1. Floods
I. Mayer, L.   II. Nash, D.   III. Series 551.48′9   GB1399
ISBN 0-04-551142-X

Printed in Great Britain by
Biddles Ltd., Guildford, Surrey

Preface

This volume contains a collection of papers presented at the 18th Annual Geomorphology "Binghamton" Symposium, held at Miami University, September 26-27, 1987, on the topic of *Catastrophic Flooding*. The term catastrophic flooding may be applied to flooding of high magnitude and low frequency, devastating floods, or floods that result in significant changes in stream channel or stream valley characteristics. Catastrophic floods may result from unusual or persistant climate regimes, and failure of natural or man-made dams.

The topic of Catastrophic flooding has generated a great deal of interest for several reasons. Catastrophic processes have been traditionally considered to be mutually exclusive of uniform processes, a legacy from the 17th and 18th century debates with diluvianism and catastrophism. Uniformitarianism has, since those debates, generally reigned as a geological paradigm. As has been the case in paleontology, the type of process operating, gradualism or punctuated equilibrium, must be determined from data. This volume will argue neither for the importance of catastrophic processes, nor for the importance of gradual and more uniform processes. Rather, the papers here presented, make a case for the careful collection and interpretation of data from which the importance and effects of catastrophic flooding may be deduced.

In planning this volume, we have identified several important topics and questions about catastrophic flooding. Topics we wanted to cover included the flow dynamics of catastrophic floods, the climatological settings that lead to floods, identification of catastrophic floods in the geological record, statistical treatment of catastrophic flood recurrence, and planetary analogs to catastrophic flooding. Specific questions included: What are the causes of catastrophic flooding? What are the effects of catastrophic flooding? How large are catastrophic floods and which parameters should be used to measure the size of a catastrophic flood? What is the sedimentology of catastrophic flood deposits and how do they differ from normal stream deposits? What effect do catastrophic floods have on erosional and depositional landforms? How can various paleoflood indicators be used to estimate hydrologic parameters of past floods? Can statistical techniques be used to describe frequency or design frequency of catastrophic floods? Can modeling be used to predict the flow dynamics of catastrophic flows?

In organizing the volume, we have tried to include those papers which address the above questions. Baker and Costa begin by presenting an overview of catastrophic flooding and providing important insights and data that describe the worldwide occurrence of catastrophic flooding. The relations among climatic variables and flooding in the United States are reviewed by Hirschboeck. Waylen and Caviedes concentrate on the relations between El Niño and flooding in coastal Peru. Several papers deal with glacial-lake outbursts. Sturm, Beget and Benson describe the floods from Strandline lake in Alaska. These data are important because they will provide conceptual frameworks within which to interpret the mechanisms of the draining of glacial-lakes. On the mid-continent Kehew and Lord, Teller and Thorliefson, and Clayton and Attig, describe catastrophic flooding at the end of the Wisconsin Glaciation. The erosional and sedimentary record of these floods may be useful in estimating the effect of large floods.

Knox discusses a method for using sedimentological studies of overbank deposits in the Upper Mississippi Valley to extend the short historical record of large floods. Carling and Glaister document the use of dam-break modeling as a tool to reconstruct a flood resulting from morraine-dam failure using a computer program designed to predict the results of man-made dam failures. Their methods may apply to flooding from other natural-dam failures. Osterkamp and Costa document the results of a large flood on sand-bed stream channel and discuss the time required for recovery. Brookes investigates the evidence for and possible causes of a medieval catastrophic flood in west central Iran. Webb notes the contrasts between catastrophic streamflow floods and debris flows in northern Arizona and southern Utah. Thomas presents the U. S. Geological Survey's methods for estimating flood frequency and dicusses their performance. Rossbacher and Rhodes show features of catastrophic flooding on other planets and examine analogs for catastrophic flooding on Earth. Differences in planetary and Earth features are discussed in the context of differeng physical constants. Craig, using the Missoula Flood, presents a method for computer modeling of flow dynamics for catastrophic floods. Inbar presents results of a catastrophic flood in the Jordan River basin and notes the importance of catastrophic fluvial processes on landform change in that Mediterranean climatic setting. Clark and others document the sedimentological and erosional changes accompanying a high magnitude flood in the humid temperate climate of Virginia. Schick examines the effects of high magnitude flooding in the hyper-arid climate of the Sinai.

In retrospect, we feel certain that the papers herein presented answered the questions we posed and provided a convincing case for the importance of field studies, statistical analysis, and modeling as tools to understand catastrophic flooding.

Larry Mayer
Miami University

David Nash
University of Cincinnati

Contents

Acknowledgements

We are grateful for the support given to the Symposium from the National Science Foundation though grant ECE-8700277. Additional support was provided by the College of Arts and Science, Miami University, the Department of Geology, Miami University, the Alumni of Miami University, the College of Arts and Science and Graduate College, University of Cincinnati, and the Department of Geology, University of Cincinnati.

We also wish to thank those editors of previous Geomorphology Symposia whose advice and assistance made this task less difficult. The assistance of Roberta Crain, Kelley Halstead, and Constance Papouras, is gratefully acknowledged.

L. M. thanks Mika and Maya for being themselves.

1

Flood power

Victor R. Baker and John E. Costa

"When the melted snows of many mountains of Tibet course toward the sea, and when, riding the crest of those thaws, the run-off of spring rains that have fallen on half a million square miles of Chinese hills flow too down the Great River, its power becomes unimaginable, even to a hopeful young hydraulic engineer." (Hersey, 1956, describing the floods in the Three Gorges of the Chang Jiang (Yangtze River), Peoples Republic of China).

ABSTRACT

Channel boundary shear stress and stream power per unit boundary area are very useful concepts in assessing the role of rare, great floods in producing major geomorphic responses in fluvial systems. These variables were determined for the largest known flash floods in small drainage basins and for six historic dam-failure floods, predominantly in the United States. The largest values (shear stress exceeding 2×10^3 Nm^{-2} and power per unit area exceeding 1×10^4 Wm^{-2}) occurred in narrow, deep flows in exceptionally steep bedrock streams. An optimum drainage area of approximately 10 to 50 km^2 seems to be associated with the most powerful flash floods.

Even more powerful floods can be analyzed by paleohydraulic procedures. The Pleistocene Missoula floods and floods on the planet Mars approached ultimate limits of flood power defined by the onset of cavitation in fluvial flows. The geomorphic effectiveness of floods seems to be linked directly, not to their magnitude (discharge) or frequency (recurrence interval), but to the shear stress and stream power per unit boundary area relative to the resistance of the channel to erosion.

INTRODUCTION

Rare, great floods introduce immense discharges into channel-valley systems. In some cases these floods produce surprisingly little geomorphic response (Costa, 1974), and in other cases spectacular effects are observed (Baker, 1977; Gupta, 1983). Nearly all previous investigations of stream- channel and valley response to extraordinary floods, and the resulting conclusions about the spatial and temporal variations in channel and valley evolution, have focused on morphologic and sedimentologic evidence (Wolman and Gerson, 1978), or precipitation variations (Newson, 1980).

Unfortunately, little quantitative hydraulic data on cataclysmic floods have been presented to assess disparities in landform response. In this paper we present a data set on cataclysmic floods, emphasizing their flow dynamics. This is a preliminary step in elucidating the relationships among flow hydraulics, sediment transport, valley morphology, and flood sedimentology.

STRESS, POWER, AND RESISTANCE

The ability of rivers to erode and transport sediment is related to channel boundary shear stress, τ, expressed in Newtons per square meter (Nm^{-2}), according to the formula

$$\tau_r = \gamma RS \tag{1}$$

where τ_r is boundary shear stress computed with hydraulic radius, γ is the specific weight of the fluid ($9800\ Nm^{-3}$ for clear water), R is the hydraulic radius (m), and S is the energy slope. In two dimensions it may be possible to substitute flow depth D (m) for R, yielding the following expression for local bed shear stress:

$$\tau_D = \gamma DS \tag{2}$$

where τ_D is boundary shear stress computed with flow depth. Following extraordinary floods, the accurate computation of shear stress is severely constrained by the ability to reconstruct the appropriate hydraulic variables. Hydraulic radius (R) is probably the least difficult to compute if approximate high-water marks can be found following the flood. The differences between Equations 1 and 2 depend on channel shape. For high width-depth ratios (w/D, where w is width (m)) (10 or greater) the distinction is negligible. For exceptionally small width-depth ratios it becomes more important, such that at w/D = 2.0, R = D/2. This results in a two-fold range of shear stress values for the same flood. Another source of variation in these formulae is γ. Highly sediment-charged water may have a specific weight which is twice that of clear water (Costa, in press a). The computation of energy slope requires the use of roughness coefficients and expansion-contraction coefficients to calculate specific energy at various points along the surveyed reach. In true slope-area indirect discharge estimates, this is actually done (Dalrymple and Benson, 1967). Frequently, however, the slope of the water-surface profile, or the channel gradient is used by geomorphologists as an expedient surrogate for energy slope. During catastrophic flash-floods on small watersheds, water-surface and channel slopes can vary as much as 96 to 104% from the true energy slope (Costa, in press c).

Energy developed per unit time, or power is another important concept for sediment transport (Bagnold, 1966). Total power per unit length of stream, Ω, expressed in watts (W) per meter (Wm^{-1}) is given by the formula

$$\Omega = \gamma QS \qquad (3)$$

where Q is discharge (m^3 s^{-1}). The concept of total stream power is tied to concepts of equilibrium in alluvial rivers. Chang (1980) and Song and Yang (1980) argue that alluvial rivers subject to prevailing water and sediment discharges tend to adjust their channel geometries to minimize Ω.

Total stream power per unit length (equation 3) varies with the discharge (size) of a river. Therefore, another useful measure of power is its value per unit area of bed, expressed in Wm^{-2} given by the formula

$$\omega_1 = \gamma QS/w \qquad (4)$$

where w is the water-surface width (m). This term is often called unit stream power.

By definition,

$$Q = wDv \qquad (5)$$

where v is mean flow velocity in ms^{-1}. Combining equations 2, 4 and 5 yields

$$\omega_1 = \tau_D v \qquad (6)$$

To distribute stream power along the flow boundary, hydraulic radius is substituted for mean depth, and ω_2 or power per unit boundary area, is expressed as

$$\omega_2 = \tau_r v \qquad (7)$$

The concept of stream power is intimately tied to sediment transport capability (Bagnold, 1966, 1977, 1980). It can also be thought of as the driving term in a net erosional threshold criterion that is balanced by resisting power (Bull, 1979). In a rare great flood an unusually large discharge is imposed on the channel and valley floor. This discharge creates a driving stream power that can also be compared to resisting factors, which can also be expressed in units of power (Bull, 1979). The resisting power in an alluvial channel is determined by factors such as hydraulic roughness, sediment load, and sediment size (Bull, in press).

UNITED STATES FLASH-FLOOD MAXIMA

The U.S. Geological Survey has a great store of unpublished hydraulic information about floods from all over the country. Typical information includes surveys of channel cross-sections, surveys of high-water marks, estimates of roughness coefficients, channel descriptions, original field notes, plane maps, conveyance and discharge computations, and stereo slides of the channel.

Because of the dangers associated with direct flood measurements, many flood-peak discharges are computed by indirect methods such as the slope-area method (Dalyrmple and Benson, 1967). These indirect discharge data were collected by hydrologists and engineers following the floods, and as a standard policy were reviewed by at least two other qualified engineers. The records provide the most accurate, consistent, and comprehensive view possible of the characteristics of extraordinary flash floods. Data from the largest floods ever docmumented for a given size drainage basin (Costa, in press b), and other large flash-floods in different physiographic areas of the United States are summarized in Table 1. Although subject to more error than direct measurements, these records provide a valuable data base of historic hydraulic conditions and characteristics. Such information can be used as guidelines when reconstructing paleoflood characteristics from other kinds of indirect evidence.

Alluvial and Bedrock Channels

From the reconstructed hydraulic data, shear stress and unit stream power for each flood can be calculated (Table 1). Values reported here employ hydraulic radius (Equation 1), true energy slope, with two exceptions, not water-surface or channel slope, and assume clear water with a specific weight of 9,800 (Nm^{-3}). Table 1 includes the largest shear stresses and stream powers per unit area computed for 28 streams draining less than approximately 1,000 square kilometers (km^2) in area, and one river, the Eel River at Scotia, California, with a drainage area of about 3,100 km^2. These data include the largest rainfall-runoff floods measured by indirect methods on drainage basins less than about 3,100 km^2 in the conterminous United States (Costa, in press b). All sites are in the United States except the Ouaieme River, New Caledonia flood of 1981, which stands out as the lone high-outlier of maximum floods per unit drainage area when compared with those recorded anywhere else in the world (Costa, in press b). Also included are data from five historic constructed-dam failure floods in the United States, and from one landslide-dam failure in Taiwan. The floods are listed in order of decreasing shear stress.

Table 1. Shear stress and unit stream power for some large flash-floods.

	Drainage Area (km²)	Discharge (m³s⁻¹)	Shear Stress (Nm⁻²)	Power per unit area (Wm⁻²)	Channel Type	Mean Depth (m)	Mean Velocity (ms⁻¹)	Energy Slope	Date of Flood
1. Ousel Creek, MT	7.56	118	2,632	18,582	Bedrock	1.8	7.1	0.2050	June 8, 1964
2. Street Creek, MT	15.5	163	2,043	14,750	Bedrock	2.6	7.2	0.1241	June 8, 1964
3. So. Fk. Pine Canyon, WA	13.99	708	1,696	15,179	Bedrock	4.3	9.0	0.0560	May 6, 1948
4. Dark Gulch, CO	2.59	204	1,477	11,698	Bedrock	1.5	7.9	0.1125	July 31, 1976
5. Ouaieme River New Caledonia	330	10,400	913	6,300	Alluvial/ bedrock	14.0	6.9	0.00807#	Dec. 24, 1981
6. Humboldt River trib., NV	2.20	251	858	8,160	Bedrock	1.5	9.5	0.0639	May 31, 1973
*7. Teton River, ID	2,204	63,944	819	10,713	Bedrock	16.5	13.1	0.0053	June 5, 1976
*8. Roaring River, CO	7.0	430	779	2,680	Bedrock	4.2	3.4	0.0240	July 15, 1982
9. Rocky Canyon, NV	10.5	407	713	6,118	Bedrock	2.0	8.6	0.0385	May 31, 1973
10. Little Pinto Creek. trib. UT	0.78	74.5	626	3,399	Bedrock	0.8	5.4	0.0799	Aug. 11, 1964
11. Maynard Gulch, ID	5.83	270	616	4,263	?	1.4	6.9	0.0462	Aug. 20, 1959
12. Big Thompson River, CO	490	799	597	3,943	Bedrock	2.5	6.6	0.0254	July 31, 1976
13. Bronco Creek, AZ	49.2	2,080	595	4,189	Alluvial	2.2	7.0	0.0193	Aug. 18, 1971
14. Big Thompson trib., CO	3.55	246	582	4,359	Bedrock	1.7	7.5	0.0401	July 31, 1976
15. Eldorado Canyon, NV	59.3	2,152	573	5,684	Bedrock	2.3	9.9	0.0267	Sept 14, 1974
16. Lane Canyon, OR	13.1	807	564	4,743	Bedrock	2.1	8.4	0.0285	July 26, 1965
17. East Br. Naugatuck Rv., CN	26.4	176	529	2,090	Bedrock	1.6	4.0	0.0367	Aug. 19, 1955
18. East Br. Salmon Brk at N. Granby CN	34.2	405	482	1,971	Bedrock	3.5	4.1	0.0164	Aug. 19, 1955
19. Sandy Run, PA	19.7	397	479	2,007	?	1.3	4.2	0.0376	July 20, 1977
20. Lahontan Reser. Trib., #3, NV	0.57	47.6	478	2,141	Bedrock	0.8	4.5	0.0650	July 20, 1971
21. Wenatchee River trib., WA	0.39	25.6	443	1,878	Bedrock	0.6	4.2	0.0766	Aug. 25, 1956

Table 1. continued

	Drainage Area (km^2)	Discharge (m^{3}s^{-1})	Shear Stress (Nm^{-2})	Power per unit area (Wm^{-2})	Channel Type	Mean Depth (m)	Mean Velocity (ms^{-1})	Energy Slope	Date of Flood
*22. Buffalo Ck.,WV	15.7	1,416	429	2,445	Bedrock	2.8	5.7	0.0159	Feb. 26, 1972
23. Meyers Can., OR	32.9	1,540	418	2,700	Bedrock	2.4	6.5	0.0193	July 13, 1956
*24. Cho-Shui River, Taiwan	259	7,860	400	2,778	Bedrock	5.5	7.0	0.00731	Aug. 24, 1979
25. Knapp Coulee trib., WA	0.73	53	376	2,162	?	1.1	5.8	0.0412	Aug. 15, 1956
*26. Kelley Barnes Dam, GA	11.9	651	373	1,462	Bedrock	3.2	3.9	0.0124	Nov. 6, 1977
27. Seco Creek, TX	368	6,510	338	1,812	Bedrock	4.9	5.4	0.00735	May 31, 1935
28. Cleghorn Can., Rapid City, SD	18	357	326	1,669	Bedrock	2.3	5.1	0.0156	June 9, 1972
*29. Laurel Run, PA	28.5	1,048	322	1,848	?	4.3	5.7	0.0090	July 20, 1977
30. Whistle Creek, VA	16.7	326	273	1,294	Bedrock	2.9	4.7	0.0134	Sept 10, 1950
31. West Nueces River, TX	1,041	16,426	189	1,421	Bedrock	10.8	7.5	0.00181	June 14, 1935
32. Trumansburg Creek, NY	33.9	503	184	887	Alluvial/bedrock	1.3	4.8	0.0150	July, 1935
33. Eel River, CA	3,113	21,300	156	719	Bedrock	14.0	4.6	0.00114#	Dec. 24, 1964
34. Mailtrail Creek, TX	195	4,810	108	451	Bedrock	2.7	4.2	0.00410	June 24, 1948
35. Newell Creek, PA	5.54	147	101	251	?	0.8	2.5	0.0132	July 18, 1942

* = dam-failure flood
# = water-surface slope

During extraordinary floods on alluvial (non-bedrock) channels, shear stress and power per unit boundary area are constrained to relatively small values by the ability of drastic width increases by overbank flow and bank erosion to accommodate most of the increase in discharge. The importance of width in determining the amount of shear stress and power per unit boundary area that can be produced at a constant discharge is clearly evident in Equation 4. An example of the effect is provided by Jimmy Camp Creek, a sand- bed alluvial channel draining the South Platte-Arkansas River divide in east- central Colorado. In 1965 this stream experienced the largest flood ever measured ($3,470 \text{ m}^3\text{s}^{-1}$) for a stream of its drainage-basin size (141 km^2) anywhere in the United States (Costa, in press b). However this flood only produced a shear stress of 61 Nm^{-2} and stream power per unit boundary area of 212 Wm^{-2}. These values are lower by about an order of magnitude than those that occurred in predominately bedrock channels which experienced lower unit discharges (Table 1).

Floods that have produced the greatest shear stresses and unit stream powers occurred predominately in bedrock channels (Table 1). Bedrock outcrops can constrain the depth of scour, as well as the water-surface width during floods. Increases in discharge are accomplished primarily by increasing depth and velocity to greater values than a width-adjustable alluvial stream allows. Velocity may increase as rising stages drown out roughness elements. Flash floods in bedrock channels have produced values of shear stress in excess of 2,000 Nm^{-2}, and as high as 2,632 Nm^{-2}, and stream powers per unit boundary area in excess of 10,000 Wm^{-2}, and as high as about 18,600 Wm^{-2}. The only alluvial channel known to us that experienced shear stress and unit stream power of the same order of magnitude as bedrock channels is Bronco Creek, Arizona (Aldridge, 1978). On July 18, 1971, a flood of 2,080 m^3s^{-1} produced a shear stress of about 600 Nm^{-2} and unit stream power of about 4,200 Wm^{-2}. We have no explanation for the supposed uniqueness of this flood.

Energy Diagrams

Energy diagrams for two of the flash floods listed in Table 1 are shown in Figure 1. Figure 1A represents the energy diagram for the flood of June 8, 1964 on Ousel Creek near West Glacier, Montana (Fig. 2) (Boner and Stermitz, 1967, p. B223). This flood produced the greatest calculated shear stress (2,600 Nm^{-2}) and unit stream power (18,600 Wm^{-2}) of any flood ever measured in the United States by the U.S. Geological Survey. The flood was supercritical, with Froude numbers (F) ranging from 1.60 to 1.62. At the beginning of the surveyed reach, total head (height above the datum, water depth, and velocity head) amounted to 12.5 m. Water depth at the downstream end of the measured reach was 3.2 m and velocity head was 2.9 m. These values represent a specific-energy head of 6.1 m. Water and energy slopes were nearly parallel, but noticeably less than channel slope. The second total energy diagram (Fig. 1B) is for the flood of August 19, 1955 on the East Branch of the Naugatuck River near Torrington, Connecticut (Fig. 3) (Water Re-

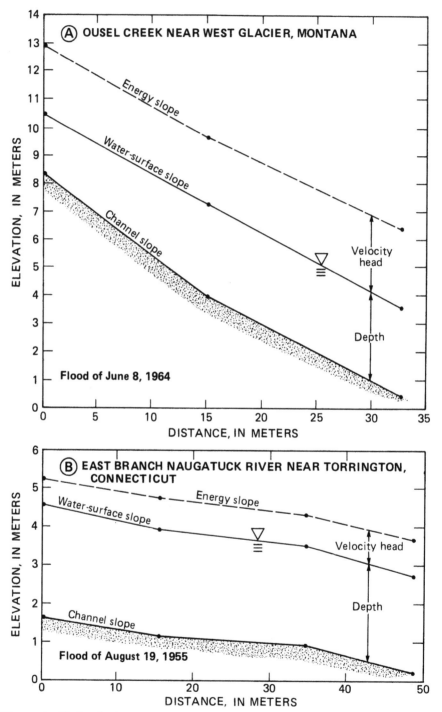

Figure 1. Hydraulic flow profiles for two exceptionally powerful floods. A. Ousel Creek near West Glacier, Montana. B. East Branch of Naugatuck River near Torrington, Connecticut. See Table 1 for other flow parameters.

Figure 2. Reach of Ousel Creek, near West Glacier, Montana, analyzed in Figure 1A.

sources Division, 1956). Water velocity was about half that in Ousel Creek, but sufficient to sustain critical flow at the peak discharge in this natural channel throughout the surveyed reach. Water, channel, and energy slopes were nearly parallel. Total energy head at the beginning of the study reach was 5.1 m, and specific energy at the end of the reach was 3.5 m. Shear stress was about 530 Nm^{-2}, and stream power per unit boundary area was about 2,100 Wm^{-2}. These values are smaller than for Ousel Creek mainly because energy slope is appreciably less, even though water depths are similar.

CATACLYSMIC FLOOD POWER

The ability of a drainage basin to maximize the shear stress and power per unit boundary area of a rare flood depends on an ideal combination of gradient, depth, and velocity defined by Equations, 1, 4, and 7. There is also an influence of size of

Figure 3. Reach of East Branch Naugatuck River near Torrington, Connecticut, analyzed in Figure 1B.

the upstream drainage area for the data in Table 1 (Fig. 4). The smallest basins are steep, but their floods are not deep or especially swift. Larger basins generally have deeper floods and somewhat faster flood velocities, but are not as steep as smaller basins. Basins of approximately 10 to 50 km^2 seem to have the valley and hydraulic characteristics that optimize flood depth, energy slope, and velocity to maximize flood power (Figure 4). In an investigation of the spatial variations of fluvial processes in the Henry Mountains, Utah, Graf (1982, 1983) determined that the maximum stream power associated with the 10-year flood occurs in drainage basins of 50 to 100 km^2.

To produce stream power per unit boundary area greater than about 2000 Wm^{-2} on rivers draining areas larger than about 1,000 km^2, two circumstances are necessary: (1) a reach of river incised into a bedrock canyon; and (2) an upstream drainage area capable of producing extraordinary flood discharges. Four historic examples of this situation are : (a) Teton River, Idaho, which is incised into ash-flow tuff bedrock, and had a large dam-failure flood in 1976; (b) Katherine Gorge, Northern Territory, Australia, developed along intersecting joint patterns in highly resistant Precambrian sandstone; during the pronounced wet season in this tropical area, large floods have been produced in the last 400 years, and documented using paleoflood hydrology techniques (Baker and Pickup, in press); (c) Pecos River, Texas, which is incised into limestone bedrock along the Balcones Escarpment in south-central Texas, and drains one of the major flash-flood regions in the United States; in 1954 Hurricane Alice caused an enormous flood on this river (Patton and Dibble, 1982) and (d) the Three Gorges of the Chang Jiang (Yangtze River) in central China, where the river is deeply incised into limestone bedrock, and drains the snowmelt

Table 2. Comparative flow dynamics of great historic and prehistoric river floods.

River	Channel Type	Peak Discharge (m^3s^{-1})	Slope	Depth (m)	Velocity (ms^{-1})	Bed Shear Stress (Nm^{-2})	Power Per Unit Area (Wm^{-2})
A-Amazon	Alluvial	3×10^5	1×10^{-5}	60	2	6	12
M-Mississippi	Alluvial	3×10^4	5×10^{-5}	12	2	6	12
Q-Chang Jiang (Qutang Gorge)	Bedrock	1×10^5	2×10^{-4}	85	11.8	175	2×10^3
P-Pecos River (Texas)	Bedrock	2.7×10^4	2×10^{-3}	30	12	600	7×10^3
K-Katherine Gorge (Australia)	Bedrock	6×10^3	3×10^{-3}	45	7.5	1.5×10^3	1×10^4
B-Bonneville Flood	Bedrock	9×10^5	4×10^{-3}	63	10-26	2.5×10^3	7.5×10^4
MF-Missoula Flood (Average)	Bedrock	1×10^7	3×10^{-3}	70	10	2×10^3	2×10^4
S-Missoula Flood (Soap Lake Constriction)	Bedrock	5×10^6	1×10^{-2}	100	30	1×10^4	3×10^5
W-Missoula Flood (Wilson Creek)	Bedrock	6×10^6	2×10^{-3}	80	15	1.6×10^3	2.4×10^4
R-Missoula Flood (Rathburn Prairie)	Bedrock	2×10^7	6×10^{-3}	175	25	1×10^4	2.5×10^5

and rainfall from eastern Xizang (Tibet) Province and much of western China. In 1870, the largest flood in an historical record of 1,000 years occurred on the Chang Jiang from a large snow-melt and rainstorms. In marked contrast to these large-river floods in bedrock areas, the largest historical floods on large alluvial rivers like the Mississippi and the Amazon only produces unit stream powers of about 12 Wm^{-2} (Fig. 4).

The data in Figure 4 represent the largest unit discharges and stream powers, computed from the best possible hydraulic data, known to us. The general trend represented in Figure 4 is not expected to change very much in the near future, unless a major river with a deep bedrock gorge experiences an extraordinary flood greater than those on the Teton, Pecos, or Chang Jiang. Flood events that fall

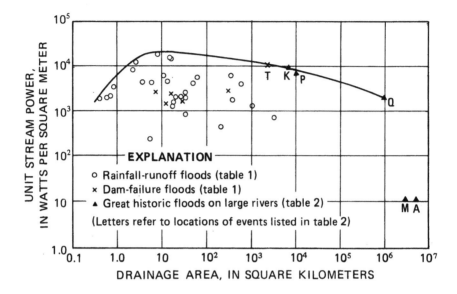

Figure 4. Plot of stream power per unit boundary area, versus drainage basin area. Rainfall-flood data from Table 1 are plotted as (o); dam-failure floods are plotted as (X); Teton dam failure is labeled "T"; five floods from Table 2 are plotted as (Δ). The curve defines the maximum known flood power in streams and rivers of different size drainage areas.

above the plotted limiting-curve should prove very interesting to investigate. The implications of Figure 4 are significant for long-term landscape evolution, and support Graf's (1982, 1983) interpretations of the spatial and temporal variability in sediment-related processes in different-sized watersheds.

Paleoflood Hydrology

The role of cataclysmic floods in fluvial geomorphology has been difficult to assess because the magnitude and infrequence of such events result in great difficulties for direct measurement. Paleoflood hydrology may be the only practical means of studying cataclysmic flood processes in many settings. For the reconstruction of paleoflood hydraulics, the most accurate procedure is to study slackwater deposits and paleostage indicators in appropriate geomorphic settings (Baker, in press; Baker and others, 1983; Kochel and Baker, 1982, in press; O'Connor and others, 1986; Stedinger and Baker, 1987).

Paleoflood hydrologic techniques can be used to expand the observational data base of Table 1. Although paleoflood data involve numerous simplifying assumptions (Baker, in press), such as the use of Equation 2 instead of 1, they do allow

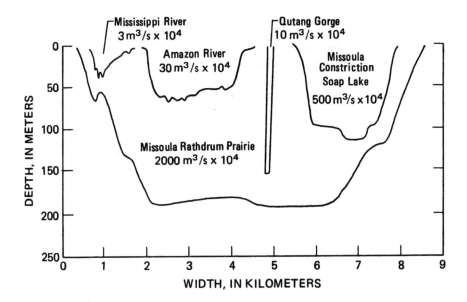

Figure 5. Comparison of channel cross sections for flood flows. The number associated with each cross section is the approximate flood discharge, in m^3/s.

some general assessments of a broader range of flows than are possible with standard historic hydrologic data sets. Reconstructed characteristics of some large floods using paleoflood techniques are included in Table 2.

Cataclysmic Flood Dynamics

Table 2 compares the flow dynamics of floods in large alluvial rivers (Amazon and Mississippi) to values from some great historic floods and paleofloods in relatively confined bedrock channels. Figure 5 illustrates some of the channel sections for these floods. Note that the combination of relatively steep slopes and narrow, deep cross sections yields very high values of both bed shear stress and steam power per unit area. Large floods on great alluvial rivers, such as the Mississippi and Amazon, yield very low shear stresses and unit powers, mainly because of the very low slopes for these rivers.

Hydraulic measurements for the 1870 flood of the Chang Jiang (Yangtze) show that very large rivers may exert very high power per unit area in the appropriate settings. At Qutang Gorge the exceptional flood of 1870 flushed a discharge of 1×10^5 $m^3 s^{-1}$ approximately 100 m wide and 85 m deep. The associated power per unit area was at least 2,000 Wm^{-2}, and in some deep bedrock pools would have been nearly twice this value. Smaller bedrock rivers can have exceptionally powerful floods. Examples include (1) Big Thompson River, Colorado, (Table 1), (Costa,

1978), (2) the Pecos River, Texas, (Kochel and Baker, 1982); and (3) the Katherine Gorge, Northern Territory, Australia, (Table 2) (Baker and Pickup, in press). The largest calculated values of boundary shear stress and stream power per unit area were achieved in the great cataclysmic floods of the Pleistocene. Table 2 shows values for an "average" Missoula flood flow, as well as for the Soap Lake constriction, Rathdrum Prairie, and Wilson Creek (Baker, 1973). In addition, a new calculation for the Bonneville Flood (Jarrett and Malde, in press) is also show.

By plotting various powerful flood flows on a velocity-depth diagram (Figure 6) it is possible to compare many of the factors discussed above. The diagram separates into two flow regimes defined by a Froude number (F) equal to 1.0, such that

$$F = \quad v/(gD)^{1/2} \tag{8}$$

where g is the acceleration of gravity (9.8 ms^{-2}). Supercritical or shooting flow occurs when F > 1.0; subcritical or tranquil flow occurs when F < 1.0. In natural channels, Froude numbers greater than 1.0, based on average channel depth and velocity, are very unusual. In alluvial channels with erodible bed and banks, supercritical flow cannot be sustained for any appreciable length of time because boundary erosion increases depth, decreases velocity, and reduces Froude numbers below 1.0. In bedrock channels, supercritical flow is more common and can be sustained for longer periods of time, but as stage increases to 10-15 m, flows are so deep that formation of supercritical flow is then supressed (Fig. 6).

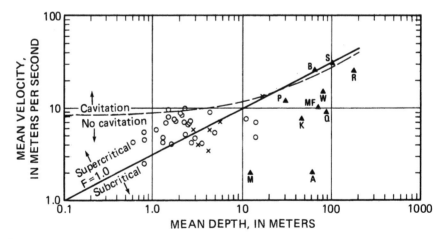

Figure 6. Depth-velocity diagram for various flood flows listed in Tables 1 and 2. The solid line separates supercritical from subcritical flow (equation 8). The dashed line is the critical condition for the inception of cavitation (equation 9). Data from Table 1 are indicated by open circles (rainfall-runoff floods), and crosses (dam-failure floods). Data from Table 2 are indicated by triangles. Letters refer to data in Table 2.

Another important threshold shown in Figure 6 is that which dictates the onset of cavitation in the flow. Cavitation is a process whereby extremely high stream velocity causes a sufficiently large reduction in pressure that the vapor pressure of water is reached, and water vapor occurs as small bubbles. If flow velocity decreases, lessening the local pressure, these bubbles become unstable and collapse with a force as great as 30,000 atmospheres (Barnes, 1956). Cavitating flows may be highly erosive, and have been attributed to the formation of bedrock potholes (Barnes, 1956) and intense bedrock erosion (Baker, 1974). The critical velocity for the inception of cavitation, V_c in ms^{-1}, is given by

$$V_c = 2.6 \, (10 + D)^{1/2}, \qquad\qquad (9)$$

where D is flow depth (m) (Barnes, 1956; Baker, 1974). Flows that exceed this threshold are unstable, since the water is transforming to vapor. It appears from Figure 6 that few, if any powerful natural flows barely exceed the conditions expressed by Equation 9. This suggests that channel adjustments produced by cavitation tend to inhibit or reduce the forces that would cause the threshold to be crossed in nature.

Figure 6 illustrates probable limiting conditions on values of power per unit boundary area in rivers. For extraordinary floods with a depth range of 1 to 10 m (Table 1), Equation 9 indicates a maximum velocity of approximately 10 ms^{-1}. For flood depths of 1 to 10 m, a velocity of 10 ms^{-1}, sediment-free water, and a slope of 0.01, Equation 6 indicates an approximate maximum of 1×10^3 to 1×10^4 Wm^{-2}. For floods deeper than about 10 m, Equation 9 shows V_c rises markedly with depth. At 100 m depth, characteristic of the Missoula Floods, V_c can be as high as about 30 ms^{-1}. For a slope of 0.01 in clear water (γ = 9800 Nm^{-3}), the approximate maximum ω_1 is 3×10^5 Wm^{-2}.

The reasons for the immense disparities in stream power per unit area for various rivers can be explained in hydraulic terms. Normal variations in river width-to-depth ratios (40 < w/D < 2) will produce a maximum factor of two variation in ω as D is substituted for R. Normal variations in γ during floods will also produce a factor of two variation in ω_1. Velocity is inversely proportional to roughness (n). The range in n for many river floods is about 0.02 < n < 0.04, especially in cases where extreme floods submerge roughness elements such as boulders and vegetation. Unlike other factors, D can vary by about two orders of magnitude (Figure 6) from approximately 1.0 to 100 m. Similarly, S can also vary by several orders of magnitude, from 1×10^{-5} for the Amazon to 1×10^{-2} for Missoula Flood constrictions.

Because slope is generally set within an approximate range by a river's longitudinal profile, it is clear that downstream, low-gradient fluvial reaches can only achieve high power by greatly increasing depth. Narrow, deep cross sections are significant because, for a given discharge, much more depth can be achieved. In addition, constrictions may substantially increase local slope, as in the Missoula Flood examples.

Cataclysmic Floods on Mars

The Missoula Floods are the largest known discharges of fresh water on Earth (Baker and Bunker, 1985). Moreover, as shown above, they probably represent the most powerful fluvial phenomena experienced on earth. However; it is likely that even greater cataclysmic floods occurred during the early history of Mars. The Martian outflow channels are a class of trough-like landforms that display evidence of large-scale fluid flow on their floors (Mars Channel Working Group, 1983). The outflow channels are immense, as much as 100 km wide and 2000 km in length. It was recognized shortly after their discovery that they possessed a suite of bedforms and morphological relationships similar to those of the Missoula Flood area (Baker and Milton, 1974). These features include streamlined uplands, longitudinal grooves and inner-channel cataracts, depositional fan complexes, anastomosis, and scour marks (Baker, 1982). Particularly important are the scour marks which must have resulted from a fluid flow possessing a free, upper surface. The entire assemblage of landforms can only be explained by the action of a fluid with physical properties similar to turbulent water. The present consensus is that water in concert with entrained debris and ice was the primary agent of outflow channel genesis (Mars Channel Working Group, 1983).

Parameters for maximum flood flows on Mars are discussed by Komar (1979), Carr (1979), and Baker (1982). In calculating power per unit area, γ is 3720 Nm^{-3} because of the lower Martian gravity. A critical cavitation velocity V_m (ms^{-1}) can be defined for a Martian flood (Baker, 1979, 1982), which reduces in the lower Martian atmospheric pressure to

$$V_m = 1.6 \ (D)^{1/2}. \qquad (10)$$

For a hypothetical Martian flood flowing 100 m deep at a gradient of 0.01, equations 10 and 6 yield a maximum ω_1 of 6×10^4 Wm^{-2}. For a flood 500 m deep at 0.01 gradient, the maximum ω_1 is 1×10^5 Wm^{-2}. Thus, maximum flood power per unit area on Mars is comparable to that of the largest paleofloods on Earth. In comparison to other processes which operate to shape the surfaces of planets, cataclysmic floods seem to be second only to high-velocity bolide impacts in terms of power generation per unit surface area (Baker, 1985).

Effects of Cataclysmic Flood Power

Stream powers of magnitudes reported in this paper can accomplish a variety of unusual sedimentologic and hydraulic phenomena, including (a) suspension of gravels, (b) erosion of bedrock, (c) movement of boulders several meters in diameter, and (d) the transport of enormous quantities of sediment. The effects of dynamic flood conditions are especially noteworthy with respect to the sizes of sediment carried in various modes of transport, (i.e., as bedload, suspension, and washload). As discussed by Komar (1980; in press), it is theoretically possible to extend the domains of these transport modes to evaluate intense floods. Komar finds that typical river floods generating shear stresses of 10 Nm^{-2} can entrain and transport sediment grains of about 1 cm diameter, while sediment finer than 1 mm moves in suspension, and the washload consists of sediments finer than 0.1 mm. Exceptionally powerful floods with shear stresses of 10^3 Nm^{-2} can transport gravel as coarse as 10 to 30 cm in suspension and sand-size diameters of 0.4 to 0.8 mm and finer in the washload. The increase in τ to 10^4 to 10^5 Nm^{-2} for the Missoula Floods implies that even boulders can be transported in suspension with sand-size grains in the washload. Estimates based on this analysis of sediment transport in clear water are likely to be conservative, especially for the Missoula Floods where erosion of the loess of the Palouse Formation could have resulted in high concentrations of washload. These high concentrations reduce the settling velocities of the coarser sediments so that they could be transported more readily in suspension. Such analyses of the modes of sediment transport confirm that the Missoula Floods had phenomenal capacity for eroding and transporting sediments.

Even more modest flow stresses and powers can initiate the transport of enormous boulders. Williams (1983) reports that a minimum power per unit area of about 1000 Wm^{-2} will move boulders with intermediate diameters of 1.5 m, and shear stresses of about 500 Nm^{-2} can move boulders with intermediate diameters of 3 m. As shown in Tables 1 and 2, dozens of floods exceeding these values of power and shear stress have been documented. Flood power is an important concept in geomorphology. Two examples can be used to demonstrate this fact. Nanson and Hean (1985) demonstrated that streams in humid coastal New South Wales (Australia) respond to cataclysmic floods in proportion to their stream power and to abundance of erodible material. Nanson (1986) noted that the character of floodplain formation also varies with stream power. When rare, great floods exceed the resistance threshold, massive erosion occurs destroying a floodplain that may have formed over centuries. Patton and Baker (1977) documented similar phenomena in central Texas.

A second example of the geomorphic role of flood power is provided by study of rapids in the Grand Canyon (Kieffer, 1985). When a debris flow from a Grand Canyon tributary forms a fan in its mouth, the Colorado River encounters a blockade of coarse debris. Initially, the river must carve a narrow-deep cross section through the end of the fan. As the channel widens by erosion, water-surface width increases; therefore, for a constant discharge, power per unit area decreases. A channel develops, with rapids at low flow, that is adjusted to the influence of flood power and the resistance imposed by the fan. Thus the Colorado River morphology at the rapids is adjusted to boundary resistance and to the rare, great floods necessary for the river to modify that resistance to the efficient transport of water and sediment.

DISCUSSION

Floods with the largest discharge values per unit drainage area do not necessarily produce the greatest stream powers or shear stresses. The first four floods listed in Table 1 are not among the largest measured in terms of discharge for the same size drainage basin (Costa, in press b). This indicates that discharge alone is not the most important factor in sediment transport or landscape modification. Stream power per unit area represents the product of a flood's specific weight (sediment concentration), mean depth, energy slope, and mean velocity. It seems to be a concept that includes most of hydraulic and sedimentologic factors that are important for flood investigations.

Flood power is intricately tied to the concept of effectiveness of geomorphic agents. Where effectiveness is defined as the modification of landforms (Wolman and Gerson, 1978), resistance thresholds must be exceeded. For extremely high resistance, such as provided by bedrock or boulders, exceptional values of power per unit area of bed (Equation 4) are required to modify channels. These can only be achieved during rare, high-magnitude floods. The concept of maximum geomorphic work (Wolman and Miller, 1960) applies to the prolonged effect of discharge on sediment transport. For streams and rivers where low to medium flows are capable of transporting nearly all of the available bed and bank sediment-sizes, a great deal of geomorphic work can be accomplished by modest flows. The morphologic impacts of outstanding floods could be relatively short-lived where low flows can transport most available sediment to rapidly reconstruct in-channel bars and channel dimensions to pre-flood conditions (Costa, 1974). In these situations, streams and rivers that develop relatively small values of power per unit boundary area (less than several hundred Wm^{-2}) during floods may accomplish a great deal of geomorphic work, but a relatively frequent or dominant discharge, not the rare flood event, may be the most "effective" process in this situation.

We hope this paper will encourage others to collect or measure appropriate quantitative characteristics of cataclysmic floods. When a rare, great flood occurs, geomorphologists should record a standard list of features from which to derive the key variables (Williams and Costa, in press). This allows accurate comparisons among different floods in different morphogenetic regions. Paleoflood hydrologic procedures can also be used, where appropriate, to derive flow characteristics for ancient cataclysms. When the appropriate cataclysmic geomorphic forces can be compared to resistances in the landscape, we will have achieved a fundamental insight into the nature of landscape change.

ACKNOWLEDGEMENTS

We would like to thank William L. Graf, Arizona State University, and Garnett P. Williams and Charles Swift III, U.S. Geological Survey, for thoughtful and constructive reviews of an early version of this manuscript. All freely shared their ideas and opinions, and greatly contributed to the content of the final version.

REFERENCES

Aldridge, B. N., 1978, Unusual hydraulic phenomena of flash floods in Arizona: Conference on flash floods: hydrometeorological aspects: American Meteorological Society, Boston, Mass., p. 117-120.

Bagnold, R. A., 1966, An approach to the sediment transport problem from general physics: U.S. Geological Survey Professional Paper 422-I, 37 p.

Bagnold, R. A., 1977, Bed load transport by natural rivers: Water Resources Research, v. 13, p. 303-312.

Bagnold, R. A., 1980, An empirical correlation of bedload transport rates in flumes and natural rivers: Proceedings of the Royal Society, v. 372A, p. 453- 473.

Baker, V. R., 1973, Paleohydrology and sedimentology of Lake Missoula flooding in eastern Washington: Geological Society of America Special Paper 144, 79 p.

Baker, V. R., 1974, Erosional forms and processes for the catastrophic Pleistocene Missoula floods in eastern Washington, in Morisawa, M., editor, Fluvial geomorphology: London, Allen and Unwin, p. 123-148.

Baker, V. R., 1977, Stream channel response to floods with examples from central Texas: Geological Society of America Bulletin, v. 88, p. 1057-1071.

Baker, V. R., 1979, Erosional processes in channelized water flows on Mars: Journal of Geophysical Research, v. 84, p. 7985-7993.

Baker, V. R., 1982, The channels of Mars: Austin, Texas, University of Texas Press, 198 p.

Baker, V. R., 1985, Relief forms on planets, in Pitty, A., editor, Themes in geomorphology: London, Croom Helm, p. 245-259.

Baker, V. R., in press, Paleoflood hydrology and extraordinary flood events: Journal of Hydrology.

Baker, V. R., and Bunker, R. C., 1985, Cataclysmic late Pleistocene flooding from glacial Lake Missoula: a review: Quaternary Science Reviews, v. 4, p. 1-41.

Baker, V. R., Kochel, R. C., Patton, P. C., and Pickup, G., 1983, Palaeohydologic analysis of Holocene flood slack-water sediments, in Collinson, J. D., and Lewin, J., editors, Modern and ancient fluvial systems: International Association of Sedimentologists Special Publication 6, p. 229-239.

Baker, V. R., and Milton, D. J., 1974, Erosion by catastrophic floods on Mars and Earth: Icarus, v. 23, p. 27-41.

Baker, V. R., and Pickup, G., in press, Flood geomorphology of the Katherine Gorge, Northern Territory, Australia: Geological Society of America Bulletin.

Barnes, H. L., 1956, Cavitation as a geological event: American Journal of Science, v. 254, p. 493-505.

Boner, F. C. and Stermitz, Frank, 1967, Floods of June 1964 in Northeastern Montana: U.S. Geological Survey Water-Supply Paper 1840-B, 242 p.

Bull, W. B., 1979, Threshold of critical power in streams: Geological Society of America Bulletin, v. 90, p. 453-464.

Bull, W. B., in press, Floods -- degradation and aggradation, in Baker, V. R., Kochel, R. C., and Patton, P. C., editors, Flood geomorphology: N.Y., John Wiley and Sons.

Carr, M. H., 1979, Formation of Martian flood features by release of water from confined aquifers: Journal of Geophysical Research, v. 84, p. 2995- 3007.

Chang, H. H., 1980, Geometry of gravel streams: Journal of the Hydraulics Division, American Society of Civil Engineers, v. 106, no. HY9, p. 1443- 1456.

Costa, J. E., 1974, Response and recovery of a Piedmont watershed from Tropical Storm Agnes, June 1972: Water Resources Research, v. 10, p. 106-112.

Costa, J. E., 1978, Colorado Big Thompson flood: geologic evidence of a rare hydrologic event: Geology, v. 6, p. 617-620.

Costa, J. E., in press a, Rheologic, geomorphic, and sedimentologic differentiation of water floods, hyperconcentrated flows, and debris flows: in Baker, V. R., Kochel, R. C. and Patton, P. C., editors, Flood Geomorphology, N.Y., John Wiley and Sons.

Costa, J. E., in press b, A comparison of the largest rainfall-runoff floods in the United States with those of the Peoples Republic of China, and the World: Journal of Hydrology.

Costa, J. E., in press c, Hydraulics and basin morphology of the largest flash floods in the conterminous United States: Journal of Hydrology.

Dalrymple, Tate, and Benson, M .A., 1967, Measurement of peak discharge by the slope-area method: U.S. Geological Survey Techniques of Water Resources Investigations, Book 3, Chapter A2, 12 p.

Graf, W. L., 1982, Spatial variation of fluvial processes in semi-arid lands: in Thorne, C. R., editor, Space and time in geomorphology, George Allen and Unwin, London, p. 193-217.

Graf, W. L., 1983, Downstream changes in stream power in the Henry Mountains, Utah: Annals, Association of American Geographers, v. 73, p. 373-387.

Gupta, A., 1983, High-magnitude floods and stream channel response, in Collinson, J. D., and Lewin, J., editors, Modern and ancient fluvial systems: International Association of Sedimentologists Special Publication 6, p. 219-227.

Hersey, John, 1956, A single pebble: Alfred A. Knopf, New York, N.Y., 181 p. 22

Jarrett, R. D., and Malde, H. E., in press, Paleodischarge of the late Pleistocene Bonneville Flood, Snake River, Idaho, computed from new evidence: Geological Society of America Bulletin.

Kieffer, S. W., 1985, The 1983 hydraulic jump in Crystal Rapid: implications for river-running and geomorphic evolution in the Grand Canyon: Journal of Geology, v. 93, p. 385-406.

Kochel, R. C., and Baker, V. R., 1982, Paleoflood hydrology: Science, v. 215, p. 353-361.

Kochel, R. C., and Baker, V. R., in press, Paleoflood analysis using slackwater deposits, in Baker, V. R., Kochel, R. C., and Patton, P. C., editors, Flood geomorphology: N.Y., John Wiley and Sons.

Komar, P. D., 1979, Comparison of the hydraulics of water flows in Martian outflow channels with flows of similar scale on Earth: Icarus, v. 37, p. 156-181.

Komar, P. D., 1980, Modes of sediment transport in channelized water flows with ramifications to the erosion of Martian outflow channels: Icarus, v. 43, p. 317-329.

Komar, P. D., in press, Sediment transport by floods, in Baker, V. R., Kochel, R. C., and Patton, P. C., editors, Flood geomorphology: N.Y., John Wiley and Sons.

Mars Channel Working Group, 1983, Channels and valleys on Mars: Geological Society of America Bulletin, v. 95, p. 1035-1054.

Nanson, G. C., 1986, Episodes of vertical accretion and catastrophic stripping: a model of disequilibrium flood-plain development: Geological Society of America Bulletin, v. 97, p. 1467-1475.

Nanson, G. G., and Hean, D., 1985, The relative importance of catastrophic events in modifying channels and transporting bedload: Abstracts of Papers for the First International Conference on Geomorphology, School of Geography, Univ. of Manchester, Manchester, England, p. 437.

Newson, M. G., 1980, The geomorphological effectiveness of floods-a contribution stimulated by two recent events in mid-Wales: Earth Surface Processes, v.5, p. 1-16.

O'Connor, J. E., Webb, R. H., and Baker, V. R., 1986, Paleohydrology of pool- riffle pattern development, Boulder Creek, Utah: Geological Society of America Bulletin, v. 97, p. 410-420.

Patton, P. C., and Baker, V. R., 1977, Geomorphic response of central Texas stream channels to catastrophic rainfall and runoff, in Doehring, D. O., editor, Geomorphology in arid regions: Binghamton, N.Y., State Univ. of N.Y. Publications in Geomorphology, p. 189-217.

Patton, P. C. and Dibble, D. S., 1982, Archeologic and geomorphic evidence for the paleohydrologic record of the Pecos River in west Texas: American Journal of Science, v. 282, p. 97-121.

Song, C.C.S., and Yang, C. T., 1980, Minimum stream power: theory: Journal of the Hydraulics, Division American Society of Civil Engineers, v. 106, no. HY6, p. 1477-1487.

Stedinger, J. R., and Baker, V. R., 1987, Surface water hydrology: historical and paleoflood information: Reviews of Geophysics, v. 25, p. 119-124.

Williams, G. P., 1983, Paleohydrological methods and some examples from Swedish fluvial environments, I -- cobble and boulder deposits: Geografiska Annaler, v. 65A, p. 227-243.

Williams, G. P., and Costa, J. E., in press, Geomorphic measurements after a flood, in Baker, V. R., Kochel, R. C., and Patton, P. C., editors, Flood geomorphology: N.Y., John Wiley and Sons.

Wolman, M. G., and Gerson, R., 1978, Relative scales of time and effectiveness of climate in watershed geomorphology: Earth Surface Processes, v. 3, p. 189-208.

Wolman, M. G., and Miller, J. P., 1960, Magnitude and frequency of forces in geomorphic processes: Journal of Geology, v. 68, p. 54-74.

2

Catastrophic flooding and atmospheric circulation anomalies

Katherine K. Hirschboeck

ABSTRACT

An analysis of the atmospheric circulation patterns associated with twenty-one catastrophic floods in the conterminous United States demonstrates that each flood can be linked to anomalous patterns of circulation. Extreme regional floods over broad areas evolve from different types of large-scale anomalous behavior: uncommon locations of typical circulation features, unusual combinations of atmospheric processes, rare configurations in circulation patterns, and exceptional persistence of the same circulation pattern. Extreme local flash floods over small drainage areas can be classified into synoptic categories of existing flash-flood forecasting schemes, and in addition, these small-scale catastrophic events exhibit sensitivity to large-scale circulation anomalies. Blocking configurations in the upper-air flow pattern are important features during catastrophic flooding episodes. A clustering of catastrophic events in time is evident and may be related to the frequency of blocking or the existence of alternate states of equilibria in the atmosphere. This episodic behavior has important implications for geomorphology, especially in terms of recovery times between recurring catastrophic events, and the probability of occurrence of channel-forming sequences of extreme floods.

INTRODUCTION

With the exception of dam failures, ice jams, jökulhlaups and tsunamis, nearly all incidents of extreme flooding evolve from definable large-scale atmospheric circulation patterns and specific meteorological features embedded within them. Certain floods -- due to their great magnitude, sudden occurrence, or extreme destructiveness -- have been labeled "catastrophic." The extreme nature and rarity of these hydrologic events raises the question of whether or not they have been spawned from equally extreme and unusual atmospheric conditions.

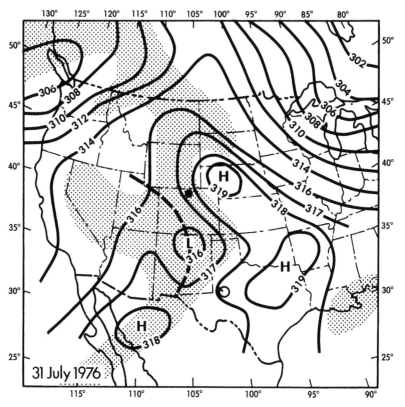

Figure 1. Upper air weather map just before the catastrophic Big Thompson Canyon flood. This detailed 700 mb map for 1200 GMT, 31 July 1976, shows a short-wave low pressure trough (dashed line) moving through a broad upper air ridge of high pressure. Height contours of the 700 mb surface are given in dekameters. (Modified from Maddox et al., 1977).

After the recent catastrophic floods of Rapid City, South Dakota (June, 1972), Big Thompson Canyon, Colorado (July, 1976), and Johnstown, Pennsylvania (July, 1977), detailed analyses were undertaken of observed precipitation patterns, radar displays, satellite imagery, and synoptic weather charts to better understand the evolution of the storms that generated these momentous flash floods (Dennis et al., 1973; Maddox et al., 1977; Hixit et al., 1982). The key meteorological features for each of these extreme events were relatively small-scale atmospheric phenomena such as intense thunderstorms, squall lines, and short-wave troughs. These in turn developed within, and were influenced by, the larger synoptic-scale wave pattern of upper-level steering winds.

For example, in the Big Thompson flash flood, supercell thunderstorms developed to great heights in anomalously moist unstable air, aided by upslope winds, and convergence associated with a short-wave trough (Fig. 1). A similar wave pat-

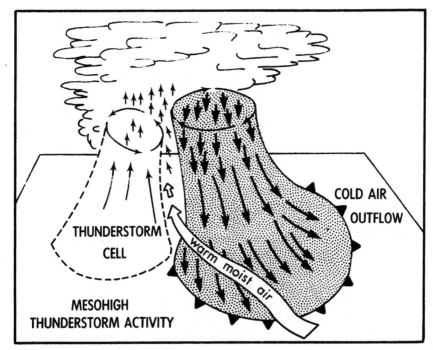

Figure 2. Schematic diagram of thunderstorm activity associated with mesohighs. Cold air outflow from the rainy area of a large thunderstorm complex spreads out ahead of the storm, acting like a localized cold front, and having some of the attributes of a small high pressure center. The mesohigh activity induces subsequent intense convective activity by forcing warm, moist air to rise over the outflow boundary. (Modified from Eagleman, 1983)

tern produced the Rapid City flood. Precipitation totals from these storms were excessive because winds aloft were weak and the thunderstorm cells remained nearly stationary for several hours (Maddox et al., 1977). Similarly, in the Johnstown flash flood, a short-wave trough, topographic relief, and anomalously high values of precipitable water vapor in the atmosphere played a role in generating the flood. Furthermore, in the Johnstown flood, *mesohigh* features associated with quasi-stationary thunderstorm activity (see Fig. 2) were key factors in concentrating heavy rains over relatively small areas (Hoxit et al., 1982).

These case studies illustrate how a variety of meteorological elements at various scales can combine to produce the anomalous atmospheric conditions that result in a catastrophic flood. Due to their small spatial scales and short durations, many of the localized meteorological features important for flash flood generation in small catchments are not easily identified on large-scale weather maps, nor do they exhibit readily discernable anomaly patterns in the broad scale atmospheric circulation. In contrast, extensive regional catastrophic flooding in larger drainage basins tends to be produced by quite obvious large-scale synoptic features that are either persistent or recurrent over a given area.

In this paper, I will examine the atmospheric circulation features responsible for generating the largest maximum observed floods in the United States in drainage areas ranging from less than 0.5 km^2 to 3,000,000 km^2. My purpose is to assess how well catastrophic flood events can be linked to various atmospheric circulation anomalies. I will then discuss the geomorphic implications of atmospheric circulation anomalies in terms of response times and recovery periods following catastrophic flooding episodes.

CATASTROPHIC FLOODING IN THE UNITED STATES

Although many floods that have occurred in the conterminous United States have been described as "catastrophes," there are no standard criteria for classifying a given flood as a catastrophic event. I will consider as catastrophic the maximum floodflows that have been observed and officially recorded for various drainage basin areas. My selection of catastrophic flood peaks is taken from the compilations of Crippen and Bue (1977) and Rodier and Roche (1984). In the catalog by Crippen and Bue (1977), United States maximum floodflows that occurred through September 1974 in areas less than 25,900 km^2 are listed and plotted against drainage area to form an envelope curve. Rodier and Roche's (1980) World Catalogue of Maximum Observed Floods presents a more recent listing (through 1980) for a smaller sample of U. S. floods.

Figure 3 depicts an envelope curve for twenty-one selected catastrophic peak flow events in the United States, plotted against drainage area. These twenty-one floods were compiled as follows. For small basins (less than 370 km^2), I reviewed the largest floods from Crippen and Bue's nationwide envelope curve and selected from these on the basis of Costa's (1985) identification of the twelve largest rainfall-runoff floods in the United States (solid circles in Fig. 3). Costa's twelve floods all occurred in relatively small catchments in the western states or in Texas, so for a wider regional representation, I included four floods (#4, 6, 14, 15) from the Colorado Front Range and central Great Plains regions that appear as maxima on the Crippen and Bue nationwide curve, but were not included by Costa because they plotted slightly below the curve defined by his twelve maximum floods. To represent intermediate and larger drainage areas, I included five floods (#16, 17, 18, 19, 20) from basins larger than 500 km^2 that are listed in the Rodier and Roche catalog. Finally, to reflect an "upper limit" for drainage area on the envelope curve, I added an entry (#21) for one of the more extreme flood events on the Mississippi River (Chin et al., 1975). (Because of diversions, levees, and changing stage-discharge relations (see Belt, 1975), the "maximum observed" flood peak on the Mississippi is highly debatable.) In order to emphasize catastrophic rainfall-runoff relationships, my selection of floods excluded events due to nonclimatic fac-

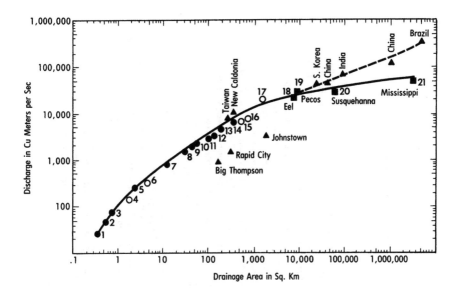

Figure 3. Envelope curve for catastrophic peak flow events in the United States. See text for sources. Flash floods are indicated by circles; large regional floods by squares. Solid   circles indicate the twelve largest rainfall-runoff floods in the United States identified by Costa (1985). Triangles indicate other floods of interest. The dashed line represents the world curve as defined by Rodier and Roche (1984).

tors such as dam breaks, as well as floods arising from snowmelt or ice jams.

Due to the varied sources from which my data have been selected, I cannot claim that the resulting curve represents a definitive upper limit of flooding in the conterminous United State. Nevertheless the portion of the curve defined by the Costa data is highly representative and the overall curve is a fair approximation of the extreme upper limits of United States flooding through 1980. The floods defining the envelope curve are listed in Table 1 and their geographic locations are given in Figure 4. Also included in the table and plotted on the figures are three of the more noteworthy United States catastrophic flash floods: Big Thompson, Rapid City, and Johnstown, Pennsylvania. In addition, some of the world's largest floods (selected from Rodier and Roche) have also been plotted on Figure 3. The curve defining the largest flood peaks in the world coincides fairly well with the United States curve for basins up to about 10,000 km^2, and then departs dramatically (dashed line).   Those extreme global floods that exceed events on the U.S. curve for the same drainage area are in tropical and monsoon climates that are not represented in the conterminous United States.

Table 1 and Figure 3 reflect the relative higher frequency of localized catastrophic flash flooding in small and medium size basins, in comparison with the more rare large-scale catastrophic flooding that encompasses broad regions. Most

Table 1. United States catastrophic floods and atmospheric circulation anomalies

ID#	Station Name	Area (sq km)	Peak Q (cms)	Date	Circulation Classification	Additional Anomalous Circulation Behavior
1	Wenatchee R. Trib at Monitor, WA	.39	25.6	8-2-56	Western III	Intense cutoff low, blocking ridge in central Pacific
2	Lahontan Res Trib nr Silver Springs, NV	.57	47.6	7-20-71	Western I	Abrupt change from deep Aleutian trough to strong ridge.
3	Little Pinto Ck trib nr Old Irontown, UT	.78	74.5	8-11-64	Western II	High latitude flow exhibited strong blocking pattern.
4	Boney Draw at Rockport, MO	1.97	144.0	7-18-65	Frontal	Hot, humid conditions prevailed beneath central U.S. ridge.
5	Humboldt R. Trib nr Rye Patch, NY	2.20	251.0	5-31-73	Western III	Strong cutoff low and deep trough.
6	Stratton Ck nr Washta, IA	4.92	311.0	8-09-61	Mesohigh	Deep trough over SE U.S. displaced cold fronts to south.
7	Lane Canyon nr Molin, OR	13.10	807.0	7-26-65	Western I	Extremely deep Aleutian low in unusual location.
8	Meyers Canyon nr Mitchell, OR	32.90	1,540.0	7-13-56	Western I	Abnormal southward displacement of westerlies; split flow.
9	Bronco Ck nr Wikieup, AZ	49.20	2,080.0	8-18-71	Western IV	Subtropical High ridge extended far westward from Atlantic.
10	Eldorado Canyon at Nelson Landing, NV	59.30	2,152.0	9-14-74	Western II	Upper level closed low; thunderstorm cell moved down valley.
11	North Fork Hubbard Ck nr Albany, TX	102.00	2,920.0	8-04-78	(combination)	TS Amelia; cold front; dry layer; Balcones uplift
12	Jimmy Camp Ck nr Fountain, CO	141.00	3,510.0	6-17-65	Synoptic	Very strong amplification of Rossby waves; deep cutoff low.
13	Mailtrail Ck nr Loma Alta, TX	195.00	4,810.0	6-24-48	(combination)	Cold front; Balcones uplift; easterly wave?
14	Seco Ck above D'Hanis, TX	368.00	6,510.0	5-31-35	(combination)	Balcones uplift; strong thunderstorm; MCC?
15	Little Nemaha R nr Syracuse, NE	549.00	6,370.0	5-09-50	Synoptic	Stronger than normal southerly flow at surface and aloft.
16	East Bijou Ck at Deer Trail, CO	782.00	7,760.0	6-17-65	Synoptic	Same as Jimmy Camp Creek (#12).
17	West Nueces R nr Brackettville, TX	1,813.00	15,600.0	6-14-35	(combination)	Balcones uplift; strong thunderstorm; MCC?
18	Eel R at Scotia, CA	8,063.00	21,300.0	12-23-64	Large-Scale I	Strong blocking ridge in Pacific; slipt flow; persistent jet.
19	Pecos R at Comstock, TX	9,300.00	27,440.0	6-28-54	Large-Scale II	TS Alice remnant stalls; 500 mb wave; Balcones uplift.
20	Susquehanna R at Harrisburg, PA	62,400.00	28,900.0	6-24-72	Large-Scale III	Unusual path of TS Agnes; anomalous SSIs; Atlantic blocking.
21	Mississippi R at Vicksburg, MS	2,964,300.00	55,600.0	5-12-73	Large-Scale IV	Repeated development of trough in roughly same position.
	(peak of record at Vicksburg)		58,900.0	2-17-37		
	Big Thompson Canyon at mouth, CO	155.00	884.0	7-31-76	Western I	Stationary supercell thunderstorm; high precipitable water.
	Rapid Ck at Rapid City, SD	236.00	1,433.0	6-09-72	Western I	Stationary supecell thunderstorm; high precipitable water.
	Johnstown (Conemaugh R at Seward, PA)	1,850.00	3,260.0	7-20-77	Mesohigh	Mesohighs; large and long-lived MCC.

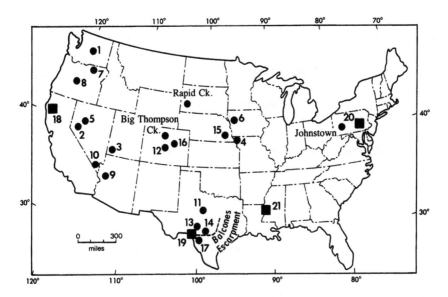

Figure 4. Location map of catastrophic events. Small-basin flash floods are indicated by circles; large regional floods by squares.

extreme flash flood events have been concentrated in western United States, although the eastern slopes of the Colorado Front Range, the central Great Plains, and central Texas have also experienced small-basin catastrophic floods. This geographic clustering appears to be related to both physiographic and climatic factors.

ATMOSPHERIC CIRCULATION ANOMALIES

Highly unusual atmospheric circulation patterns might be considered the atmospheric equivalents of "catastrophic" events. Anomalous atmospheric activity can occur over a wide spectrum of space and time scales. Figure 5 depicts a simplified version of the standard subdivision of scales for atmospheric processes proposed by Orlanski (1975), with some illustrations of potentially anomalous activity at each scale. At the *meso-β* scale, areas of unusually deep convection, intense squall lines, and persistent and severe "supercell" thunderstorms might be viewed as anomalous circulations that concentrate part of the enormous energy and moisture of the atmosphere into relatively small areas. At larger *meso-α synoptic* scales of activity, anomalous circulation features tend to be exhibited in particularly strong or persistent fronts, unusually intense or atypically-located tropical cyclonic storms, mesoscale convective complexes (MCCs) (Maddox, 1980; Rodgers et al., 1983), and short-wave troughs that migrate through larger-scale quasi-stationary upper air waves, triggering convection and intense precipitation.

The term "atmospheric circulation anomaly" is most commonly used to refer to unusual configurations or persistence in the *macroscale* features of the upper atmosphere. The *macro-β scale* is characterized by long (Rossby) waves, their ridges and troughs, and their attendant cyclonic and anticyclonic eddies; while the *macro-α scale* includes the ultra-long thermally- and topographically-anchored planetary waves that reflect the mean state of the circumpolar vortex. Current research on macroscale anomalous atmospheric flows has focused on: blocking, persistent anomalies, the possibility of multiple equilibria in the atmospheric circulation, and forcing of the anomalies by either internal or external factors. Each of these topics has some relevance for the question of how catastrophic floods are related to atmospheric circulation anomalies.

Blocking

Blocking is a large-scale perturbation in the typical zonal (west-east) movement of high and low pressure systems that takes the form of a quasi-stationary long wave. It is characterized by high-amplitude meridional flow aloft and the presence of high-latitude anticyclonic and low-latitude cyclonic eddies, known also as cutoff (omega) highs and cutoff lows (Fig. 6). The blocking generally persists for one to two weeks and exhibits an abrupt transition from zonal to meridional flow that oc-

Ts / Ls	1 MONTH	1 DAY	1 HOUR	1 MINUTE	1 SECOND		
KM / 10,000	Mean stationary planetary waves / Ultra-long waves					MACRO α SCALE	
2,000		Rossby waves / Ridges & troughs / Cyclones & anticyclones				MACRO β SCALE	
200			Fronts / Hurricanes / MCCs / Short-wave troughs			MESO α SCALE	
20				Squall lines / Supercell / Thunder storms		MESO β SCALE	
	CLIMATOLOGICAL SCALE	SYNOPTIC PLANETARY SCALE	MESO SCALE	MICRO-SCALE		DEFINITION	

Figure 5. Standard subdivision of scales for atmospheric processes, modified from Orlanski (1975). The time scale (Ts) refers to the typical period between events. The length scale (Ls) refers to the typical wavelength between features; e.g., the distance from one ridge axis to the next ridge axis. Only the meso-β through macro-α scales are shown.

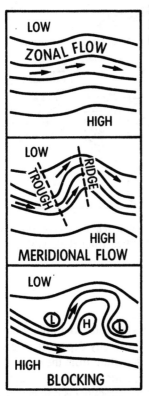

Figure 6. Schematic diagram of blocking in relation to more typical zonal and meridional flow. (Modified from Oliver and Fairbridge, 1987).

curs where the basic westerly jet stream splits into two branches (Rex, 1950a, b). Blocking occurs most often over the northeastern Atlantic and northeastern Pacific Oceans, with the latter location having a particularly strong influence on the weather of western United States. It is noteworthy that the occurrence of Pacific blocking has been linked to variations in tropical sea surface temperatures, El Niño, and the Southern Oscillation (White and Clark, 1975; Horel and Wallace, 1981). Blocking in the Atlantic Ocean can also influence the weather of the United States, as demonstrated by Resio and Hayden (1975) who linked Atlantic blocking to increased extratropical storm activity along the east coast. Blocking plays a major role in generating episodes of extreme flooding because of the persistent and anomalously-located storm tracks that tend to be associated with the phenomenon. The Northern California floods of February 11-19, 1986 resulted from a blocking pattern in the North Pacific ocean that shifted the main branch of the winter jet stream to a more southerly location for longer than a week. This circulation anomaly repeatedly supported the development of massive low pressure systems off the California coast, and steered a succession of devastating storms into the region over a nine-day period (Fig. 7).

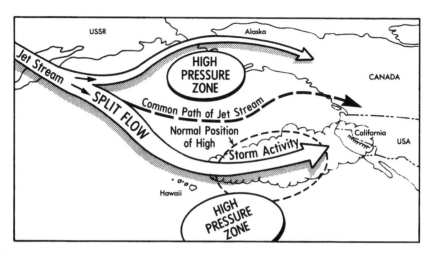

Figure 7. Diagram depicting the blocking action that led to severe flooding in Northern California during 11-19 February 1986. (Modified from Hart, 1986).

Persistent Anomalies

In climatology, the term "anomaly" does not always refer to a highly unusual atmospheric event, since it is used frequently to refer to any departure of a value from the long-term climatic mean. If however, the departure from the mean is sufficiently great and remains so for an unusually long period of time, the event is often referred to as a "persistent anomaly" (Dole, 1986). Blocking is one mechanism that is likely to generate persistent anomalies, and most of the current research in this area has revolved around the persistence of blocking over the North Pacific and North Atlantic oceans.

Over the continents, various configurations of the Rossby waves that are similar to blocking can exhibit anomalous persistence by either remaining quasi-stationary for extended periods of time or by repeatedly developing in roughly the same location. These preferred continental ridge and trough positions may or may not be linked to blocking over the ocean, but they have a substantial influence on the climate of specific areas -- often bringing droughts to regions underlying persistent high pressure ridges, and floods to regions underlying the southeastern sectors of troughs (where frontal activity and storm tracks are concentrated).

Multiple Equilibria

Ever since the earliest observations of variations in the mid-latitude westerlies, meteorologists have noted a tendency toward two different states of large-scale motion in the atmosphere. Under a zonal flow regime (see Fig. 6), motion is strongly west-to-east, reflected in low amplitude Rossby waves. Under meridional flow con-

ditions, winds are steered in more northerly and southerly trajectories and amplitudes in the Rossby waves are high. Blocking is an example of an extreme high-
amplitude condition, which will eventually break down when cutoffs occur and zonal flow is resumed. Episodes of dominance by zonal and meridional flow conditions have been successfully linked to twentieth century climatic variations (e.g.
Dzerdzeevskii, 1969; Lamb, 1977), and specifically to flood variability (Knox et
al., 1975; Knox, 1984).

Recent theoretical work and modeling have suggested the possibility of multiple equilibria in the atmospheric circulation, with one stable steady state that corresponds to relatively strong zonal flow in the ultra-long planetary waves, and another that corresponds to a higher-amplitude, more meridional state with an apparent tendency toward more frequent blocking (Charney and DeVore, 1979; Hansen,
1986; Sutera, 1986). The importance of this new research, as suggested by Sutera,
is that if "more than one regime occurs, then the mean state of the atmosphere
would not be adequately represented by its long-term average, but rather by the set
of average states obtained by considering, separately, each individual regime."
(Sutera, 1986, p. 227). The possibility of two modes of operation in the planetary
waves -- with one mode perhaps more prone to anomalous blocking activity -- has
important implications for flood variability and could be reflected in a temporal
clustering of catastrophic events during high-amplitude circulation regimes.

Internal and External Forcing

An area of active debate among dynamic and synoptic meteorologists centers
around the importance of various forcing mechanisms for large-scale wave activity.
It is widely agreed that external forcing by orographic drag and land-sea thermal
contrasts is an important mechanism for developing the long-term, macro-α scale
mean planetary waves (Wallace, 1983). However, internal, as well as external, factors have been advanced as the dominant forcing mechanisms for anomalous wave
patterns, and blocking, at other macro-α and macro-ß scales. Those who promote
internal forcing mechanisms have demonstrated that interactions between ultra-long
planetary waves and synoptic-scale traveling systems can form and sustain large-
scale blocking patterns in a process referred to as eddy forcing (e.g., Shutts, 1986;
Pierrehumbert, 1986; Egger et al., 1986). Similarly, Lindzen (1986) and Reinhold
(1987) suggest tha interference and interaction between waves and flow configurations of different scale can produce planetary-wave anomalies by a mechanism that
is purely internal to the dynamics of the atmosphere. Furthermore, Lindzen (1986)
maintains that these interactions alone can explain the observed evidence of multiple equilibria.

An alternative view is expressed by those who have promoted external forcing
mechanisms as explanations for atmospheric circulation anomalies and blocking
wave patterns. This perspective is presented most notably in the works of Jerome
Namias, from 1944 to the present. Self-enhancing feedback between continental
surfaces and the atmosphere, due to persistent soil moisture conditions or snow

cover, is one external mechanism that has been linked to climatic anomalies (Namias, 1962; 1963). Of much greater significance is the interaction between the atmosphere and the ocean, especially as it is manifested in teleconnections between sea surface temperature (SST) anomalies and corresponding atmospheric circulation anomalies. The major El Niño/Southern Oscillation (ENSO) event of 1982/83 has been linked to persistent global circulation anomalies (Chen, 1983; Quiroz, 1983; Yarnal, 1985) and episodes of major flooding at diverse locations in the western hemisphere, including Peru, Bolivia, and Ecuador in South America, the Pacific coast of North America from Oregon to Baja California, the Lower Mississippi Valley, and coastal areas of the Gulf of Mexico (Quiroz, 1983). On the basis of varying modeling results and the strong evidence for internal control mechanisms, the overriding influence of SSTs on the thermal forcing of anomalous wave configurations has been questioned (Lindzen, 1986; Reinhold, 1987). Nevertheless, success in long-range forecasting and strong correlations with regional climatic anomalies have made SSTs and teleconnections a primary research area with important implications for catastrophic flooding.

Classification of Large-Scale Anomaly Patterns

Most of the recent theoretical and observational research on circulation anomalies has focused specifically on blocking, but other forms of large-scale anomalous behavior are also possible. Inspection of surface and upper air synoptic charts during unusual climatic episodes reveals that anomalous atmospheric behavior can manifest itself in a variety of ways -- some related to true blocking regimes, and others that are not necessarily tied to blocking. I have classified large-scale circulation patterns into four types of anomalous behavior. These types are not based on internal atmospheric dynamics, but on how a point on the earth's surface would experience the unusual circulation pattern. This kind of response-based classification can better define the ways in which rare atmospheric patterns result in catastrophic flooding events.

The four large-scale anomaly types are: an anomalous location or unseasonal occurrence of an otherwise typical circulation mechanism (*Large-Scale I*), an unusual combination of several common meteorological mechanisms occurring simultaneously (*Large-Scale II*), an extremely rare configuration in the upper air pattern itself (*Large-Scale III*), and the unusual persistence in space and time of a specific wave configuration (*Large-Scale IV*). While Large-Scale II might occur during any kind of circulation regime, due purely to random factors, Large-Scale I, III, and IV are more likely to occur during either true blocking regimes, or during meridional regimes characterized by high-amplitude, quasi-stationary waves.

Classification of Small-Scale Anomalies

Due to their small spatial scale and short duration, mesoscale events that produce flash floods do not exhibit anomaly patterns that are as obvious, and as persis-

tent, as larger-scale anomalies, such as blocking. However, the conditions necessary to transform a "typical," mesoscale storm into one with catastrophic effects are
often linked to larger-scale features such as thick and laterally extensive layers of
unusually high precipitable water vapor, and specific upper-air flow patterns that affect the path, strength, and persistence of severe storms.

Recently, efforts by R. A. Maddox and his colleagues to improve flash flood
forecasting by using synoptic observational techniques have produced an extremely
useful classification scheme for meteorological events that generate flash floods.
Maddox and Chappell (1978) and Maddox et al (1979) introduced a four-fold classification for circulation features that are associated with flash floods, dividing them
into *Synoptic events, Frontal events, Mesohigh events, and Western events.* Maddox et al (1980) subsequently refined the classification by subdividing the Western
flash flood category into four additional types (*Western I-IV*). The diagnostic characteristics that define each category are related to local properties of temperature,
moisture, instability, wind speed, and wind direction at various levels in the atmosphere. In addition, each type is associated with a particular synoptic-scale pattern
of both surface and upper-level circulation features. The patterns are distinguished
primarily on the basis of how short-wave troughs move through specific configurations of longer-wave ridges and troughs. An advantage of the scheme is that it
bridges small-scale and large-scale atmospheric activity, ultimately tying a local
flash flood event into the broader regional pattern that is linked to macro-scale circulation features.

The diagnostic details of the classification for central and eastern United States
flash flood events can be found in Maddox et al (1979), and for western events in
Maddox et al (1980). The Maddox classification is based on the detailed analysis of
at least 180 flash flood events of varying sizes that occurred during 1973-78. Due
in part to the time period sampled, only a few of these floods are among the catastrophic floodflows being considered in this paper. An initial stage in the Maddox
analysis was a pilot study of twenty-one "significant" flash flood events (Maddox
and Chappell, 1978). These included the catastrophic floods of Rapid City (1972);
Eldorado Canyon, Nevada (1974); Big Thompson Canyon (1976); and Johnstown,
Pennsylvania (1977) all of which are plotted on Figures 3 and 4 and listed in Table
1. However, of these, only the Eldorado flood (#10) falls on the envelope curve of
maximum discharge versus drainage area shown in Figure 3.

In a sense, any flash flood occurrence can be considered "catastrophic" because
of the suddenness of the event. Therefore the synoptic patterns of the Maddox classification can be viewed as circulation "mini-anomaly" patterns that produce small-
scale catastrophic flood events. The maximum floodflows that define the envelope
curve of Figure 3 raise an interesting question, however. Do the atmospheric circulation patterns associated with these extreme events also fit the Maddox classification scheme, and if so, is their catastrophic nature linked to any additional anomalous atmospheric behavior, such as that seen during larger-scale catastrophic flood
events? In the following section I will explore these questions as I discuss the specific circulation anomalies responsible for the 21 events plotted on Figure 3.

CIRCULATION CLASSIFICATION
FOR UNITED STATES
MAXIMUM OBSERVED FLOODS

Classification Procedure

My circulation classification for the twenty-one maximum events that define the envelope curve in Figure 3 is based on interpretation of: surface and 500 mb height daily weather maps (NOAA Daily Weather Maps - Weekly Series); monthly and weekly mean 700 mb height and departure-from-normal 700 mb height maps (Monthly Weather Review, weather and circulation summaries by month); and a variety of additional references that addressed specific floods. The original Maddox classification -- designed for forecasting purposes -- was based on detailed interpretation of hourly synoptic charts, including analysis of wind speeds, temperature, dewpoints, precipitable water, and stability indices. For the non-forecasting purposes of this study, each of the flash flood events was assigned to one of the Maddox categories on the basis of information found on daily weather maps alone. The large-scale flood events were treated individually and not tied into the Maddox classification.

Thirteen of the twenty-one events were classified as one of the standard types described in Maddox et al (1979) and Maddox et al (1980). The four flash flood events in Texas (#11, 13, 14 and 17) did not conform as well to the Maddox prototypes, but this was not unexpected because events associatioed with weather systems of tropical origin were specifically excluded from the Maddox et al (1979) and Maddox et al (1980) samples. The remaining large floods, on the Eel, Pecos, Susquehanna, and Mississippi rivers, were each associated with major large-scale circulation anomaly patterns (Large-Scale I-IV). In the following sections I will briefly describe the Maddox flash flood types and the catastrophic events from Figure 3 that are associated with each type.

Western Catastrophic Flash Floods

Western Type I Events. The Western Type I event occurs when a weak 500 mb short-wave trough moves northward along the western side of a long-wave ridge (Figure 8a). The Big Thompson Canyon flood depicted in Figure 1, and the Rapid City flood, were both Type I events (Maddox et al., 1980). In this type, the short-wave trough often originates from a stationary long-wave trough or cutoff low, situated just off the west coast. Heavy precipitation occurs in the warm air just ahead of the short wave, where instability and moisture values are high.

This pattern was associated with the extreme flash floods of the Lahontan Reservoir tributary in western Nevada (#2), Lane Canyon in northeastern Oregon (#7), and Meyers Canyon in central Oregon (#8). All three events occurred in July and each was associated, in the macro-scale, with an upper-level low pressure system

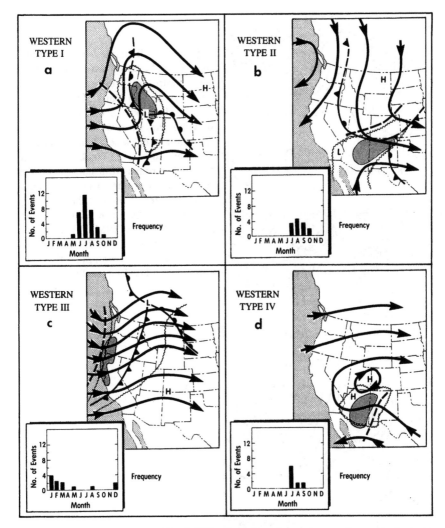

Figure 8. Generalized combined 500 mb and surface patterns for typical western flash flood events. Heavy dashed line shows location of 500 mb short-wave trough. Shaded area shows region of potential for heavy precipitation and flash flooding. Also shown are monthly frequencies for each type, based on the 1973-78 sample of Maddox et al (1980). (Modified from Maddox et al., 1980).

off the west coast that was located unusually far south for this time of year. During the week of the Lane Canyon flood, blocking in Asia and the Arctic was apparently responsible for an extremely deep Aleutian Low that was situated in a rarely-observed location for July (Andrews, 1965) (Figure 9a and 9b). During the time of the Meyers Canyon flood, split flow in the westerlies resulted in their abnormal southward displacement and the development of a trough in the eastern Pacific in a

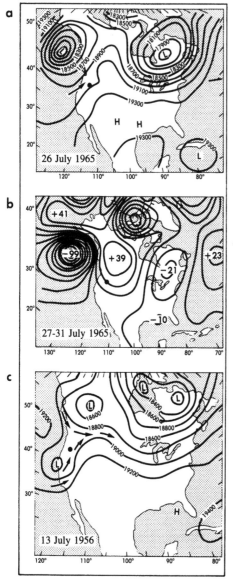

Figure 9. Examples of large-scale anomalous behavior associated with two catastrophic Western Type I flash flood events. (a) Daily 500 mb weather map for the Lane Canyon flood, showing the short-wave trough moving counterclockwise out of the deep Aleutian low. (b) Weekly 700 mb height departures from normal, in tens of feet, showing the extremely anomalous character of the circulation during the time of the Lane Canyon flood. (Source: Andrews, 1965) (c) Daily 500 mb height weather map for the Meyers Canyon flood, showing split flow and a trough displaced unusually far south.

region usually dominated by ridges (Krueger, 1956) (Figure 9c). The Lahontan tributary flood was associated with somewhat different macro-scale features, but did occur soon after a major mid-month upheaval in the global circulation (Wagner, 1971). Each of these cases illustrates that, in addition to nicely fitting into the Maddox classification, the catastrophic nature of these extreme flash floods also can be linked to major global circulation anomalies.

Western Type II Events.   These events are associated with a 500 mb short-wave trough that moves southward along the eastern side of a long-wave ridge (Fig. 8b). Heavy rains occur in the unstable air ahead of the short wave, especially when moist air from the Southwest "summer monsoon" is available at low levels.

The catastrophic Eldorado Canyon flash flood of 14 September 1974 (#10) was a Type II event (Maddox et al., 1980). Its short-wave trough was unusually deep, and formed an upper-level closed low over Nevada during the time of the flood. An important factor in the severity of this event was the eastward track of the major thunderstorm cell as it moved down the length of the canyon, toward its mouth. The storm's movement caused intense rainfall and runoff to be superimposed on previously generated flood waves coming from upstream (Glancy and Harmsen, 1975). It is important to emphasize, therefore, the well-known fact that many small-scale meteorological factors and local basin characteristics can have a major effect on the catastrophic nature of a flash flood, and may therefore be more significant than the presence of any large-scale circulation anomalies.

Western Type III Events.   Heavy precipitation from this type affects large areas and is associated with a strong synoptic weather system that moves into the west coast from the Pacific Ocean. The type is characterized by a surface front and a 500 mb trough aloft (Figure 8c). Occasionally the trough will develop into an intense upper-level cutoff low. Local flash flooding occurs in response to this pattern when mountainous terrain triggers thunderstorm activity in the moist air that is being steered in at low levels by the synoptic system.

The floods of the Wenatchee River tributary in central Washington (#1) and the Humboldt River tributary in northwestern Nevada (#5) each occurred in conjunction with Type III features -- a deep upper-level cutoff low, and the passage of a surface cold front. In the Wenatchee flood the center of the upper-level low moved directly across Washington state with westerly flow circulating counterclockwise around it bringing moist oceanic air into central Washington from the southwest ahead of the front. (Not unlike a higher- amplitude version of Figure 8c, which shows conditions after the rain-producing front passed.) The Humboldt tributary circulation pattern also was characterized by higher-amplitude waves in the westerlies than are indicated in Figure 8c, and eventually developed into a blocking-like configuration at higher latitudes, with an omega high situated between two deep upper-level lows.

Western Type IV Events.   These events occur when a weak short-wave trough moves through zonal flow, and a strong east-west 500 mb high pressure ridge is situated to the north or south of the short-wave's path. When the ridge is to the north, the zonal flow will be from the east and the short wave is called an easterly wave. This type tends to be associated with the Southwest "summer monsoon" season.

The catastrophic flash floods of Little Pinto Creek tributary in southwestern Utah (#3) and Bronco Creek in central western Arizona (#9) were each associated with a broad west-to-east trending 500 mb high-pressure ridge over southwestern United States. A weak short-wave trough moved over south- western Utah from the west, around the north side of the elongated ridge, to produce the Little Pinto flood; and a weak easterly short-wave moved across Arizona to the south of the ridge, in a manner similar to that depicted in Figure 8d, to generate the Bronco Creek flash flood. Both of these events corroborated the statement in Maddox et al., 1980) that Type IV events tend to have weak and ill-defined surface patterns. Their anomalous nature therefore can best be evaluated by a more detailed analysis of moisture-layer thicknesses and instability indices. For both of these cases, the amount of available moisture was significantly above the mean. On the large scale, the elongate upper-level high-pressure ridge was stronger than normal for both events, suggesting intense surface convectional heating and a thick, moist atmosphere during these monsoon-season floods.

East Slope and Central Plains Catastrophic Flash Floods

Synoptic Events. Flash flood events of this type tend to develop with an intense synoptic-scale cyclone and a quasi-stationary front at the surface, in conjunction with a major trough aloft at the 500 mb level (Figure 10a). A strong short-wave trough often moves through the larger trough, increasing instability and triggering convection. Storms and heavy rains are concentrated in the warm, moist air ahead of the front. Synoptic events can be fairly widespread and long-lived, affecting several states and lasting two to three days in some cases. General widespread flooding may occur, but local flash flooding is associated with convective storms which develop repeatedly over the same general area, delivering heavy rains. The pattern is most frequent in spring, early summer, and fall during periods of adjustment in the global circulation.

Two of the maximum flash floods on the envelope curve (#12 Jimmy Camp Creek, and #16 East Bijou Creek in east central Colorado) were produced by the same Synoptic type event in June of 1965. The upper-air pattern associated with this event (Fig. 11) shows a highly-anomalous 500 mb configuration, dominated by an intense cutoff low over the west that steered warm, unstable air northward into eastern Colorado. This blocking pattern caused the attendant surface cold front to become stationary and remain in roughly the same position for at least three days, causing extreme rains. The situation closely resembled an especially persistent version of Figure 10a. According to Posey (1965a) the extreme amplification of the upper-air ridges and troughs occurred when three separate Rossby wave-trains came into phase during the third week of June.

The circulation pattern for the 1950 flood event in the Little Nemaha River in southeastern Nebraska (#15) was much less dramatic, but was characterized by an upper-air trough and a surface cold front, fitting the Maddox Synoptic event description. In addition, a short-wave trough moved over the area and triggered intense thunderstorm activity ahead of the front.

Frontal Events. This type is distinguished by a stationary or slowly-moving front at the surface that is usually oriented west to east. The upper-air pattern is characterized by a broad ridge, through which a weak, meso-scale short-wave trough often moves (Fig. 10b). Storms and heavy rains are triggered on the cool side of the surface warm front when warm unstable air flows over the frontal zone. The July 1965 catastrophic flash flood in Boney Draw in northwestern Missouri (#4) best resembles the Frontal type of event. During the week of the flood, a broad upper-level ridge prevailed over the central plains, allowing hot humid conditions to build and promote intense convective activity. A weak short-wave trough aloft and a west-to-east trending stationary front at the surface were the triggers for the heavy rainfall that produced the flash flood.

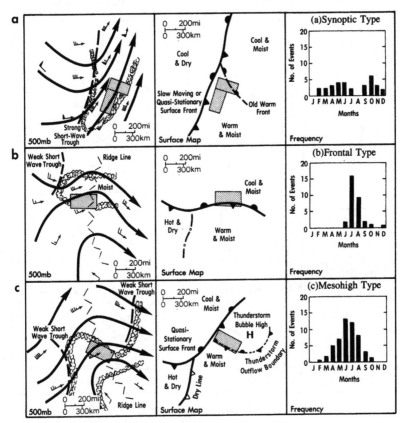

Figure 10. Generalized 500 mb and surface patterns for typical East Slope and Central Plains flash flood events. Heavy dashed line shows location of 500 mb short-wave trough. Zig-zag line shows axis of 500 mb ridge. Shaded and boxed areas show regions of potential for heavy precipitation and flash flooding. Also shown are monthly frequencies for each type, based on the 1973-77 sample of Maddox et al. (1979). (Modified from Maddox et al., 1979).

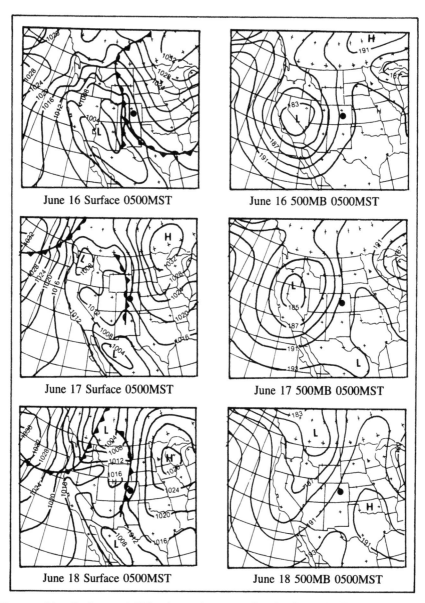

June 16 Surface 0500MST

June 16 500MB 0500MST

June 17 Surface 0500MST

June 17 500MB 0500MST

June 18 Surface 0500MST

June 18 500MB 0500MST

Figure 11. Surface and 500 mb weather maps for the Synoptic Type catastrophic flash floods of Jimmy Camp Creek (#12) and East Bijou Creek (#16), showing an intense "blocking" cutoff low and associated stationary surface front. (Source: Miller et al., 1984).

Mesohigh Events. The distinctive characteristic of a Mesohigh event is the heavy precipitation that occurs due to instability and convection along a nearly stationary cold-air outflow boundary (e.g. Fig. 2). This mesohigh outflow boundary is usually generated by thunderstorm activity occurring earlier in the afternoon or evening. Although conditions are similar to the Frontal event type, an identifiable front may or may not be in the vicinity. The corresponding upper-air pattern consists of a large-scale ridge over the area of heaviest rain, with weak short-wave troughs moving through the ridge, helping to trigger and focus storm activity (Fig. 10c).

It is likely that the Stratton Creek flash flood in western Iowa (#6) was caused by mesohigh-induced convective activity. At the time of the flood, no frontal activity was evident, but the event occurred not long after a cold front had passed through the region, inducing thunderstorms that could have provided the necessary antecedent conditions for mesohigh activity.

Although not detectable on daily weather maps, another possible cause for the flooding in Stratton Creek and the other central Plains drainages is the mesoscale convective complex (MCC). MCCs are huge (50,000 - 100,000 km^2) organized, convectively-driven systems characterized by numerous cold outflow boundaries and mesohighs that generally produce widespread and locally-intense rains (Maddox, 1980). The systems are so large that they can interact with and modify their larger-scale environment, rendering traditional ways of forecasting from the macroscale less effective. A large and long-lived MCC played an important role in the catastrophic Johnstown flood of 1977 (Bosart and Sanders, 1981). Maps of typical MCC tracks show a strong concentration of the systems in the vicinity of the three central Plains catastrophic floods (Maddox, 1980; Rodgers and Howard, 1983).

Texas Catastrophic Flash Floods

Four Texas flash flood events help to define the maximum floodflow curve of Figure 3: (#11) North Fork Hubbard Creek, (#13) Mailtrail Creek, (#14) Seco Creek, and (#17) West Nueces River. As noted earlier, the circulation patterns from which these floods were generated did not fit exactly into the Maddox classification scheme. Texas has its own unique "combination" flood regime because of a variety of factors, most notably its proximity to the Gulf of Mexico moisture source, and the effects of tropical storms and easterly waves. Another important element is the exceptionally strong orographic uplift of moist Gulf air masses along the Balcones Escarpment, which has produced some of the highest rainfall intensities in the world (Patton and Baker, 1977). Mesoscale processes, such as intense mesohigh activity and MCCs, also contribute to extreme precipitation events, in conjunction with, or independent of, the escarpment. In fall, winter, and spring, deep troughs and surface cold fronts traverse the state, resulting in great airmass contrasts, uplift, and widespread precipitation. Under certain circulation patterns, upper air flow brings in a layer of dry air that enhances instability aloft. All of these activities can be exacerbated if a still-organized tropical system moves inland, or if remnants of the system enter at upper levels. This complexity of flood-generating circulation features makes classification of the Texas events quite difficult, requiring detailed information for an accurate assessment of all the factors involved.

The Texas Hill Country flash floods of 2 August 1978 were analyzed in detail by Carcena and Fritsch (1983) and they found that a variety of distinct mechanisms interacted to focus and anchor the large stationary thunderstorm complex that produced the floods. Among these were an elevated dry layer, a mesohigh remnant of tropical storm Amelia, and forced uplift along the Balcones Escarpment. The North Fork Hubbard Creek flood (#11) occurred two days later, about 300 km to the north, and was produced by many of the same features, with a slow- moving cold front replacing the mesohigh as a lifting mechanism (Carcena and Fritsch, 1983). In the large-scale, a stronger-than- normal Atlantic subtropical high enhanced the easterly and southeasterly flow of Amelia into Texas and at the same time, brought in the dry layer at upper levels.

Examination of the somewhat rudimentary weather maps of May-June 1935 and June 1948, without additional information for enhancement or re-analysis, did not allow me to assess the Mailtrail Creek (#13), Seco Creek (#14), and West Nueces (#17) flood events in any detail. Each of these floods occurred in close proximity to the Balcones Escarpment, and each was associated with strong thunderstorm activity and southerly and southeasterly surface flow. Fronts were present in northern Texas at the times of the Mailtrail and Seco Creek floods and may have induced some large-scale instability. Low-level flow was from the east just prior to the West Nueces flood, suggesting an easterly wave, which would also explain the deep westward penetration of moist air into Texas. It is certainly possible that some or all of the Texas flood- producing mechanisms mentioned above were present during these early catastrophic events as well.

Large-Scale Catastrophic Floods

The storms that generate a typical regional flood event can usually be traced to an easily-identified synoptic weather pattern. Catastrophic regional floods, however, tend to be distinguished from more common regional floods by the anomalous behavior of the macroscale circulation patterns that drive and steer flood-generating synoptic weather systems. The largest floods on Figure 3 each evolved from different types of large-scale anomalous behavior.

Eel River Flood of December 1964 -- (Large-Scale I). Heavy rains throughout the northwestern states produced record-breaking and severely damaging floods over an unusually large area during December 1964 and January 1965. Among these was the catastrophic flood peak recorded on the Eel River at Scotia (#18) on 23 December 1964, during which the Eel transported more than ten times its maximum sediment load of record (Waananen et al., 1971). Numerous small catchments in the Eel and other northern California basins experienced dramatic geomorphic changes in response to this event (e.g. Stewart and LaMarche, 1967; Helley and LaMarche, 1968). The circulation pattern that produced this catastrophic flood was characterized by intense blocking action over the Aleutians. A split westerly airstream, similar to that of February 1986 (Fig. 7), directed the flood-producing storms into the northwest along a jet stream track that was located unusually far south, coming from the vicinity of Hawaii (Posey, 1965b). This pattern, an example of an anomalous location for an otherwise typical storm track (Large-Scale I type), also occurred, with less intensity, in December 1955 and produced major floods in the same areas.

Pecos River Flood of June 1954 -- (Large-Scale II). The storm period of June 24-28 in southwestern Texas was a direct result of the movement of Hurricane Alice from the Gulf of Mexico up the Rio Grande Valley. The hurricane in itself was not anomalous in size or strength; in fact, in the lower Rio Grande Valley the rains were only moderate for a hurricane, and it had lost most of its surface identity by the time it reached the upper valley. Rather, it was an unusual combination of events (Large-Scale II type) involving the hurricane that triggered the catastrophic Pecos River flood on the 28th, the largest flood on record in Texas and estimated as having a recurrence interval in excess of 2000 years (Patton and Baker, 1977). The key factor was that the upper-level remnant of the hurricane stalled in the vicinity of the uplift-enhancing Balcones Escarpment. At the same time, the system interacted with a weak 500 mb wave in the westerlies and transformed from a warm core tropical cyclonic system to a cold core extratropical cyclonic system (Miller et al., 1984). The resulting vigorous vertical motion, in what was still a very deep layer of moist air, produced intense thunderstorm activity and anomalously heavy rains.

Susquehanna River Flood of June 1972 -- (Large-Scale III). The flooding from Hurricane Agnes that devastated the East Coast of the United States has been called "the greatest natural disaster ever to befall the Nation" (U. S. Department of Commerce, 1973, p. 1). The outstanding aspect of this event was the great areal extent of flooding, which resulted in many very large drainage basins experiencing record-breaking flows. Peaks having recurrence intervals in excess of 100 years were recorded throughout the length of the Susquehanna River basin, which is more than 70,000 km^2 in area (Bailey and Patterson, 24th). The Susquehanna's peak at Harrisburg on June 24th is believed to be the largest for any basin of comparable size in the United States.

Although not an unusual storm at its outset, the area covered by Agnes was exceptionally large, and its slow development and movement permitted large amounts of moisture to be entrained into the system from the tropics. However, it was the influence of a highly-abnormal blocking configuration over the North Atlantic ocean (Large-Scale III type) that steered Agnes' unusual path and fed large amounts of moisture into the storm during its latter stages, as it merged with a trough in the westerlies and stagnated (Fig. 12). Anomalously warm sea surface temperatures in the western North Atlantic, that had been developing since February or March,

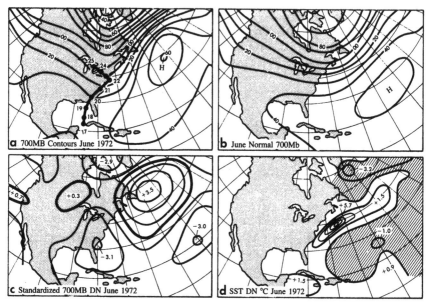

Figure 12. Anomalous circulation during the Agnes floods of June 1972. (a) June 1972 700 mb height pattern showing path of Agnes and blocking high over Atlantic. (b) Normal June 700 mb pattern based on the mean of Junes from 1947-63. (Contours are labeled in tens of feet with hundreds omitted,) (c) Deviations of the June 1972 700 mb heights from the normal pattern. (Isopleths are in standard deviations with a contour interval of 1). (d) Sea surface temperature departures from normal for 1 - 26 June. (Source: Namias, 1973).

probably played a major role in both sustaining the Atlantic blocking pattern through external forcing and positive feedback, and in directing the hurricane's path to unusually high latitudes (Namias, 1973). Furthermore, the anomalous nature of both the SSTs, and the abnormal macroscale wave configuration, may have been linked to the 1972/73 El Niño event (Namias, 1973).

Mississippi River Flood of Spring 1973 -- (Large-Scale IV). Although catastrophic, neither the flood stages nor the peak discharge volumes that occurred on the Mississippi during the great flood of 1973 were the largest ever recorded. However, the duration of this flood was unprecedented, and new records for consecutive days above flood stage were set for most of the main-stem gaging stations on the Mississippi from southern Iowa to Louisiana: St. Louis, Missouri, 77 days; Chester, Illinois, 97 days; Memphis, Tennessee, 63 days; and Vicksburg, Mississippi, 88 days (Chin et al., 1975). In fact, the cumulative runoff at Vicksburg for the first nine months of water-year 1973 was greater than in any other recorded flood year. On this basis, and because flood stage and discharge relationships have changed significantly over time, I have selected the 1973 flood at Vicksburg as an example of a catastrophic flood of extremely long duration occurring in an exceptionally large basin.

The duration and persistence of the large-scale atmospheric circulation pattern that produced this flood was equally anomalous (Large-Scale IV type). Throughout March and April of 1973, the repeated development of a trough over the south-

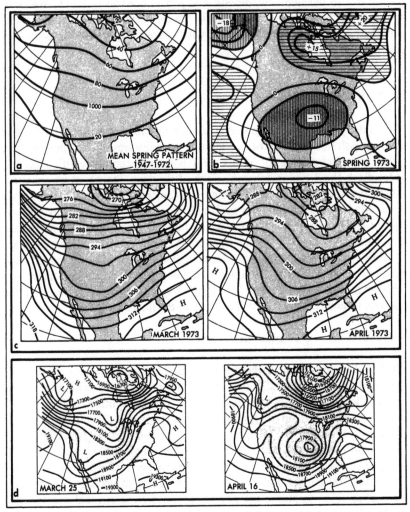

Figure 13. Anomalous circulation during the Mississippi River flood of spring 1973. (a) Mean spring 700 mb height pattern in tens of feet. (Based on March-May over the period 1947-72.) (b) Spring 1973 departure pattern from the 26-year mean, in tens of feet. Contour interval 50 ft. (Source for (a) and (b): Namias, 1979). (c) Mean monthly 700 mb height patterns for individual months of March and April 1973, in dekameters. (d) Selected daily 500 mb charts for March and April showing the position of the trough. (Source for (c) and (d): Chin et al., 1975).

ern United States steered numerous fronts and extratropical cyclones across the Mississippi basin (Fig. 13). An analysis of daily weather maps shows that the trough was positioned in roughly the same location for 60 percent of the days during March and April. Furthermore, Namias (1974) was able to trace the continuity of this remarkably persistent trough from September 1972 through August 1973 and suggested that its development, motion, and persistence was related to sea surface temperature conditions in the North Pacific On the macro-β scale, this exceptionally persistent wave pattern was associated with a strong ridge over the eastern North Pacific ocean.

DISCUSSION

From the preceding descriptions it is clear that major regional flooding, and even local flash flooding, can be tied to large-scale features of the atmospheric circulation. An important result of this study is the independent verification of the Maddox flash-flood typing scheme. Each of the (non-Texas) catastrophic flash floods fit into one of the Maddox categories, demonstrating that, as a forecasting tool, the Maddox classification is applicable to catastrophic events, as well as to the set of smaller flash floods upon which the categories were originally based. During the course of assigning each flood to a Maddox categroy, I observed that, in most cases various large-scale anomalous circulation features accompanied the occurrence of a catastrophic flash flood, even though such anomalous features were usually not an essential part of the Maddox prototype descriptions. These additional anomalous features have been listed for each of the floods in Table 1.

When Is An Anomaly "Anomalous?"

It should be stressed that what I have labeled "anomalous" behavior is subjectively based on my own familiarity with the typical circulation patterns at different geographic locations, and on the views expressed in the syntheses of monthly circulation patterns published in Monthly Weather Review. In each of the flood cases, the relevant "centers of action," (e.g. Aleutian Low, North Pacific High, Icelandic Low, and North Atlantic High, identified by Namias, 1981) usually showed monthly or seasonal departures well above or below the monthly or seasonal mean height of the 700 mb surface. Yet due to the variety of ways in which unusual flood-producing atmospheric activity is exhibited (i.e., unusual combinations, anamolous locations, repeated patterns) a single set of threshold criteria for 700 mb height departures would be difficult to objectively establish, the existence, strength, or persistence of, flood- producing anomalies.

The most general conclusion that can be drawn about the flood-producing anomalies is that various forms of blocking were instrumental in setting up most of the catastrophic events: both small-scale flash floods and large-scale regional floods. The exact location of the blocking was, of course, quite important for specific floods, but even distant blocking may have had an effect. Teleconnection studies have shown that anomalous activity at remote centers of action can significantly affect the circulation at downstream locations (e.g. Namias, 1981; Wallace and Gutzler, 1981). The importance of blocking -- an attribute of meridional as opposed to zonal long-wave configurations -- corroborates earlier claims by Knox et al (1975) and Knox (1976, 1984), that meridional circulation periods are more likely to experience large floods in North America. Moreover, the sample of flash flood events upon which this conclusion is based underscores the importance of large-scale circulation patterns for creating a favorable atmospheric environment for flooding -- even in small watersheds where flash floods have usually been attributed to the random occurrence of local thunderstorms.

Blocking is currently the center of attention for much of the theoretical, observational, and numerical modeling in progress on anomalous atmospheric flows. Blocking in itself, is a fairly normal stage in the weekly progression of circulation in the westerlies, especially in winter. The determination of when a situation becomes truly "anomalous" lies in the strength, frequency, and persistence of blocking, and these are the factors that must be examined more closely in relation to flooding.

Of particular interest to catastrophic flooding is the possibility of multiple equilibria in the large-scale circulation, ie., a blocking and a nonblocking mode (Hansen, 1986; Sutera, 1986). Figure 14 shows a time series plot of the catastrophic floods listed in Table 1. The distinct clustering of events may well be reflecting alternate modes of operation in the atmosphere.

Observational studies have shown similar episodic behavior in blocking, but most of this work has focused on winter months when the action is typically much stronger. A study of blocking during summer, when most extreme floods tend to occur, would be an important test of this hypothesis.

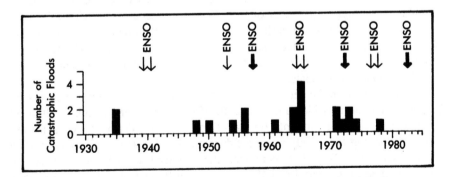

Figure 14. Time distribution of the twenty-one catastrophic floods discussed in this paper. Years of strong (heavy arrow) and moderate (thin arrow) ENSO events are also indicated. The mid-60s and early 70s seem to have been especially prone to the occurrence of extreme events. Various synoptic studies have shown that meridional types of circulation have been more frequent since 1950 (see Knox, 1984).

Strong and moderate ENSO events, as summarized by Yarnal (1985), are also indicated on Figure 14. The two main clusters of catastrophic flooding occurred in association with ENSO-related SST anomalies, but the correlation is not consistent for the other events. It appears that the complexities of the relationship between atmospheric circulation anomalies and catastrophic flooding will have to be further unraveled by a better understanding of internal and external forcing mechanisms for anomalous atmospheric activity, as well as a better discrimination of "catastrophic" floods (according to geographic location, basin size, physiography, season, and type of event).

Geomorphic Implications of Anomalous Atmospheric Circulations

One of the critical factors in determining the geomorphic effectiveness of a catastrophic flood, ie., its ability to make either a permanent or a temporary change in the landscape, is the length of the "intervening interval" until the next catastrophic event (Gupta and Fox, 1974; Wolman and Gerson, 1978; Beven, 1981). In humid regions, intervening moderate events can return a channel to its pre-flood condition in a matter of months (Costa, 1974), whereas recovery times in arid and semi-arid environments may be much longer (Wolman and Gerson, 1978; Harvey, 1984). Basin factors such as geology, topography, vegetation cover, slopes vs. streams, and bedrock vs. alluvial channels, may further complicate the effectiveness and recovery time associated with a given catastrophic flood (Costa, 1974; Patton and Baker, 1977; Newson, 1980).

The results of this study suggest that there may be an episodic tendency in the likelihood of occurrence of catastrophic events, due to variations in blocking and possible multiple modes of anomalous atmospheric behavior. The actual recurrence of another catastrophic event of equal magnitude in the same basin will always be an extremely rare event. However, during blocking episodes, it seems probable that a higher frequency of large floods will occur, even within the same watershed. Sequences of large floods can be an important factor in major channel changes and geomorphic effectiveness (Burkham, 1972; Beven, 1981), and these, too, are more likely to occur during periods dominated by blocking-related atmospheric circulation anomalies.

Another factor in the geomorphic response to catastrophic flood events is the length of time during which a river is at flood stage or out of its banks. Although this may be of minor importance for flash floods, in larger drainage basins, the duration of flooding can greatly affect the degree of erosion and overbank deposition (e.g., Kesel et al., 1974). Persistent anomalous atmospheric behavior, such as that which resulted in the long-lived 1973 Mississippi River flood, would be a major generator of this kind of geomorphic activity.

Finally, on an applied note, the geographic areas influenced by different types of circulation anomalies play a role in assessing a region's susceptibility to catastrophic flood hazards. In Texas, hydrogeomorphic and remote sensing techniques have been successfully used to delineate areas that have a high flood potential

(Baker, 1976). This circulation study has shown that, in addition to hydrogeomorphic factors, central and western Texas are especially prone to a unique type of anomalous atmospheric behavior that manifests itself in unusual combinations of meteorological processes operating together. In other parts of the country, different types of anomalies may dominate. The high occurrence of catastrophic flooding in the western states (Fig. 4) appears to be related to the strength of blocking in the North Pacific ocean, a major center of action for the circulation of the whole Northern Hemisphere. Great Plains catastrophic flash floods may be more attuned to the occurrence of mesoscale convective complexes, but extreme regional flooding in the upper and lower Mississippi Valley appears to be linked to the preferred development and anomalous persistence of a trough in the southern and western United States. Catastrophic flooding along the eastern seaboard is more likely to be affected by unusual Atlantic blocking patterns and/or sea surface temperature anomalies that influence both tropical and extratropical storm tracks.

This study has shown that some type of anomalous circulation pattern appears to be necessary to produce a catastrophic flood, however it must be emphasized that catastrophic floods do not always occur when circulation anomalies are present. Many other factors, such as basin physiography, thunderstorm cell movement, and antecedent soil moisture conditions, will ultimately determine whether or not a catastrophic flood will occur in response to an anomalous circulation pattern.

CONCLUDING REMARKS

The rare occurrences of both local and regional catastrophic flood events can be linked to anomalous atmospheric circulation patterns. These atmospheric circulation anomalies appear to develop in preferred locations, and to exhibit episodic behavior due to zonal and meridional tendencies in the large-scale atmospheric flow. The geomorphic effectiveness of catastrophic flooding may be significantly related, in both space and time, to variations in large-scale anomalous atmospheric activity. Nevertheless, it is obvious that even a very unusual anomaly pattern will not always produce a catastrophic flood. Furthermore, in the case of flash floods, the flooding response to a large-scale anomaly may be very localized, occurring in one basin and not another. The challenge awaits both hydroclimatologists and geomorphologists to continue to sort out the numerous variables which control the occurrence and the impact of catastrophic flooding.

ACKNOWLEDGEMENTS

I wish to thank John Costa for the use of his maximum flood data and for his helpful comments and suggestions. I am also most grateful for the constructive review comments provided by P. J. Barlien, W. M. Wendland, R. A. Muller, and J. M. Grymes. Mary Lee Eggart of the LSU Cartographic Section prepared all the figures.

REFERENCES

Andrews, J. F., 1965, The weather and circulation of July 1965: Monthly Weather Review, v. 93, p. 647-54.

Bailey, J. F., and Patterson, J. L., 1975, Hurricane Agnes rainfall and floods, June-July 1972: U. S. Geological Survey Professional Paper 924, 403 p.

Baker, V. R., 1976, Hydrogeomorphic methods for the regional evaluation of flood hazards: Environmental Geology, v. 1, p. 161-81.

Belt, C. B., Jr., 1975, The 1973 flood and man's constriction of the Mississippi River: Science, v. 189, p. 681-84.

Beven, Keith, 1981, The effect of ordering on the geomorphic effectiveness of hydrologic events, in Erosion and sediment transport in Pacific Rim steeplands: International Association of Hydrological Sciences (IAHS) Publication No. 132, p. 510-26.

Bosart, L. F., and Sanders, F., 1981, The Johnstown flood of July 1977: a long-lived convective system: Journal of the Atmospheric Sciences, v. 38, p. 1616-42.

Burkham, D. E., 1972, Channel changes of the Gila River in Safford valley, Arizona, 1846-1970: U. S. Geological Survey Professional Paper 655G, 24 p.

Carcena, F. and Fritsch, J. M., 1983, Focusing mechanisms in the Texas Hill Country flash floods of 1978: Monthly Weather Review, v. 111, p. 2319-32.

Charney, J. G., and DeVore, J. G., 1979, Multiple flow equilibria in the atmosphere and blocking: Journal of the Atmospheric Sciences, v. 36, p. 1205-16.

Chen, W. Y., 1983, The climate of spring 1983 -- a season with persistent global anomalies associated with El Niño: Monthly Weather Review, v. 111, p. 371-84.

Chin, E. H., Skelton, J., and Guy, H. P., 1975, The 1973 Mississippi River basin flood: Compilation and analyses of meteorologic, streamflow, and sediment data: U. S. Geological Survey Professional Paper 937, 137 p.

Costa, J. E., 1974, Response and recovery of a Piedmont watershed from tropical storm Agnes, June 1972: Water Resources Research, v. 10 p.106-12.

Costa, J. E., 1985, Interpretation of the largest rainfall-runoff floods measured by indirect methods on small drainage basins in the conterminous United States: Paper presented at the U.S.- China Bilateral Symposium on the Analysis of Extraordinary Flood Events, Nanjinj, China, October 1985.

Crippen, J. R., and Bue, C. D., 1977, Maximum floodflows in the conterminous United States: U. S. Geological Survey Water Supply Paper 1887, 52 p.

Dennis, A. S., Schleusener, R. A., Hirsch, J. H., and Koscielski, A., 1973, Meteorology of the Black Hills flood of 1972: Institute of Atmospheric Sciences Report 73-4, South Dakota School of Mines and Technology, Rapid City, 41 p.

Dole, R. D., 1986, The life cycles of persistent anomalies and blocking over the North Pacific, in Benzi, R., Saltzman, B., and Wiin-Nielsen, A, C., editors, Anomalous atmospheric flows and blocking: Advances in Geophysics, v. 29, p. 31-69.

Dzerdzeevskii, B. L., 1969, Climatic epochs in the twentieth century and some comments on the analysis of past climates, in Wright, H. E., editor, Quaternary geology and climate, v. 16 of the proceedings of the 7th congress, International Quaternary Association (INQUA): National Academy of Sciences publication 1701, Washington D. C., p. 49-60.

Eagleman, J. R., 1983, Severe and unusual weather: New York, Van Nostrand Reinhold Company, 372 p.

Egger, J., Metz, W. and Muller, G., 1986, Forcing of planetary-scale blocking anticyclones by synoptic-scale eddies, in Benzi, R., Saltzman, B., and Wiin-Nielsen, A, C., editors, Anomalous atmospheric flows and blocking: Advances in Geophysics, v. 29, p. 183-98.

Glancy, P. A., and Harmsen, L., 1975, A hydrologic assessment of the September 14, 1974, flood in Eldorado Canyon, Nevada: U. S. Geological Survey Professional Paper 930, 28 p.

Gupta, A. and Fox, H., 1974, Effects of high-magnitude floods on channel form: a case study in Maryland Piedmont: Water Resources Research, v. 10, p. 499-509.

Hansen, A. R., 1986, Observational characteristics of atmospheric planetary waves with bimodal amplitude distributions, in Benzi, R., Saltzman, B., and Wiin-Nielsen, A, C., editors, Anomalous atmospheric flows and blocking: Advances in Geophysics, v. 29, p. 101-33.

Hart, Steven, 1986, "Errant jet stream flings volley of storms at west coast": illustration appearing in The New York Times, 20 February 1986.

Harvey, A. M., 1984, Geomorphological response to an extreme flood: a case from southeast Spain: Earth Surface Processes and Landforms, v. 9, p. 267-79.

Helley, E. J., and LaMarche, V. C., Jr., 1968, December 1964, a 400-year flood in northern California: U. S. Geological Survey Professional Paper 600-D, p. D34-D37.

Horel, J. D., and Wallace, J. M., 1981, Planetary scale atmospheric phenomena associated with the Southern Oscillation: Monthly Weather Review, v. 109, p. 813-29.

Hoxit, L. R., Maddox, R. A., Chappell, C. F., and Brua, S. A., 1982, Johnstown-Western Pennsylvania storm and floods of July 19-29, 1977: U. S. Geological Survey Professional Paper 1211, 68 p.

Kesel, R. H., Dunne, K. C., McDonald, R. C., Allison, K. R., and Spicer, B. E., 1974, Lateral erosion and overbank deposition on the Mississippi River in Louisiana caused by 1973 flooding: Geology, v. 2, p. 461-64.

Knox, J. C., 1976, Concept of the graded stream, in Melhorn, W.N. and Flemal, R. C., editors, Theories of landform development: Binghamton, New York, Publications in Geomorphology, p. 170-98.

Knox, J. C., 1984, Fluvial responses to small scale climatic changes, in Costa, J. E., and Fleisher, P. J., Developments and applications of geomorphology: Berlin Heidelberg, Springer-Verlag, p. 318-42.

Knox, J. C., Bartlein, P. J., Hirschboeck, K. K. and Muckenhirn, R. J., 1975, The response of floods and sediment yields to climatic variation and landuse in the Upper Mississippi Valley: University of Wisconsin, Institute of Environmental Studies Report 52, 76 p.

Krueger, A. F., 1956, The weather and circulation of July 1965: Monthly Weather Review, v. 84, p.271-76.

Lamb, H. H., 1977, Climate: present, past and future: Climatic history and the future, v. 2: London, Methuen, 835 p.

Lindzen, R. S., 1986, Stationary planetary waves, blocking, and interannual variability, *in* Benzi, R., Saltzman, B., and Wiin-Nielsen, A, C., editors, Anomalous atmospheric flows and blocking: Advances in Geophysics, v. 29, p. 251-73.

Maddox, R. A., 1980, Mesoscale convective complexes: Bulletin of the American Meteorological Society, v. 61, p. 1374-87.

Maddox, R. A., Canova, F., and Hoxit, L. R., 1980, Meteorological characteristics of flash flood events over the western United States: Monthly Weather Review, v. 108, p. 1866-77.

Maddox, R. A., Carcena, F., Hoxit, L. R., and Chappell, C. F., 1977, Meteorological aspects of the Big Thompson flash flood of 31 July 1976: National Oceanic and Atmospheric Administration (NOAA) Technical Report ERL 388-APCL 41, 83 p.

Maddox, R. A., and Chappell, C. F., 1978, Meteorological aspects of twenty significant flash flood events: Preprints, Conference on flash floods: hydrometeorological aspects, American Meteorological Society, p. 1-9.

Maddox, R. A., Chappell, C. F., and Hoxit, L. R., 1979, Synoptic and meso-scale aspects of flash flood events: Bulletin of the American Meteorological Society, v. 60, p. 115-23.

Miller, J. F., Hansen, E. M., and Fenn, D. D., 1984, Probable maximum precipitation estimates -- United States between the continental divide and the 103rd meridian: National Oceanic and Atmospheric Adminis- tration (NOAA) Hydrometeorological Report No. 55, 245 p.

Namias, Jerome, 1962, Influences of abnormal surface heat sources and sinks on atmospheric behavior: The proceedings of the international symposium on numerical weather prediction in Tokyo, November 7-13, 1960, Meteorological Society of Japan, p. 615-27.

Namias, Jerome, 1963, Surface-atmosphere interactions as fundamental causes of drought and other climatic fluctuations: Arid Zone Research XX, Changes of climate, proceedings of Rome symposium, UNESCO and WMO, p. 345-59.

Namias, Jerome, 1973, Hurricane Agnes -- an event shaped by large- scale air-sea systems generated during antecedent months: Quarterly Journal of the Royal Meteorological Society, v. 99, p. 506-19.

Namias, Jerome, 1974, Longevity of a coupled air-sea continent system: Monthly Weather Review, v. 102, p. 638-48.

Namias, Jerome, 1979, Northern hemisphere seasonal 700 mb height and anomaly charts, 1947-1978, and associated North Pacific sea surface temperature anomalies: California Cooperative Oceanic Fisheries Investigations (CALCOFI) Atlas No. 27, Marine Life Research Group, Scripps Institution of Oceanography, LaJolla, California, 275 p.

Namias, Jerome, 1981, Teleconnections of 700 mb height anomalies for the Northern Hemisphere: California Cooperative Oceanic Fisheries Investigations (CALCOFI) Atlas No. 29, Marine Life Research Group, Scripps Institution of Oceanography, LaJolla, California, 265 p.

Newson, Malcolm, 1980, The geomorphological effectiveness of floods -- a contribution stimulated by two recent events in mid-Wales: Earth Processes, v. 5, p. 1-16.

Oliver, J. E. and Fairbridge, R. W., 1987, The encyclopedia of climatology: New

York, Van Nostrand Reinhold Company, 986 p.

Orlanksi, Isidoro, 1975, A rational subdivision of scales for atmospheric processes: Bulletin of the American Meteorological Society, v. 56, p. 527-30.

Patton, P. C. and Baker, V. R., 1977, Geomorphic response of central Texas stream channels to catastrophic rainfall and runoff, *in* Doehring, D. O., ed., Geomorphology in arid regions: Fort Collins, Colorado, Publications in Geomorphology, p. 189-217.

Pierrehumbert, R. T., 1986, The effect of local baroclinic instability on zonal inhomogeneities of vorticity and temperature, *in* Benzi, R., Saltzman, B., and Wiin-Nielsen, A, C., editors, Anomalous atmospheric flows and blocking: Advances in Geophysics, v. 29, p. 165-82.

Posey, J. W., 1965a, The weather and circulation of June 1965: Monthly Weather Review, v. 93, p. 573-78.

Posey, J. W., 1965b, The weather and circulation of December 1964: Monthly Weather Review, v. 93, p. 189-94.

Quiroz, R. S., 1983, The climate of the "El Niño" winter of 1982-83 -- a season of extraordinary climatic anomalies: Monthly Weather Review, v. 111, p. 1685-1706.

Reinhold, Brian, 1987, Weather regimes: the challenge in extended- range forecasting: Science, v. 235, p. 437-41.

Resio, D. T., and Hayden, B. P., 1975, Recent secular variations in mid-Atlantic winter extratropical storm climate: Journal of Applied Meteorology, v. 14, p. 1223-34.

Rex, D. F., 1950a, Blocking action in the middle troposphere and its effect upon regional climate, I. An aerological study of blocking action: Tellus, v.2, p. 196-211.

Rex, D. F., 1950b, Blocking action in the middle troposphere and its effect upon regional climate, II. The climatology of blocking action: Tellus, v. 2, p. 275-301.

Rodgers, D. M., Howard, K. W., and Johnston, E. C., 1983, Mesoscale convective complexes over the United States during 1982: Monthly Weather Review, v. 111, p. 2363-69.

Rodier, J. A., and Roche, M., 1984, World catalogue of maximum observed floods: International Association of Hydrological Sciences (IAHS-AISH) Publication No. 143, 354 p.

Shutts, G. J., 1986, A case study of eddy forcing during an Atlantic blocking episode, *in* Benzi, R., Saltzman, B., and Wiin-Nielsen, A, C., editors, Anomalous atmospheric flows and blocking: Advances in Geophysics, v. 29, p. 135-62.

Stewart, J. H. and LaMarche, V. C., Jr., 1967, Erosion and deposition produced by the flood of December 1964 on Coffee Creek, Trinity County, California: U. S. Geological Survey Professional Paper 422-K, 22 p.

Sutera, Alfonso, 1986, Probability density distribution of large-scale atmospheric flow, *in* Benzi, R., Saltzman, B., and Wiin-Nielsen, A, C., editors, Anomalous atmospheric flows and blocking: Advances in Geophysics, v. 29, p. 227-49.

U. S. Department of Commerce, 1973, Final report of the disaster team on the events of Agnes: National Oceanic and Atmospheric Administration (NOAA) Natural

Disaster Survey Report 73-1, 45 p.

Waananen, A. O., Harris, D. D. and Williams, R. C., 1971, Floods of December 1964 and January 1965 in the far western states: U. S. Geological Survey Water-Supply Paper 1866-A, 265 p.

Wagner, A. J., 1971, Weather and circulation of July 1971: Monthly Weather Review, v. 99, p. 800-06.

Wallace, J. M., 1983, The climatological mean stationary waves: observational evidence, in Hoskins, B., and Pearce, R., editors, Large-scale dynamical processes in the atmosphere: London, Academic Press, p. 27-53.

Wallace, J. M., and Gutzler, D. S., 1981, Teleconnections in the geopotential height field during the Northern Hemisphere winter: Monthly Weather Review, v. 109, p. 785-812.

White, W. B. and Clark, N. E., 1975, On the development of blocking ridge activity over the central North Pacific: Journal of the Atmospheric Sciences, v. 23, p. 489-502.

Wolman, M. G., and Gerson, R., 1978, Relative scales of time and effectiveness of climate in watershed geomorphology: Earth Surface Processes, v. 3, p. 189-208.

Yarnal, Brent, 1985, Extratropical teleconnections with El Niño/Southern Oscillation (ENSO) events: Progress in Physical Geography, v. 9, p. 315-52.

3
El Niño and annual floods in coastal Peru

Peter R. Waylen and César N. Caviedes

ABSTRACT

A three component mixed Gumbel distribution is found to satisfactorily model the observed annual flood frequencies of rivers in coastal Peru which display highly variable annual peak flood characteristics corresponding to three sets of ocean-atmosphere conditions. Exceptionally warm waters (El Niño) cause extensive heavy rains particularly in the northern lowlands; unusually cold waters (anti-El Niño) restrict both the quantity and distribution of precipitation everywhere. Model parameters reveal marked spatial trends in the severity and extent of flooding during any set of offshore conditions and conspicuous alterations in the pattern of regional flood characteristics with elevation. The analysis suggests a strong negative correlation of rainfall-runoff relationships between northern and southern Peruvian rivers.

INTRODUCTION

El Niño-Southern Oscillation (ENSO) events have been extensively documented in recent years. Disturbed meteorological conditions resulting in heavy rains (Mugica, 1984) and leading to extensive flooding (Caviedes, 1984) in arid coastal Peru are coupled with warm waters of El Niño and the unusual invasion of humid equatorial air masses. The consequences of these events on the surface hydrology of coastal Peruvian rivers and the probabilities of occurrence of extreme discharges were analyzed in a previous paper by Waylen and Caviedes (1986) who suggested that the effect of basin hypsometry on the discharges measured at low-lying gauging stations be examined closely. Transmission losses in the arid coastal environment are thought to be high, although unmeasurable, because the real inputs in the higher segments of the river basin were unknown. Utilizing government and supplementary unpublished data, this paper reviews the significance of ENSO events to the pattern of regional catastrophic flooding and the influence of hypsometry in the distribution of precipitation and associated runoff for all coastal rivers of Peru.

Castrophic flooding in this region has been extensively documented in historic accounts since the arrival of the Spaniards in Peru in 1521 (Hamilton and Garcia, 1986). There is also strong archaeological evidence that floods affected Chimu agricultural communities along coastal Northern Peru, triggering a succession of cultural collapses and abandonnment of some settlements, agricultural lands and hydraulic structures (Nials et al., 1979).

STUDY AREA AND DATA

The drainage pattern of coastal Peru is characterized by numerous streams which flow sub-parallel west-southwest towards the Pacific Ocean from the Cordillera Negra, the western- most range of the Andes. These streams originate in interior basins or at the foothills of the partially glaciated Cordillera Blanca (Fig. 1). The rivers are extremely allogenic as the contributions from their lower courses is minimal. Most rivers rise at elevations of over 3000 m a.s.l., or occasionally over 5000 m, then descend rapidly to the coastal plain below 100 m. The coastal plain is widest (200 km) in northern Peru, diminishing southward to approximately 30 km at the latitude of Trujillo (8°S). From Trujillo to the southern boundary of Peru, the coastal plain is so narrow that steep cliffs become the most common feature of the Peruvian coast. In a few places the cliffs are interrupted by narrow alluvial fans or by marine terraces cut on Tertiary deposits such as the Tablazo de Pica, one of the few flat coastal areas south of Lima.

The cold Peru Current induces stable conditions in the lower atmosphere resulting in an inversion layer between 600 m and 900 m elevation along the central and southern coast of Peru. This causes high humidity and constant fog in the air between the ocean surface and the inversion layer, but inhibits rain formation along the coast. Thus, short-lived rain episodes are possible only when,

1) an extreme ocean warming breaks the stabilizing effects of the Peru Current during the summer months of years with El Niño, or
2) when powerful frontal systems from the subtropical Pacific are able to penetrate as far north as the latitude of Lima (12°S) during the winter time.

Along the western slopes of the Cordillera Negra, annual precipitation depends strongly on elevation, ranging from less than 10 mm on the coast to over 1500 mm in the northern Andes, (Servicio de Agrometeorologia e Hidrologia [SENAMHI], 1962; Francou and Pizarro, 1985). In the Peruvian Andes bordering southern Ecuador, the isohyets exhibit a clear north-south trend as the influence of equatorial air masses increase, whereas to the south the distribution of precipitation follows the altimetric levels of the Andes (i.e., arranged in belts parallel to the

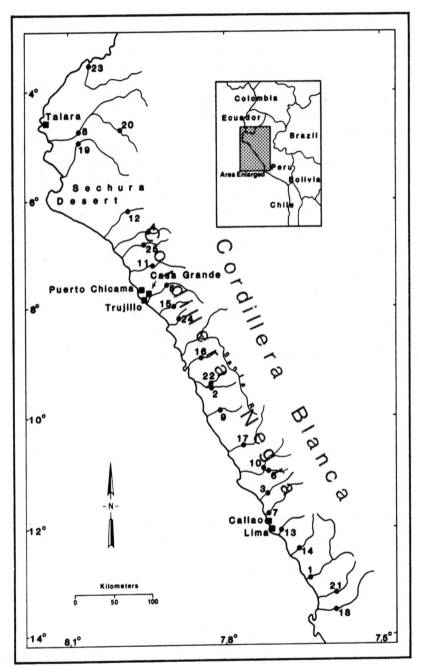

Figure 1. Major rivers in the study area of coastal northern and central Peru. Also identified are streamflow gauging stations for the study area which are are listed in Table 1.

coast line). In the northern rivers flooding is highly seasonal and the result of summer (December-May) rainfall, which may vary considerably from year to year. Annual streamflow is augmented by small snowmelt contributions in only a few small basins located in the interior of the Cordilleras. South of the Rio Santa (9°S) the scarce summer rains occur more frequently between January and April, and streamflow exhibits a marked March maximum. South of the Rio San Juan (13.5°S) the annual range in discharge becomes less pronounced, as a result of the increased contribution of winter frontal rains to discharge (Lettau and Lettau, 1978).

Due to their proximity to the major fluctuations in ocean-atmosphere conditions and associated El Niño events, the northern rivers suffer the maximum intensity of catastrophic flooding as revealed by the events of 1982-83 (Caviedes, 1984). The rivers of the south exhibit less pronounced flood levels during corresponding abnormal years. The annual flood series of twenty-five rivers, listed in Table 1, draining areas of between 800 and 13000 km^2 and located between 3°S and 14°S are analyzed. The selected stations provide more than 900 station-years of historic record which are available from the Peruvian Government (SENAHMI, 1962) and from local water management offices. There may be errors in flood measurements, particularly during the most intense events, but such errors are unavoidable even with the most modern technology and suitable hydraulic conditions (Potter and Walker, 1981). The majority of the stations are properly established at or near major urban centers and data were published under the auspices of the United Nations until 1960. Records affected by recent flow regulation schemes have been excluded from the analysis.

METHODS

Annual flood is defined as the largest daily discharge, Q', observed in a year:

$$Q' = \max |Q_i| \quad i = 1, 2, \ldots 365 \tag{1}$$

where Q is daily discharge and i denotes dates in Julian days. The magnitude of annual floods has traditionally been modelled by the Gumbel distribution, with a probability function:

$$F_T(Q' \geq x) = 1 - \exp(-\exp[-\alpha\{x - \beta\}]) \tag{2}$$

where, by the method of moments

$$\alpha = 1.281/\sigma \tag{3}$$

and

$$\beta = \mu - 0.45\sigma \tag{4}$$

μ and σ^2 are the population mean and variance respectively (Lowery and Nash, 1970). The Gumbel distribution generally fits the observed data satisfactorily when

Table 1. Locations and sizes of drainage basins in north coastal Peru.

No.	River	Gauge Location	Lat. (°.'S)	Long. (°.'W)	Area (km^2)	Years of Record
1.	Cañete	Toma Imperial	13.00	79.09	4900	33
2.	Casma	Puente Casma	09.28	78.17	2082	29
3.	Chancay	Santo Domingo	11.17	76.58	1360	39
4.	Chancay	Carhuaqero	06.38	79.29	2860	59
5.	Chicama	Salinar	07.04	78.58	4087	62
6.	Chico	Sayan	11.09	77.12	750	18
7.	Chillon	Magdalena	11.48	77.00	1849	40
8.	Chira	Sullana	04.53	80.42	12712	22
9.	Huarmey	Puente Huarmey	10.04	78.10	2046	28
10.	Huaura	Sayan	11.08	77.12	2630	31
11.	Jequetepeque	Ventanillas	07.17	79.17	4809	52
12.	Leche	Puchaca	06.24	79.30	914	51
13.	Lurin	Manchay	12.08	76.51	1332	22
14.	Mala	La Capilla	12.25	76.24	1934	22
15.	Moche	Quirihuac	08.02	78.50	1873	60
16.	Nepeña	San Jacinto	09.10	78.15	1335	25
17.	Pativilca	Alpas	10.37	77.31	4050	24
18.	Pisco	Letrayoc	13.40	75.50	3487	34
19.	Piura	Piura	05.13	80.38	7836	53
20.	Quiroz	Paraje Grande	04.26	80.15	1793	25
21.	San Juan	Conta	13.27	75.58	2966	35
22.	Sechin	Puente Casma	09.29	78.17	776	15
23.	Tumbes	Tumbes	03.34	80.28	4565	12
24.	Viru	Huacapongo	08.25	78.40	803	34
25.	Zaña	El Batan	06.50	79.18	845	54

the annual floods are produced by a single hydrologic process (Todorovic, 1978). It performs less well when two or more processes are responsible for the generation of annual floods. Waylen and Caviedes (1986) found that a three component mixed Gumbel model of the form

$$F_T(Q' \geq x) = \rho_1(1- \exp(-\exp[- \alpha_1\{x-\beta_1\}])) +$$
$$\rho_2(1- \exp(-\exp[- \alpha_2\{x- \beta_2\}])) +$$
$$\rho_3(1- \exp(-\exp[- \alpha_3\{x- \beta_3\}])) \qquad (5)$$

performed adequately in this region. The subscripts refer to the three distinct flood generating conditions, the ρ's are estimated as the relative frequencies of each annual set of conditions and the parameters α and β are estimated from the appropriate sub-sample using Equations (3) and (4). The annual at-a-station variability may then be summarized by the mean and variance of floods generated by each process and the relative frequency with which each process is observed.

This paper will proceed by the following steps.

(a). Identify years in which El Niño, normal and anti-El Niño conditions prevail in the southeastern Pacific.

(b). Extract flood sub-samples for each of the appropriate conditions from the annual flood series of rivers in the region.

(c). Estimate statistics and parameters of each sub-population.

(d). Derive a mixture distribution using parameter estimates from (c) and relative frequencies from (a) and compare to the observed data.

(e). Examine the spatial variability of the model parameters.

RESULTS

Identification of Processes

Conceptually there should exist a strong link between sea-surface temperatures, annual precipitation and flood discharges in this region. Examination of the synchronous period of deviations of sea surface temperatures at Puerto Chicama and hydrometeorological data at Casa Grande and the Chicama River (Fig. 2) indicates that the relationship is not as consistent as might be expected. Because the interactions at the ocean-atmosphere/land interfaces are not known or fully understood in this region, the sea surface temperatures do not provide a good *a priori* basis for the classification of annual floods by generating process (Cunnane, 1985). Not even the discharges themselves can be used as indicators of the generating processes involved. Figure 3 depicts the empirical estimates of the return periods of annual floods in the study area of four years, widely considered to be influenced by El Niño conditions (Schweigger, 1941; Merriman, 1955; Posner, 1954; Wyrkti et al., 1976). It can be seen that there is considerable variability in the severity of flooding between contiguous basins.

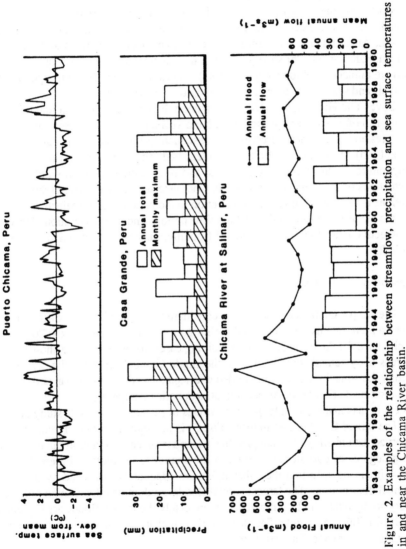

Figure 2. Examples of the relationship between streamflow, precipitation and sea surface temperatures in and near the Chicama River basin.

In the absence of any acceptable single classificatory hydrologic variable the annual series are divided on the basis of oceanic-atmospheric anomalies derived from a search of the pertinent literature. Schweigger (1959) singles out El Niño years as those which concurrently produce heavy summer rains in Northern Peru, abnormally high sea surface temperatures at Puerto Chicama and Callao, the depletion of fish stocks and the starvation of sea birds. Quinn and Burt (1970) utilize barometric and pluviometric criteria to distinguish El Niño years. If summer pressures at Darwin, Australia rise and rainfall on several equatorial Pacific islands (Canton, Malden, Fanning and Christmas) increase, then anomalous ocean-atmosphere conditions are induced. Bellido Delgado (1983) made similar classifications upon the basis of air and sea surface temperatures along the Peruvian coast.

Covey and Hastenrath (1978) were among the first to attempt an identification of anti-El Niño years, as opposed to El Niño years, on the basis of sea surface temperatures and atmospheric pressures. Subsequently Barnett (1984), Van Loon and Shea (1985) and Philander (1985) have identified "cold episodes" in the sea surface temperatures of the southeastern Pacific and related them to zonal winds and the fluctuations of the Southern Oscillation. Complete accordance between these classifications is very difficult given the variation in the natural criteria selected, the geographic locations of the studies and the time periods under consideration. However an objective classification, in terms of flooding, may be extracted and is listed in Table 2.

Probability distributions

The simple Gumbel distribution is considered to be significantly different from the observed data using a modified Kolmogorov-Smirnov test at the 0.20 level of significance for three rivers. In all remaining cases the observed data show varying degrees of systematic deviation from the single Gumbel function (Fig. 4). Subsequently the annual series are subdivided and separate estimates of mean and standard deviation made (Table 3). As oceanic-atmospheric changes are at so large a spatial scale it is reasonable to apply the same values of ρ_i across the whole study area.

The mixed Gumbel distribution (Fig. 5) is not significantly different from the observed data in any of the sampled basins, and is successful in reproducing the varying degrees of curvature observed in the annual series plots in Figure 4. In all cases, it can be seen that the subsamples are adequately described by the single Gumbel distribution. The largest annual floods are dominated, although not exclusively so, by events in El Niño years. Conversely the lowest floods are dominated by anti-El Niño years and reflect their statistical characteristics on the lower end of the plot of the mixed Gumbel distribution.

Spatial variability

When fitted with the mixed Gumbel distribution the annual flood series in this region show some strong spatial trends. In order to make spatial comparisons it is

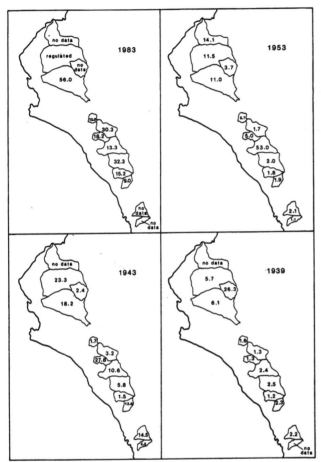

Figure 3. Observed return periods (years) of annual floods over the northern portion of the study area during four years influenced by El Niño.

necessary to standardize the annual flood parameters. The flow data are expressed as specific runoff ($m^3s^{-1}km^{-2}$) and listed in Table 3. In general flood discharges increase towards the north and east (Fig. 6) corresponding to the increased influence of equatorial conditions and orographic enhancement of precipitation. As might be expected the pattern of annual flood size is controlled by elevation in normal and anti-El Niño years. In El Niño years the pattern becomes more strongly influenced by latitude, as the equatorial air masses related to the Inter-tropical Convergence Zone (ITCZ) and the Equatorial Counter-current move further South.

The varying curvatures of the annual flood series plots (Fig. 5) is a measure of the proportional differences in the flood statistics between each set of generating conditions. Figure 7 depicts the ratio of the mean annual flood in El Niño years relative to those of anti-El Niño years. The greatest change in conditions appears to be associated with basins with centroids in northern coastal locations and diminishes in basins whose centroids lie southward and into the mountains. The dimi-

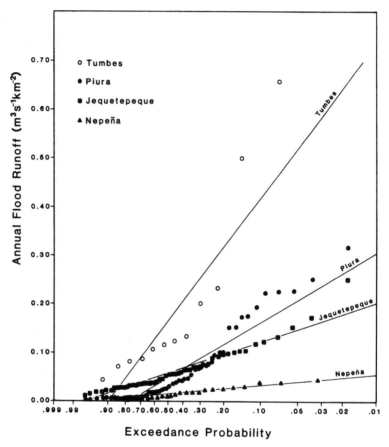

Figure 4. Examples of the observed annual flood frequencies of four rivers and the estimated Gumbel funtion. Data are standardized to units of specific discharge $(m^3 s^{-1} km^{-2})$ for comparison purposes.

nishing change of flood characteristics with elevation is consistent with Mugica's (1984) observations of precipitation during the 1982-83 El Niño event. A comparison of precipitation totals in the first six months of 1983 to mean annual precipitation at stations in the Piura (19) and Chira (8) River basins revealed a 40- to 60-fold increase in coastal regions compared to a doubling in the mountains. At northern locations, orographic precipitation will be induced at lower elevations in the warmer air masses, while the southerly movement of the ITCZ will greatly enhance convective activity.

The difference in mean annual flood size between years of various oceanic-atmospheric conditions may also be a function of groundwater transmission losses

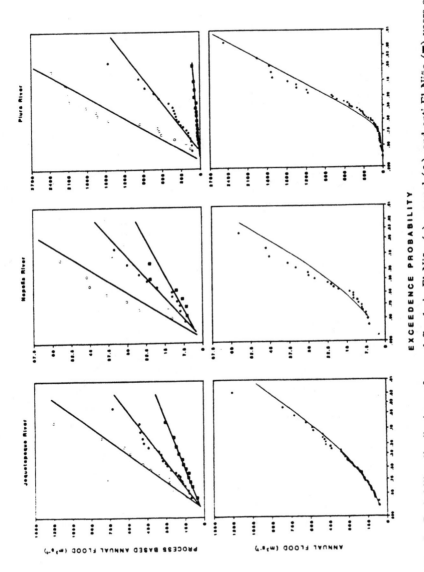

Figure 5. Probability distributions of annual floods in El Niño (o), normal (•), and anti-El Niño (■) years on three sample rivers. The corresponding mixture models are shown on the bottom row.

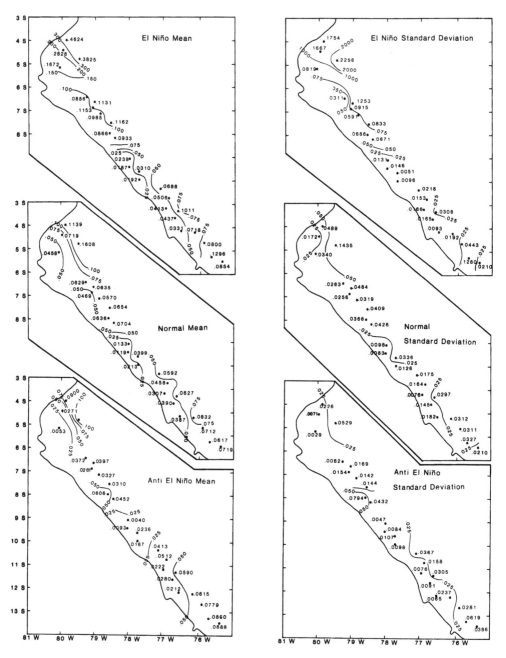

Figure 6. Areal variation of mean and standard deviation of annual floods $(m^3s^{-1}km^{-2})$ due to the three sets of ocean-atmosphere conditions in northern Peru.

Table 2. Annual classification of prevailing oceanic-atmospheric conditions (1911-1983).

El Niño	Normal		Anti-El Niño	
1911	1915	1940	1912	1967
1925	1916	1942	1913	1968
1926	1917	1944	1914	1970
1932	1919	1946	1924	1979
1933	1919	1946	1924	1979
1939	1920	1949	1930	1980
1941	1922	1952	1937	
1943	1923	1955	1947	
1953	1927	1956	1948	
1957	1928	1959	1950	
1965	1929	1969	1951	
1972	1931	1971	1954	
1973	1934	1975	1960	
1977	1935	1976	1963	
1978	1936	1981	1964	
1983	1938	1982	1966	

along the arid coastal plain. The larger northern basins (notably the Piura and Chira) have a greater proportion of their basin areas on the coastal plain (Fig. 9). During normal and anti-El Niño years, when summer precipitation on these regions is low, transmission losses will be high, while during El Niño years coastal rainfall will reduce, or reverse, the direction of groundwater flow. In the south, the absence of a coastal plain does not permit the collection of excessively large quantities of precipitation and the magnitudes of floods under each set of conditions do not vary markedly.

Figure 8 illustrates that there is a fairly consistent power function relationship between mean annual flood and drainage basin area in El Niño years. Two basins stand out as positive residuals; the Tumbes (23) and Quiroz (21) to the north of the study area. The influence of equatorial and mountainous conditions on these northern basins has already been mentioned, in this case they reveal higher flood runoffs than their more southerly counterparts. There also appears to be a group of basins, Casma (2), Huarmey (9), Nepeña (16), Sechin (22), Chico (6) and Lurin (13), with mean flood discharges below those which might be expected from the regional relationship. A possible explanation may be found from an examination of Figure 1. which reveals that these basins do not extend as far into the central Andes as the

Table 3. Estimates of the mean and standard deviation of annual floods under differing ocean-atmospheric conditions.

No. River	El Niño		Normal		Anti-El Niño	
	Mean	St. Dev.	Mean	St. Dev.	Mean	St. Dev.
1. Cañete	.0800	.0443	.0712	.0311	.0779	.0281
2. Casma	.0310	.0051	.0399	.0336	.0236	.0107
3. Chancay	.1011	.0308	.0627	.0297	.0590	.0305
4. Chancay	.1131	.1253	.0635	.0263	.0397	.0169
5. Chicama	.1162	.0833	.0654	.0409	.0311	.0144
6. Chico	.0413	.0167	.0307	.0076	.0222	.0076
7. Chillon	.0437	.0165	.0390	.0145	.0280	.0091
8. Chira	.2828	.1667	.0719	.0172	.0271	.0074
9. Huarmey	.0192	.0096	.0213	.0124	.0167	.0098
10. Huaura	.0506	.0153	.0458	.0164	.0412	.0158
11. Jequetepeque	.0984	.0591	.0570	.0319	.0327	.0142
12. Leche	.0886	.0311	.0629	.0484	.0373	.0082
13. Lurin	.0331	.0093	.0387	.0182	.0212	.0649
14. Mala	.0718	.0192	.0832	.0312	.0615	.0237
15. Moche	.0866	.0686	.0636	.0366	.0607	.0793
16. Nepeña	.0238	.0131	.0133	.0074	.0090	.0047
17. Pativilca	.0688	.0218	.0591	.0175	.0512	.0367
18. Pisco	.0854	.0210	.0719	.0210	.0688	.0386
19. Piura	.1672	.0819	.0458	.0340	.0053	.0029
20. Quiroz	.3825	.2256	.1608	.1435	.1112	.0529
21. San Juan	.1296	.1280	.0617	.0327	.0860	.0619
22. Sechin	.0187	.0146	.0195	.0083	.0093	.0084
23. Tumbes	.4624	.1754	.1139	.0489	.0899	.0226
24. Viru	.0933	.0671	.0704	.0426	.0452	.0432
25. Zaña	.1153	.0916	.0469	.0256	.0261	.0154

*All data are in units of specific runoff ($m^3 s^{-1} km^{-2}$).

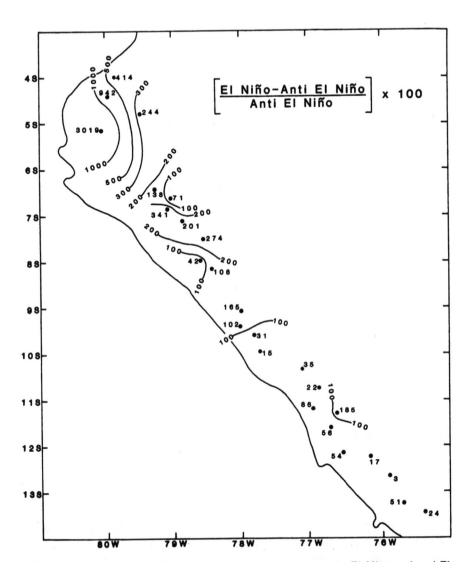

Figure 7. Areal variation of the ratio of mean annual floods in El Niño and anti-El Niño years.

other basins and that this western most branch of the Andes, the Cordillera Negra, is the driest of all. In the central region of the study area (9-10°S) a large portion of the Andes is drained by the Santa River which flows parallel (southeast-northwest) to the major mountain ridges and orthogonal to the regional drainage.

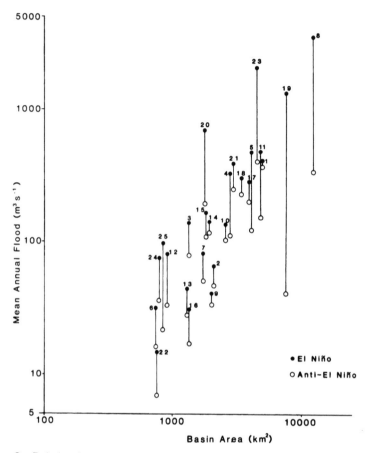

Figure 8. Relation between basin area and mean annual flood size in El Niño and anti-El Niño years.

The relative importance of orographic precipitation during El Niño years is therefore considerably reduced in the Sechin, Huarmey, Nepeña and Casma Rivers, and non-existent during normal or anti-El Niño years as they drain only the western slopes of the Cordillera Negra. A similar explanation may be proposed for the anamalous behavior of the Chico River which is "beheaded" by the main trunk stream of the Huaura (10), and for the Lurin which drains the region below the headwaters of both the Mala (14) and Cañete (1).

When mean annual floods in anti-El Niño years are plotted on the same graph a very marked and consistent regional shift in basin area/runoff relationship is observed. Tumbes and Quiroz fall along the line described by the majority of basins in El Niño years. These in turn appear to adopt the El Niño characteristics of the drier "beheaded" basins. Such an observation suggests extensive shifts in regional flood hydrology result from upset oceanic-atmospheric conditions. The Piura River (19) shows the most dramatic change in conditions and is the major exception to this generalization. In anti-El Niño years its mean annual flood is more in keeping with that of the "beheaded" basins. The Piura basin occupies a very large area of arid coastal plain and the headwaters correspond to one of the lowest segments of the Andes, seldom exceeding 3000 m in elevation. Within each of the broad bands described in Figure 8, there exists individual variability in the degree of change in mean annual flood size from one basin to another. The Chira (8), in El Niño years, approaches the characteristics of its northern neighbors, but during anti-El Niño years shows a considerably diminished mean due to a large proportion of low lying land. In contrast the Huarmey (9) exhibits minimal change between conditions and the Cañete (1) hardly any at all. These observations have important implications for the regional hydrology of Peru, namely that the increased pluviosity associated with ENSO events does not effect all regions of coastal Peru equally, but that it, and therefore flooding, is dependent upon both latitude and elevation.

Mugica (1984) observed a negative correlation between elevation and increases in precipitation in El Niño years and average conditions in the Departamento de Piura and locally this appears to be borne out by the examination of flood characteristics. In order to evaluate the role of elevation in determining the degree of

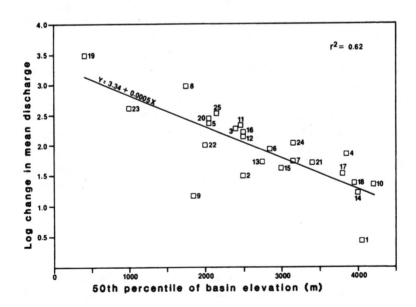

Figure 9. Linear regression of median basin elevation and the log of change in mean annual flood ([El Niño mean - anti-El Niño mean]/anti-El Niño * 100 %)

change in flood conditions caused by upset atmosphere-ocean circulation the median elevation of each basin was calculated from hypsometric curves and regressed against the percentage change in mean annual flood (Fig. 9). Given the crudity of the measures and the areal scope of the study a remarkably strong functional relationship exists (r^2 = 0.62), whose form

$$\text{Log(Change)} = 3.34 - .0005 \text{ MEDIAN (m)} \qquad (6)$$

is suggestive of the observed negative physical correlation reported by Mugica (1984).

There are two very pronounced residuals from the overall relationship, the Cañete (1) and Huarmey (9). Figure 1. shows that the northern headwaters of the Cañete drain a much larger area of the Andean interior than any other sampled basin. The unexpectedly small change in flood conditions in the Cañete would indicate that in the most southerly portions of the study area the headwaters of basins may be experiencing droughts while only the lower, coastal, portions of the same basin may be receiving excessive rains. Caviedes (1982) noted that annual precipitation along the northern coast of Peru correlated negatively with records in the highlands of Southern Peru and the Bolivian Altiplano. Further support for this hypothesis can be found in an examination of the record of the Santa River, which had been excluded from any previous analysis due to its lack of representativeness of coastal conditions. Comparison of El Niño and Anti-El Niño conditions reveals lower mean annual floods in the former years than the latter. Richey and Dunne (1986) have intimated that such a correlation between coastal Peru and the interior of the continent is also present in the stage record of the Amazon River at Manaus.

Other independent hydrometeorological records support the use of latitudes 10-12°S as an approximate southern limit to extensive flooding in response to ENSO events, although there will still be increased local rainfall in coastal regions. Short records of stage of the Chicrin River (Fig. 10) at 3000 m in the headwaters of the Chancay (3) (provided by I. Lausent, Centre National de Recherche Scientifique CNRS-CEGET, Bourdeaux) and precipitation at Chuquibambilla (3971 m; 14° 47'S), in the central Peruvian Andes (courtesy of B. Francou, CNRS-CENES, Caen) show strong negative correlation with marked El Niño years and large increases in anti-El Niño years.

CONCLUSIONS

The proposed mixture model provides a good fit to the observed distributions of annual floods on rivers in the study area, portions of which may annually experience highly variable flood conditions and extremely large "outliers". The improvement in fit of a nine parameter distribution over the standard Gumbel distribution is not surprising. However, there is very strong independent physical evidence for subdividing the annual flood series into three sub-samples. Subjectivity in classifi-

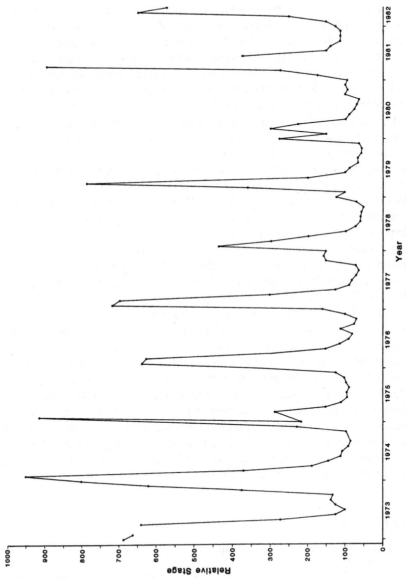

Figure 10. Time Series of stage on the Chicrin River (lower).

cation is removed, and the need for parameter estimation at each station is considerably reduced, by the reasonable assumption that changes in atmosphere/ocean conditions are regional and that the proportions of the mixtures are therefore constant.

Examination of floods at the regional level reveals temporal and spatial patterns of flooding that are otherwise obscured on an individual basis. The majority of basins, particularly in the north, reveal marked differences between flooding in each of the sub-samples, however the probabilistic approach allows the incorporation of such considerable variation within each sample. The variability diminishes south of a latitude of about 11°S. In the northern regions the changing spatial nature of the between-sample characteristics clearly reflects the physical atmosphere/ocean processes dominant at the time. In any year the effects of El Niño reveal a very patchy distribution from one basin to the next. However, when considered together each process reveal very pronounced spatial trends.

This may be a consequence of the fact that El Niño events are not the result of massive southward invasions of warm water but rather the result of regionally restricted intrusions of warm waters offshore Peru in isolated pockets of various numbers and size depending on the severity of the El Niño event (Smith, 1983). This hypothesis is in keeping with this study's observations of the magnitude of flooding and supports the adoption of a stochastic approach.

The importance of elevation in controlling the degree of variability in flooding experienced between El Niño and anti-El Niño conditions has been identified and crudely quantified by the techniques employed. There are deviations from the proposed relationship notably in the south portion of the study area that support the hypothesis of an inverse hypsometric relationship between conditions along the north coast and those above an elevation of 3000 m in the south.

These conclusions have immediate implications for future damage reduction programs, land use planning and construction projects in the areas of catastrophic flooding. The continental-scale impacts of atmosphere-ocean disturbances on flood hydrologic responses are suggested at the margins of the study area. These impacts are consistent with other local and regional hydrometeorological indices reported elsewhere. Historically there may also be archaeological significance in the explanation of abandoned coastal settlements, inter-basin transfers of water and migration of peoples to the highlands (Moseley, 1983).

REFERENCES

Barnett, T. M., 1984, Prediction of El Niño 1982-83: Monthly Weather Review, v. 112, p. 1403-1407.

Bellido Delgado, R., 1983, Metodo empirico para pronosticar la ocurrencia del fenomeno del Niño: Documenta (Lima) 11, p. 23-28.

Caviedes, C. N., 1982, On the genetic linkages of precipitation in South America. In: Fortschritte landschaftsoekologischer und klimatologischer Forschungen in den Tropen: Freiburger Geographische Hefte, No. 18, p. 55-77.

Caviedes, C. N., 1984, El Niño 1982-83: Geographical Review, v. 74, p. 267-290.

Covey, D. L. and Hastenrath, S., 1978, The Pacific El Niño phenomenon and the Atlantic circulation: Monthly Weather Review, v. 106, p. 1280-1287.

Cunnane, C., 1985, Factors affecting choice of distribution for flood series: Hydrological Sciences Journal, v. 30, p. 23-37.

Francou, B. and Pizarro, L. 1985, El Niño y la sequia en los altos Andes Centrales: Peru y Bolivia: Bulletin de l'Institute Francais d'Etudes Andines, v. 14, p. 1-18.

Hamilton, K. and Garcia, R. R., 1986, El Niño/Southern Oscillation events and their associated midlatitude teleconnections: Bulletin of the American Meteorological Society, v. 67, p. 1354-1361

Lettau, H. H. and Lettau, K., 1978, Exploring the world's driest climate: Institute for Environmental Studies, University of Wisconsin, Report 101, 264 pp.

Lowery, M. D. and Nash, J. E., 1970, A comparison of methods of fitting the double exponential distribution: Journal of Hydrology, v. 10, p. 259-275.

Merriman, D., 1955, El Niño brings rain to Peru: American Scientist, v. 43, p. 63-76.

Moseley, M. E., 1983, The good old days were better: Agrarian collapse and tectonics: American Anthropologist, v. 85, p. 773-799.

Mugica, R., 1984, Departamento de Piura rainfall in 1983: Tropical Ocean- Atmosphere Newsletter, v. 28, p. 7.

Nials, F. L. et al., 1979, El Niño: The catastrophic flooding of Coastal Peru: Field Museum of Natural History, Bulletin 50, p. 4-9.

Philander, S. G. H., 1985, El Niño and La Niña: Journal of Atmospheric Sciences, v. 42, p. 2652-2662.

Posner, G. S., 1954, The Peru Current: Scientific American, v.190, p. 66-71.

Potter, K. W. and Walker, J. W., 1981, A model of discontinuous measurement error and its effect upon the probability distribution of flood discharge measurements: Water Resources Research, v. 17, p. 1505-1509.

Quinn, W. H. and Burt, W. V., 1970, Prediction of abnormally heavy precipitation over the Equatorial Pacific dry zone: Journal of Applied Metereology, v. 6, p. 20-27.

Richey, J. E. and Dunne, T., 1986, Regional-scale variability in the hydrology of the Amazon River: Eos, v. 67, p. 935.

Schweigger, E. H., 1941, Studies of the Peru Coastal Current with reference to the extraordinary summer of 1939: Proc. of Sixth Pacific Science Congress, v. 3, p. 177-195.

Servicio Agrometereorologico e Hidrologico, 1962, Boletin de Estadistica Meteorologica e Hidrologica: Nos. 2,3,4,7,8,9, 11,12,13,14,18 and 19. Ministerio de Agricultura, Republica del Peru.

Schweigger, E., 1959, Die Westkuste Sudamerikas im Bereich des Peru-Stroms: Keyserche Verslagsbuchhandlung, Heidelberg- Munchen, West Germany.

Smith, R.L., 1983, Peru coastal currents during El Niño 1976 and 1982: Science, v. 221, p. 1397-1398.

Todorovic, P., 1978, Stochastic models of floods: Water Resources Research, v. 14, p. 345-356.

Van Loon, H. and Shea, D. J., 1985, The Southern Oscillation. Part IV: The precursors south of 15°S to the extremes of the oscillation: Monthly Weather Review, v. 113, p. 2063-2074.

Waylen, P. R. and Caviedes, C. N.,1986, El Niño and annual floods on the North Peruvian littoral: Journal of Hydrology, v. 89, p. 141-156.

Wyrkti, K. et al., 1976, Predicting and observing El Niño: Science, v. 191, p. 343-346.

4

Observations of Jökulhlaups from ice-dammed Strandline Lake, Alaska: implications for paleohydrology

Matthew Sturm, James Beget and Carl Benson

ABSTRACT

Strandline Lake, a glacier-dammed lake in south-central Alaska containing over 7.1×10^8 m^3 of water, has produced a series of jökulhlaups during at least the last 20 years. Observations, vertical aerial photography, and time lapse photography made during jökulhlaups in 1982, 1984, and 1986 indicate that the flood waters follow multiple pathways as they drain under or through, and along the margins of the Triumvirate Glacier which impounds the lake. As the flood discharge increases, the drainage changes from primarily marginal to sub- or en-glacial. Detailed measurements of the lake stage made from 35mm films taken during the 1984 jökulhlaup, coupled with the hypsometric function of lake stage and water volume, have been used to derive the discharge hydrograph for the 1984 event. The hydrograph shows five distinct episodes where discharge was abruptly reduced for short periods, superimposed on an overall trend in which discharge was increasing nearly exponentially.

The reductions in discharge are probably due to blockage or collapse of subglacial tunnels. In 1984 the blockages were temporary, but the 1986 jökulhlaup was prematurely terminated. Strandline Lake was left partially full of water as of this writing. This suggests that in some cases tunnels can become totally blocked during a flood.

Caution is needed in interpreting prehistoric sequences of flood deposits produced by jökulhlaups. The sequence of Pleistocene flood deposits in varved lake sediments in northern Washington state may contain a record of temporary reductions in discharge and premature terminations of jökulhlaups from Lake Missoula.

INTRODUCTION

Opportunities to observe jökulhlaups (outburst floods) from ice-dammed lakes are rare. These lakes are generally in remote areas and often fill and discharge without attracting notice. In the few cases where they have been observed, usually only the spectacular denouement of the cycle of filling and draining is witnessed.

Detailed observations of jökulhlaups from modern ice-dammed lakes can serve as analogs for the similar, albeit larger, events which took place during the Pleistocene Epoch. We can guide our intrepetations of prehistoric flood deposits and discharge events by observing modern ice-dammed lakes .

We were fortunate to observe the 1982 jökulhlaup of ice-dammed Strandline Lake in Alaska (Sturm and Benson, 1985), and to monitor the lake during its refilling. This recharge culminated in a jökulhlaup in September 1984, which we monitored in detail.

The recent history of Strandline Lake is characterized by irregular intervals between jökulhlaups, marked differences in the magnitudes of the jökulhlaups, and significant differences in the styles of the flood events. We believe the processes observed during these floods have implications for understanding the failure of ice-dammed lakes in the geologic past.

PREVIOUS WORK

The recent history and geometry of Strandline Lake was described by Sturm and Benson (1985). Figure 1 shows the lake and the Triumvirate Glacier which impounds it. Table 1 compiles the history of the lake through 1986.

The 1982 jökulhlaup from Strandline Lake apparently was initiated when the water level in the lake rose high enough to lift the portion of the Triumvirate Glacier which dammed the lake (Sturm and Benson, 1985). This initiated drainage of 7.1×10^8 m^3 of water, and gave rise to a flood with a peak discharge estimated to be 6×10^3 m^{3}s^{-1}, using the empirical formula of Clague and Mathews (1973).

METHODS

The 1984 jökulhlaup was studied by on-site observers, periodic fixed-wing and helicopter over-flights, and aerial photogrammetry. Two automatic 35mm cameras, designed and fabricated for studies of glacier motion by W. Harrison and others (1986), were installed overlooking Strandline Lake and the Triumvirate Glacier. Photographs were taken six times daily. Time-lapse 8mm movies were also made of the lake. In addition, we contracted to have vertical aerial photographs taken immediately before and after the jökulhlaup.

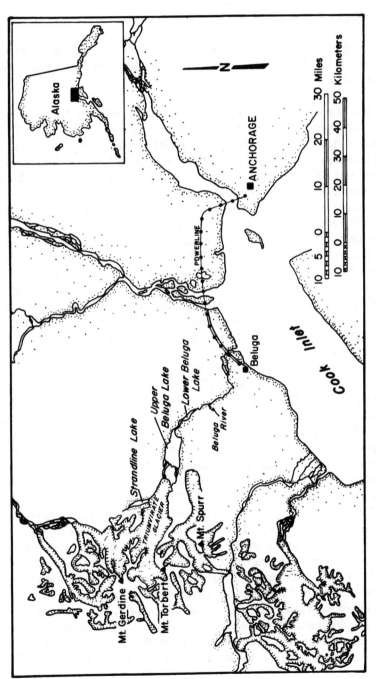

Figure 1. Location map showing glacier-dammed Strandline Lake and its relationship to the Triumvirate Glacier. Reproduced from Sturm and Benson, 1985, courtesy of the International Glaciological Society from Journal of Glaciology.

Determining the discharge from Strandline Lake during the 1984 jökulhlaup required two steps:

1.) measuring the hypsography of the lake basin from photogrammetry, and
2.) observing the change in the surface elevation of the lake from the time-lapse photography and on-site surveying.

Two sets of pre- and post-flood aerial photographs were used to construct a 1:10,000 scale topographic map of the lake basin accurate to ± 2.5 m (Fig. 2) using a stereo plotter system. The maps were digitized using a Talos digitizer to determine the lake surface area at 17 elevations between 240 m a.s.l. (lowest water, lake empty) and 400 m a.s.l. (11 m above recent maximum lake level). This defined the hypsometric curve for the lake (Fig. 3).

The recharge and discharge of the lake was determined by measuring the change in the elevation of the water surface of the lake at periodic intervals. The elevation changes were converted to water volume changes using the hypsometric curve (Fig. 3). During the initial stages of the 1984 jökulhlaup, between August 17 and September 12, all measurements were made either by leveling down to the lake water surface from a bench mark, or, for observations made over short intervals, reading off temporary staff gauges. After September 7, changes in lake elevation were also measured using 35 mm time-lapse cameras. The cameras took date- and time-stamped photographs on dimensionally stable film from which water surface displacements could be measured.

Two methods were used to reduce the photographic data. The first method consisted of making precise enlargements (7.63 X) of selected frames (Fig. 4). An overlay tracing was made from these enlargements of three grounded bergs and the location of the water line on the sides of the bergs between September 7th and 15th. The change in the position of the water line on the overlay was measured with calipers to ± 0.1 mm, so that the elevation of the water line could be determined to ± 0.5 m at 2500 m, the average distance of the target bergs from the camera. A similar technique was employed using the Talos digitizer on projections of the 35 mm negatives (19.4 X enlargement). This method could resolve distances to ± 0.025 mm on the enlargement, corresponding to ± 0.05 m elevation change, but the identification of the water line was more difficult than on the higher resolution photographs. To improve the reliability of the measurements, a second set of digitized measurements were made on only those frames on which the water surface could be identified unambiguously. Measurements from the two photogrammetric methods were combined with survey measurements to produce the hydrograph shown in Figure 5. Where direct comparison was possible, lake surface elevations determined by different methods agreed to better than 0.75 m. Errors in estimating discharge vary with the deviance from actual water elevation and the time interval between measurements, but the maximum error is estimated to be less than 10%.

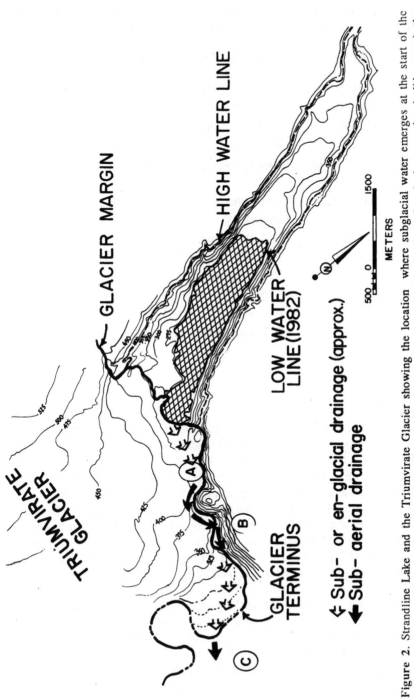

Figure 2. Strandline Lake and the Triumvirate Glacier showing the location where subglacial water emerges at the start of the sub-aerial portion of the flood path (A), the visible section of the flood path along the glacier margin (solid arrows), the inferred location of the subglacial flood paths (open arrows), the marginal pool (B), and the point at the glacier terminus where flood waters emerge (C).

Table 1. Known jökulhlaups from Strandline Lake. Prior to 1974 the record is incomplete, and it is likely that some events have been omitted.

DATE	EVENT	SOURCES and NOTES
1940	FLOOD	"Beluga Joe" Sneller as told to E. P. Whittemore
1954	LAKE FULL	USGS Photographs
1958	FLOOD	W. Shultz, Anchorage Times, 15 July 1979
1960	LAKE LOW	Aerial photographs, A. Post, USGS
1970	LAKE FULL	Aerial photographs, A. Post, USGS
1970	FLOOD (?)	
1974	Lake 3/4 FULL	Aerial photographs, North Pacific Aerial Survey, Inc (NPAS)
1974	FLOOD	E. P. Whittemore
1979	FLOOD	E. P. Whittemore, S. Shomper
1980	FLOOD	E. P. Whittemore, S. Shomper
1982	FLOOD	Authors
1984	FLOOD	Authors
1986	FLOOD	G. Nibler, M. Arline

RESULTS

The 1984 Jökulhlaup

On August 17, 1984, Strandline Lake had filled to within 5 m of the high water level (380 m a.s.l.) that triggered the 1982 jökulhlaup. The lake crested at 389 m.a.s.l. and began to drain on the 29th or 30th of August. On August 31, a small discharge was observed along 1.5 km of the glacier margin below Strandline Lake, though this flow was subglacial or englacial for a kilometer below the lake (Fig 2). Marginal drainage occurred at the contact between ice and the bedrock cliffs which line the north side of the glacier. For several days the marginal drainage ponded in a large pool (B on Fig 2). When this pool had filled to an approximate depth of 120 m on or about September 1, subglacial drainage was initiated and water began to flow out of the glacier terminus (C on Fig. 2). Terminus discharge was initially about 20 $m^3 s^{-1}$.

During the next two weeks the discharge increased noticeably in the marginal channel, which deepened due to melting and/or erosion. It is possible that other englacial or subglacial routes were simultaneously in use. The connection between the marginal drainage and the lake remained completely below the ice surface, and water could be observed "boiling up" at the head of the marginal channel under hydrostatic pressure (A in Fig 2). Large fractures developed in the ice of the Triumvirate Glacier over the area where subglacial flow was occurring.

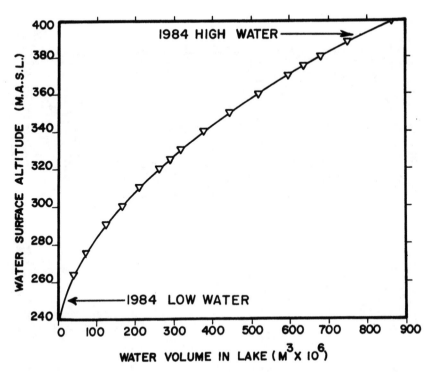

Figure 3. Hypsometric curve for Strandline Lake derived from photogrammetric maps. High and low water for 1984 are shown.

A temporary reduction in the discharge occurred between September 5 and 8, but was not recognized due to the generally low discharge. A second short reduction in discharge occurred between September 8 and 9. This reduction, and a third temporary reduction which occurred between September 9 and 11, also escaped notice.

However, a rapid increase in discharge to greater than 1000 m^3s^{-1} on September 11 caused the level of the Beluga River to rise rapidly 50 km down-stream of Strandline Lake. The rising water carried blocks of glacier ice. We anticipated that the lake would drain quickly, but an over-flight on the morning of September 12th revealed, surprisingly, that the discharge had dropped to less than 100 m^3s^{-1}. The marginal channel was still in use, but at a much reduced flow rate. A high water line 2 m above the stream below the terminus of the Triumvirate Glacier was marked by stranded blocks of ice up to several meters in diameter which lined the entire course of the channel. The waterlevel in Strandline Lake at this time was 22 m below the high water level, and the entire lake was choked with ice bergs. Many of the bergs were concentrated against the ice face of the Triumvirate Glacier, and it was impossible to observe the point where the water drained from the lake.

Figure 4. Three selected frames from the time-lapse sequence of photographs taken between September 7 and 15. Large berg in the center fore-ground is approximately 105 m high in last frame.

Two days later, on September 14, the discharge increased rapidly to greater than 3400 m^3s^{-1}, and the lake emptied within 24 hours. At some time during this period the marginal channel was abandoned in favor of en- or sub-glacial tunnel(s). This may have occurred by stoping failure (Paige, 1955) of the ice forming the floor of the marginal channel, because after the event, a collapse hole more than 10 m in diameter was visible in the ice floor of the channel near its up-stream end.

Figure 5. Hydrograph of the 1984 jökulhlaup of Strandline Lake. Five temporary reductions in discharge are marked by stars.

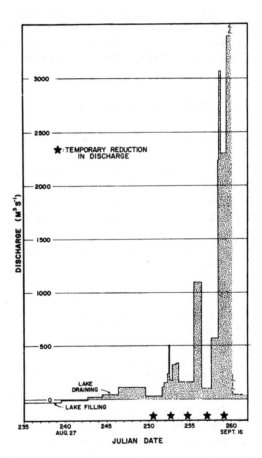

The 1984 Hydrograph

Figure 5 is the hydrograph of the 1984 jökulhlaup from Strandline Lake. Five times during the course of the jökulhlaup the discharge decreased, followed by a rapid increase. These five temporary reductions in discharge are marked by stars on the figure. The last one occurred when the discharge exceeded 3000 m^3s^{-1}.

The peak measured discharge was 3.4 x 10^3 m^3s^{-1}, but it is likely that the lake had a maximum discharge in excess of 5.0 X 10^3 m^3s^{-1}. This can be estimated as follows: at the time of our last measured discharge (3.4×10^3 m^3s^{-1}), the total water volume which had drained from the lake was 5.2×10^8 m^3. This is obtained by integrating under the hydrograph curve up to 1600 hours on September 15, 1984, when the last usable picture was taken. At that time, floating ice in the lake became so thick that it was no longer possible to distinguish the water surface on the time lapse photographs. From observations made early on September 16 it is clear that the lake was empty before noon. At least 1.87×10^8 m^3 of water had to drain in less than 12 hours. This would require a mean discharge of 3.8×10^3 m^3s^{-1}. By extrapolating the rapid increase in discharge when last measured, it is reasonable to assume a peak discharge somewhat higher than 5.0×10^3 m^3s^{-1}. This is a greater discharge than that of the Missouri River near St. Louis.

DISCUSSION OF RESULTS

Our observations of Strandline Lake have two major implications:

1.) the filling and draining of ice-dammed lakes can be irregular in periodicity and style of draining, and
2.) draining during a single jökulhlaup can be irregular, punctuated by periods of reduced flow during which the discharge can drop to virtually zero.

The first point is best illustrated by referring to Table 1. Recent jökulhlaups have occurred at intervals of two years, but as recently as 1974 to 1979, there was a five-year period during which no flood occurred. The magnitude of the five floods have also varied markedly. The flood of 1979 destroyed the bridge across the Beluga River and water rose high enough to occupy the entire flood plain. The 1980 flood did the same, but the 1982 and 1984 floods produced significantly lower water stages and little damage.

The 1980 flood occurred in winter when the lakes and rivers along the flood path were frozen, while all other known floods have occurred in late summer. It seems likely that erosion and sedimentation during this flood differed from the other floods, as substrates along the flood path were frozen. Large variation in the magnitude and consequences of jökulhlaups as a result of seasonal timing have also been observed on the Kenai River which carries the discharge from glacier-dammed Snow Lake (Post and Mayo, 1974; Chapman, 1981).

It is unlikely that the variations in period and style of flooding are the result of changes in the Triumvirate Glacier. Examination of aerial photographs of the glacier from 1970 on reveal only slight down-wasting in the area of Strandline Lake.

We hypothesize that jamming of icebergs in drainage tunnels or the collapse failure of parts of the tunnel may have caused the temporary reductions in discharge observed during the 1984 jökulhlaup. The five reductions during 1984 were short-lived and outflow from the lake was never reduced to zero. Erosion and/or melting of the partial obstruction ultimately re-opened and enlarged the tunnel. Following each period of reduced flow, the discharge increased more rapidly than predicted by Clarke's (1982) computer model of jökulhlaup discharge (Fig 6). This supports the hypothesis that an obstruction was melted or swept out of an existing tunnel, because the high discharge resumed more rapidly than could be achieved by melting a new tunnel.

A review of published jökulhlaup hydrographs shows variable discharge during catastrophic floods from other ice-dammed lakes. The 1974 hydrograph for Snow Lake (Chapman, 1981) and the 1967 hydrograph for Summit Lake (Mathews, 1973) both show periods of reduced discharge lasting up to 24 hours. The 1970 hydrograph for a jökulhlaup originating from Vatnajökull, Iceland, shows a similar period of reduced discharge which lasted 2 days (Bjornsson,1972).

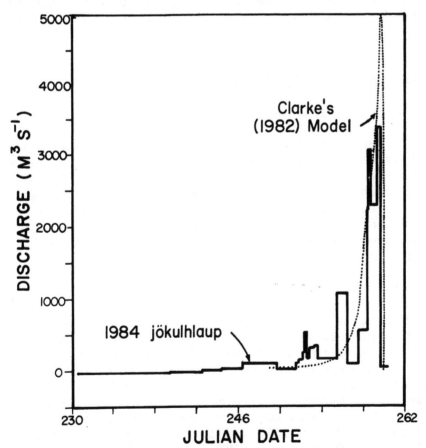

Figure 6. A comparison of the actual 1984 discharge hydrograph for Strandline Lake (solid histogram) with a theoretical hydrograph computed by Clarke's (1982) model. Note that the model does not include temporary reductions in discharge. Actual increase in discharge following the reductions is more rapid than predicted by theory.

The marked reductions in discharge during the 1984 jökulhlaup of Strandline Lake suggest the possibility that jökulhlaups can end prematurely, before the lake is completely empty. If an ice-berg jammed in a subglacial tunnel reduced the water flow to zero, then ice deformation would close and seal the tunnel. The partially drained lake would be unable to re-float the ice and unable to re-initiate the flow. The jökulhlaup would be prematurely terminated; the lake would begin to refill.

In September, 1986, Strandline Lake began to discharge in a jökulhlaup that at first appeared similar to the 1984 event. After 10 days, however, the lake stabilized at a level 80 m below the 1986 high water. Discharges as high as those in 1982 and 1984 did not occur, and the lake did not fully drain. Since September, lake water has continued to discharge at a low rate, and outflow from the lake has been roughly balanced by inflow. It seems likely that outflow will cease during the winter, and that the lake will refill and initiate a new jökulhlaup cycle in the near future.

PALEOHYDROLOGIC IMPLICATIONS

Terminations During Pleistocene Lake Missoula Jökulhlaups?

The observations at Strandline Lake have implications for paleo-hydrologic interpretation of jökulhlaup sediments. We propose that interruptions in discharge and premature terminations should be expected during a long series of jökulhlaups from ice-dammed lakes. It should be possible to identify premature terminations in the geologic record because peak flood discharge would be smaller than normal, and the duration between floods would be shorter. Here we re-examine the record of jökulhlaup deposits from Pleistocene Lake Missoula preserved in varved lake sediments in northern Washington state in light of this new model.

Premature Terminations of Lake Missoula Jökulhlaups

Missoula Flood deposits separated by varves record interruptions of Missoula Flood waters into lakes dammed by the Cordilleran Ice Sheet in eastern Washington. The varved layers between flood beds record the time interval between jökulhlaups. Atwater (1986) identified 89 flood beds in exposures and drill cores from varved lake sediments in the Columbia River Valley. Half of the varve sequences separating flood units were incomplete, but several of the remaining complete flood beds followed anomalously soon after the preceding floods, suggesting that a number of the floods from glacial Lake Missoula were prematurely terminated. About one flood bed Atwater (1986, p.11) states, "a puzzling departure from...overall trends is an apparently complete varved interval...in borehole 3 that has one-quarter the number of its neighbors' varves". The varved interval in question has 10 annual layers, while surrounding beds have about forty. We suggest this short varve sequence indicates the preceding flood terminated prematurely,

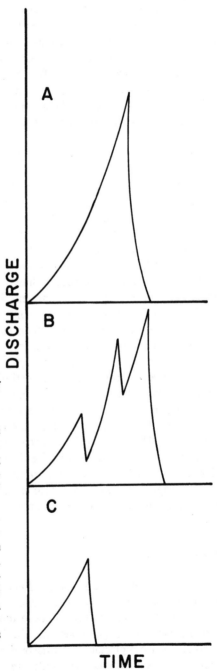

Figure 7 Idealized hydrographs and sedimentation records for a series of jökulhlaups from the same ice-dammed lake. A.) "Normal" hydrograph, characterized by uninterrupted flow, with rapidly increasing discharge followed by abrupt cessation of flow, producing a single thick graded bed. B.) Hydrograph interrupted by two periods of temporarily reduced discharge. Peak discharge is less than A, but total discharge is identical. Three distinct graded beds are produced. C.) Hydrograph which shows premature termination. Peak and total discharge are significantly less than A or B. Recurrence interval to the next flood is shorter than for A or B. A single, thin graded bed consisting of finer material than A or B is produced.

leaving glacial Lake Missoula partly (three-quarters?) filled. The next flood occurred as soon as the Missoula basin refilled. Both Waitt (1984) and Atwater (1986) have identified several other Missoula floods separated by anomolously short recurrence intervals intercalated with flood beds with longer intervals.

If several of the floods from Lake Missoula were prematurely terminated, their peak discharges would be significantly less than those produced during normal floods. We would expect large differences in flood discharge and competence to be reflected in the thickness and coarseness of the flood deposit (Fig. 7). Sedimentary sequences similar to this have been reported by Atwater (1986).

Hiatuses During Individual Jökulhlaups from Lake Missoula

In ideal cases it should be possible to recognize a geologic record of temporary reductions in the discharge of prehistoric floods. Each pulse of water would produce its own flood stage, albeit with lower peak discharge than an uninterrupted flood (Fig 7). Several of the 89 flood beds present in sections and drill core from the Columbia River Basin show at least two cycles of upward fining within a single flood bed. One flood bed shows four (Atwater, 1986). Similarly, Waitt (1980) found several flood beds with two or more cycles of upward fining at Burlingame Canyon. Perhaps these complex flood beds record temporary reductions in discharge during outburst floods from glacial Lake Missoula. The absence of varves between upward fining sequences indicates that the maximum duration of any period of reduced discharge was less than a year.

Temporary reductions in discharge and premature termination of jökulhlaups have implications for quantitative estimates of paleo-flood discharge from ice-dammed lakes. Two different but complimentary approaches have addressed the problem of jökulhlaup discharge retrodiction. A strong empirical relationship exists between initial lake size and peak discharge (Clague and Mathews, 1973; Beget, 1986). The use of such empirical regression equations, however, assumes total, uninterrupted discharge of the ice-dammed lake and therefore over-estimates discharge from a lake which only partially drains. Similarly, computer models of lake drainage through ice dams also assume total, uninterrupted drainage, and discharge estimates made in this way should also be considered maxima (Clarke et al, 1984).

SUMMARY

Our observations at Strandline Lake indicate that irregularities in jökulhlaup discharge and recurrence interval are to be expected from ice-dammed lakes. Temporary interruptions and premature terminations of jökulhlaups can occur during lake drainage. We suggest here that similar events may have occurred during outburst floods from glacial Lake Missoula as well as other glacier-dammed lakes. Premature flood terminations would result in low peak discharge and an anomalously short interval until the next flood. Temporarily interrupted discharge would produce multiple flood peaks and complex flood deposits with multiple cycles of upward grading. We have applied these ideas to explain irregularities in flood deposits ascribed to outburst floods from glacial Lake Missoula. Because large variations in recurrence interval and discharge of floods are to be expected from an ice-dammed lake, it is not necessary to hypothesize climatic fluctuations, changes in extent or thickness of ice dams, or variations in volume or depth of glacial lakes to account for variations in discharge and recurrence interval of pre-historic jökulhlaups.

ACKNOWLEDGEMENTS

We were assisted by many people in making observations on Strandline Lake. We wish to thank S. Shomper, V. Watson, E. Wittemore, D. Witte, B. Campbell, T. Clarke, G. Liston, C.Benson Jr., B. Sturm, M. Arline, C. Moore and G. Nibler for making these observations. Field support was generously provided by Chugach Electric Company and the National Weather Service in Anchorage. We wish to thank G. K. C. Clarke for running a simulation of the 1984 jökulhlaup on his computer model. This work was supported by the State of Alaska, Department of Natural Resources, Division of Geologic and Geophysical Surveys.

REFERENCES

Atwater, B. F., 1986, Pleistocene Glacial-lake Deposits of the Sanpoil River Valley, Northeastern Washington: U.S. Geological Survey Bulletin 1661, p.1-39.

Beget, J. E., 1986, Comment on "Outburst floods from glacial Lake Missoula", by G. K. C. Clarke, W. H. Mathews, and R. T. Pack: Quaternary Research, v. 25, p. 136-138.

Bjornsson, H., 1977, The cause of jökulhlaups in the Skafta' River, Vatnajökull: Jökull, 27.AR, p.71-77.

Chapman, D. L., 1981, Jökulhlaups on Snow River in Southcentral Alaska: NOAA Technical Memorandum NWS AR-31, United States Department of Commerce, p. 1-48.

Clague, J. J. and. Mathews, W. H, 1973, The magnitude of jökulhlaups: Journal of Glaciology, v. 28, p. 501-504.

Clarke, G. K. C., 1982, Glacier outburst floods from "Hazard Lake", Yukon Territory, and the problem of flood magnitude prediction: Journal of Glaciology, v. 28,p. 3-22.

Clarke, G. K. C., Mathews, W. H., and Pack, R. T., 1984, Outburst floods from glacial Lake Missoula: Quaternary Research, v.22, p. 289-299.

Harrison, W. D., Raymond, C. F., Moore, C. S., Senear, E., and MacKeith, P., 1986, Time lapse camera data from Variegated and Other Glaciers 1979-1984: Geophysical Institute Report UAG- R(304), Geophysical Institute, University of Alaska, Fairbanks, Alaska, U.S.A.

Mathews, W. H., 1973, Record of two jökulhlaups: Symposium on the Hydrology of Glaciers, Publication No. 95, International Union of Geodesy and Geophysics, International Association of Scientific Hydrology, Commission of Snow and Ice, p. 99-110.

Paige, R. A., 1955, Subglacial stoping or block caving: a type of glacier ablation: Journal of Glaciology, v.2, p. 727-729.

Post, A. and Mayo, L. R., 1971, Glacier dammed lakes and outburst floods in Alaska: U. S. Geological Survey Hydrologic Investigation Atlas HA-455.

Sturm, M. and Benson, C. S., 1985, A history of jökulhlaups from Strandline Lake: Journal of Glaciology, v. 31, p. 272-280.

Waitt, R. B., Jr., 1980, About forty last-glacial Lake Missoula jökulaups through Southern Washington: Journal of Geology, v. 86, p. 653-679.

Waitt, R. B. Jr., 1984, Periodic jökulhlaups from Pleistocene glacial Lake Missoula--new evidence from varved sediment in northern Idaho and Washington: Quaternary Research, v. 22, p. 46-58.

5

Glacial-lake outbursts along the mid-continent margins of the Laurentide ice-sheet

Alan E. Kehew and Mark L. Lord

ABSTRACT

Most of the meltwater from the Laurentide Ice Sheet was not discharged into proglacial outwash streams, but rather was ponded along the ice margin. Basins containing these proglacial lakes were the result of regional slope toward the ice sheet, crustal depression, and dams formed by accumulations of stagnant ice or previously formed ice-marginal glacial landforms. When lake levels rose and overtopped low points of their basins, or ice or debris dams failed, glacial-lake outburst floods occurred. Where lake outlets developed in glacial drift or poorly consolidated sedimentary bedrock, outburst floods rapidly downcut through the non-resistant substrate to completely drain the lake. Alternatively, where hard bedrock was encountered during downcutting, the lake level stabilized near that elevation.

Outburst floods were highly erosive and left their mark upon the landscape in the form of huge trench-shape channels with a suite of characteristic geomorphic components. These immense, distinctive channels lead away from all major glacial-lake basins at the southern margin of the Laurentide Ice Sheet. In addition, the midcontinent plains of the U. S. and Canada are crossed by hundreds of these deeply incised channels, which record the meltwater history of the retreating ice.

The volume of sediment eroded by the outburst floods was enormous. Floods from the southern outlets of Lake Agassiz and the lakes in the Great Lakes basins funneled into the Mississippi Valley and transported their sediment load to the Gulf of Mexico. These events probably dominate the Pleistocene glacial intervals recorded in the deposits of the Mississippi fan and Louisiana slope. In the western plains of the U. S. and Canada, outburst floods progressed from basin to basin along the ice margin like a line of falling dominoes; the arrival of a flood at a basin triggered the downcutting of its outlet and the draining of the lake. The coarse-sediment fraction, mostly sand, was dumped as huge fans which spread over the floors of the basins inundated by the floods. The concurrent enlargement of the outlets was so rapid that the fine fraction never had sufficient time to settle out and was washed along through downstream basins until a lake was reached that was large enough to contain the discharge without completely draining.

Meltwater flow across broad outwash plains, as traditionally envisioned, was, in fact, not typical along many segments of the Laurentide margin. The impoundment of meltwater to form glacial lakes, followed by sudden drainage and formation of glacial-lake spillways was a more significant meltwater process in most of the mid-continent area. The lasting effects of these outburst floods include the establishment of the courses of most major rivers in the area glaciated by the Laurentide Ice Sheet.

INTRODUCTION

As the Laurentide Ice Sheet advanced into areas underlain by poorly indurated rock, glacial erosion and deposition increased. Much of the Great Lakes region, as well as the western plains of the U. S. and Canada, are underlain by relatively thick sequences of till and other forms of glacial drift. Glacial advance upslope onto the Great Plains resulted in compressional flow and thick ablation-drift deposits. Ice-marginal deposits include bands of high-relief ablation drift that covered stagnant-ice remnants and ridges of ice-thrust materials emplaced in frozen-bed marginal zones. Crustal depression, in addition to regional topography and ice-marginal deposits, insured the presence of closed basins in the vicinity of the ice margin during retreat of the ice sheet. Glacial meltwater gradually filled these marginal basins rather than flowing away from the ice over outwash plains or in valleys leading away from the glacier.

As water levels rose in the proglacial lakes, ice or debris dams failed causing the incision of outlets and the release of tremendous volumes of water. Torrential discharges from the glacial-lake outbursts were diverted along the ice margin or around ridges of ice-marginal sediment in the direction of decreasing elevation. Commonly, these flow paths led to lower glacial lakes. The voluminous influx of meltwater overloaded the capacity of smaller lakes, causing their outlets to be widened and deepened. If the outlet substrate was non-resistant, complete drainage of the lake took place. The outburst then migrated to the next lower lake basin augmented by the flow from the lake just inundated (Kehew and Clayton, 1983).

The first outburst recognized in the Great Plains originated from Lake Regina in Saskatchewan (Kehew, 1982) and proceeded to flow through Lakes Souris and Hind on its path to Lake Agassiz. Recent investigations have shown that the Lake Regina outburst was not a rare, isolated event. In fact, it appears that the catastrophic partial or total drainage of glacial lakes was probably the most typical meltwater process along many segments of the Laurentide Ice Sheet. Other examples in the Lake Agassiz region were described by Clayton (1983), Matsch (1983), and Teller and Thorleifson (1983). In the eastern Great Lakes region, outbursts from Lake Maumee were reported by Vaughn and Ash (1983) and Fraser and Bleuer

(1985). An event of this type in northern Illinois and Indiana was recognized as early as 1925 by Ekblaw and Athy.

The purpose of this paper is to review the erosional and depositional effects of Laurentide glacial-lake outbursts and to present new examples of these phenomena. Reconnaissance investigations suggest that glacial-lake outbursts are so common in the area of the Laurentide Ice Sheet that they should be incorporated into all models of deglaciation. The effects of these events must figure prominently in the history of the Mississippi Valley, the evolution of the Mississippi fan and Louisiana slope in the Gulf of Mexico, and in the development of the drainage system of most of northern North America.

EROSIONAL EFFECTS

The erosional capacity of glacial-lake outbursts was truly enormous. The spillways connecting glacial-lake basins on the prairies are easily visible on landsat imagery and rank among the most pronounced topographic features of the midcontinent region (Fig. 1). The volume of sediment eroded by outburst floods is equivalent to the volume of the spillway channels. Present spillway dimensions underestimate erosion because of thick sections of Holocene fill deposited in the spillways. The sediment eroded from spillway channels leading to the Mississippi Valley was deposited in the lower Mississippi Valley and/or the Gulf of Mexico. In the western plains, material eroded from spillways was deposited in glacial-lake basins.

Perhaps even more remarkable than the amount of erosion was the intensity of erosion. Outburst floods quickly trenched through the glacial drift and continued downcutting into bedrock if the bedrock was poorly indurated. Cretaceous and Tertiary sedimentary rocks on the Great Plains were easily cut through by the floods. Studies of spillway fills in North Dakota (Boettger, 1986) indicate that there was no continuous depositional phase in the spillway channels--Holocene sediment directly overlies bedrock on the spillway bottoms. Discharges therefore were short-lived and abruptly terminated.

Investigation of spillway morphology in the Dakotas, Minnesota, Manitoba, and Saskatchewan (Fig. 2) (Kehew and Lord, 1986) led to the development of a generalized geomorphic model of spillway erosional features (Fig. 3). The most diagnostic criteria are listed in Table 1; these features are described in detail in Kehew and Lord (1986). In general, spillways are wide, deep channels that begin abruptly at glacial-lake outlets and maintain their size and shape until they end at other lake basins. A flood origin for the valleys is indicated by the lack of tributary valleys of comparable size with the exception of other spillways. Spillways commonly terminate where they meet other glacial-lake basins (Kehew and Clayton, 1983), thus implying that glacial lakes were suddenly subjected to huge inflows of water. The depositional phenomena accompanying these cataclysmic inflows will be discussed in the following section.

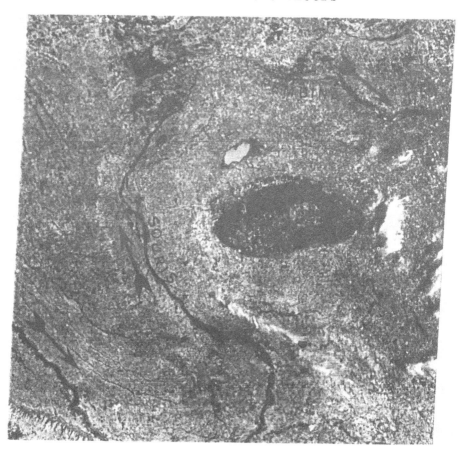

Figure 1. Landsat image of Souris and Pembina spillways in North Dakota and Manitoba. Dark areas of spillways are reservoirs. Oval, dark area at center is the forested Turtle Mountains upland. North at top. Width of image from west to east approximately 200 km. Arrows indicate direction of flow of outbursts in spillways.

Many spillways consist of two prominent components--a broad, gently sloping upper level (the outer zone), and a centrally positioned trench-like inner channel (Figs. 3 and 4). Shallow flow initially covered the outer zone if no existing drainageway lay in the path of the outburst flood. The common lack of distinct outer-zone margins suggests that stagnant ice formed the original channel bottom. As incision of the outer zone progressed, boulders from the underlying till began to accumulate as a coarse channel-bottom lag. This boulder concentration increased the resistance to flow in the broad, shallow outer-zone channel. In response, the flow. began to carve out a narrow, deep inner channel at the center of the outer zone which provided less resistance to flow. In addition to the lag-covered surface, other characteristic outer-zone features include longitudinal grooves and streamlined residual hills (Kehew and Lord, 1986). Streamlined erosional residuals are protected

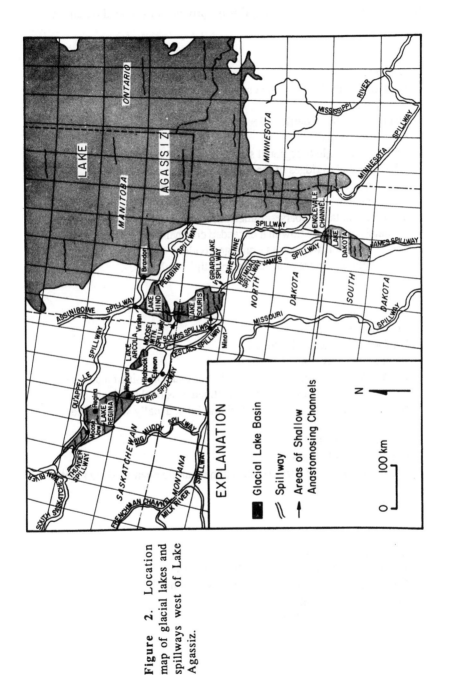

Figure 2. Location map of glacial lakes and spillways west of Lake Agassiz.

Table 1. Geomorphic features of spillways produced by glacial outbursts.

General

Lack of tributaries other than small Holocene valleys or other spillways.

Usually contain underfit Holocene streams.

Deeply entrenched.

Trend at an angle to regional slope.

Often parallel to ridges of ice-marginal deposits.

Constant size from lake outlet to termination at topographically lower lake basin or junction with other spillway.

Terminate at large, coarse-grained sediment fans in glacial-lake basins.

Channels eroded by flow that exceeded channel capacity may lead away from spillway across drainage divides.

May contain terraces representing multiple outbursts.

Inner channel

Trench-like shape.

Uniform width and side slopes.

Regular meander bends.

Occasional bifurcation to form parallel or anastomosing channels separated by linear ridges or streamlined erosional residuals.

May contain isolated erosional residual hills, usually streamlined.

1-3 km in width.

25-100 m in depth.

Lack of slip-off slopes.

May contain bars of very coarse sediment along channel sides at infrequent intervals.

Outer zone

Broad, scoured surface.

May contain shallow anastomosing channels.

Boulder lag may be developed by incision into till.

Longitudinal grooves.

Streamlined erosional hills.

from erosion during flow as they attain the streamlined shape of an airfoil. This shape, which is approximated in plan view by the lemniscate loop (Komar, 1984), minimizes flow separation and drag on the landform (Fig. 5).

The inner channel of the spillways is the most diagnostic indicator of outburst flooding. The size and shape of these channels are not consistent with an origin by gradual downcutting by small streams. The width and depth of the inner channels correlate with discharges of 10^5 m^3s^{-1} or more if bankfull flow is assumed (Kehew and Clayton, 1983; Lord, 1984). The uniform channel shape, constant valley-side slope, and lack of slipoff slopes rule out gradual erosion of the channels. A compound origin, however, including multiple episodes of outburst flooding separated by other types of fluvial activity, is very likely for some of the spillways. Multiple events are indicated by terraces underlain by outwash sediment and/or outburst flood sediment. In addition, some spillways were occupied by glacial meltwater streams after their incision.

Inner channels commonly contain erosional remnants produced by the bifurcation of the flow around obstacles or the simultaneous incision of two or more branches of the inner channel. Long, parallel inner-channel segments separated by narrow bedrock or till highs were presumably produced by concurrent erosion of two longitudinal grooves within the outer zone (Kehew and Lord, 1986). Shorter erosional remnants in the inner channels include streamlined hills which were preserved in the highly erosional flow because of their shape.

DEPOSITIONAL EFFECTS

As the capacity for erosion by glacial-lake outbursts was enormous, so was the propensity for formation of deposits reflective of the huge volumes of sediment-charged water that formed them. Eroded sediment was deposited as gravel bars within the inner channels and large, coarse-grained fans in glacial lakes recipient to the outburst discharges. Deposition of the fine-grained fraction was limited to large basins that did not drain completely during outburst inflows. In the glaciated plains west of Lake Agassiz (Fig. 2), the source sediment for the outburst deposits consists of poorly indurated till and bedrock. The till typically is composed of approximately equal parts sand, silt, and clay, with only a few percent gravel. Subjacent to the till is poorly indurated Paleocene and Cretaceous bedrock composed of silt and sand with minor amounts of fine sand and lignite. Most of the descriptions of deposits in this section are based on detailed work in the Souris and Des Lacs spillways (Fig. 2). Similar deposits are present in the Qu'Appelle, Minnesota, Sheyenne, Assiniboine, and Thunder spillways.

The discharge hydrographs of glacial-lake outbursts were probably similar to those of historical jökulhlaups, showing a steadily increasing discharge to a peak followed by a sudden cessation of flow--a pattern opposite to that of storm hydrographs (Marcus, 1960). As a consequence of the outburst discharge characteristics, flow competence within the spillways seldom decreased enough to permit deposi-

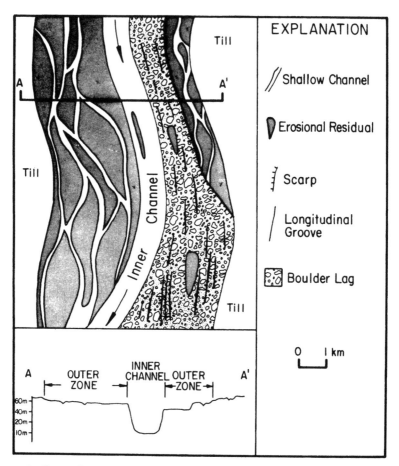

Figure 3. Generalized geomorphic model of spillway morphology. (From Kehew and Lord, 1986)

tion of any of the sediment load. Most sediment eroded by glacial-lake outbursts in the region west of lake Agassiz was dumped in glacial lakes. Though outburst deposits are relatively rare in the spillways, they are morphologically and texturally distinct from other glaciofluvial deposits.

Spillway Deposits

Glacial-lake outburst deposits in spillways consist of large scale bars. The bars occur in two depositional settings within the spillways: point bar positions and in alcoves formed by landslides during incision of the spillways (Fig. 6). The average dimensions of the bars are approximately 2 km in length, 0. 5 km in width, and 20 m in thickness (Fig. 7). These bars are primary bedforms with little or no subsequent alteration since their deposition. Bars of similar dimensions have been

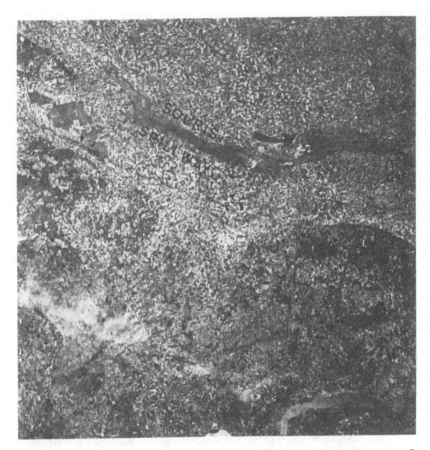

Figure 4. Landsat image of Souris spillway in southeastern Saskatchewan. Outer zone is visible as the broad, non-cultivated area (5-10 km wide) flanking the incised inner channel. Cultivation is prevented by surficial boulder lag in outer zone. North at top. Width of image approximately 200 km. Arrow indicates direction of flow in spillway.

described in association with other catastrophic floods, most notably those from the Lake Bonneville and Glacial Lake Missoula floods (Malde, 1968; Baker, 1973).

Internally, the bars consist of massive, matrix-supported, very poorly sorted cobble gravel (Fig. 8) commonly containing boulders 1 m in diameter with exceptional boulders up to 3 m in diameter. The maximum grain size in the bars decreases in the downstream direction and toward the side of the spillway. Texturally, in sharp contrast to the material eroded to form the spillways, the average composition of the bars is 2 percent clay plus silt, 17 percent sand, and 81 percent gravel (Lord, 1984). The erosive power of the outbursts can be shown by comparing the amount eroded by the discharges with the amount deposited within the spillways. For example, the Lake Regina outburst eroded an estimated 26 km^3 of sediment (mostly till) from the Souris and Des Lacs spillways (Fig. 2). By comparison, only an estimated 0. 75 km^3 of sediment, or 2. 9 percent, was redeposited within the spillways (Table 2) (Lord and Kehew, in review). The vast majority of sand and almost all of the silt and clay were conveyed through the spillway system into downstream glacial lakes.

Flow during the outburst discharges was probably hyperconcentrated. This interpretation is consistent with estimated sediment-water concentrations, the high competence required to transport the sediment load, and the intermediate character of the bars between debris-flow and clear-water deposits. It is also likely that, once material was entrained into flow, most clast sizes, including gravel, were transported in suspension. Deposition within the spillways was limited to areas of substantial flow expansion such as at bends in the spillways or alcoves created by landslides, and did not occur along the spillway bottoms. When deposition was trig-

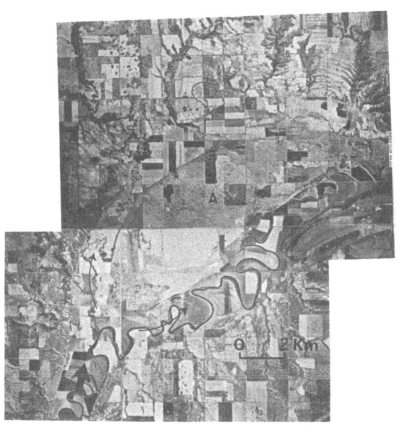

Figure 5. Air photo mosaic of large streamlined erosional residual (A) in Assiniboine spillway near Brandon, Manitoba (Fig. 2). North at top. Constructed from photos A26428-86, 87, 103; Manitoba Department of Natural Resources, Survey Mapping Branch. Arrows indicate direction of flow in spillway.

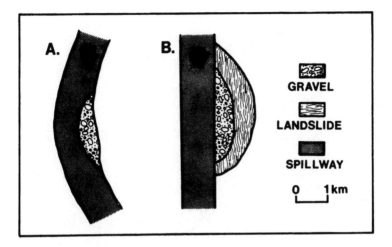

Figure 6. Schematic diagram showing two positions in which large-scale bars formed within spillways: A. Point bar position deposit, B. Alcove deposit on the spillway side of a landslide.

Table 2. Comparison of textures and volumes of material eroded versus redeposited (within spillways) by the Glacial Lake Regina outburst.

	Sediment eroded		Sediment deposited		Percent material redeposited in spillway
	Percent*	Volume (km³)	Percent**	Volume (km³)	
CLAY+SILT	67	17.4	2	0.01	0.1
SAND	28	7.3	17	0.13	1.8
GRAVEL	5	1.3	81	0.61	46.1
TOTAL	100	26.0	100	0.75	2.9

* Data from Kehew (1983)
** Data from Lord (1984)

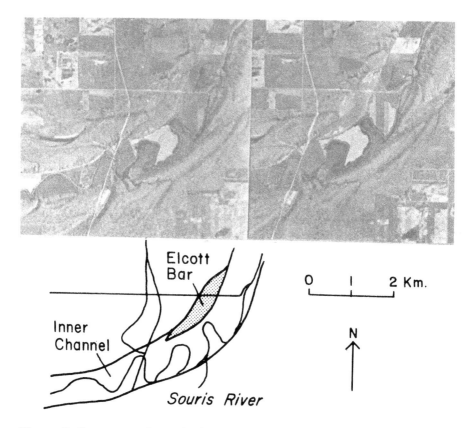

Figure 7. Stereogram of a point-bar position deposit within the Souris spillway near Elcott, Saskatchewan. North at top. Photos A21756-37, 38; Saskatchewan Department of Highways and Transportation, Central Survey and Mapping Agency.

gered by flow expansion, it occurred indiscriminate of clast size, resulting in deposits of massive, matrix-supported gravel.

Glacial Lake Deposits

The glacial lakes that received torrential discharges from the outbursts were the major sediment sinks for the eroded material. The form taken by the deposits in the lakes was largely dependent on three factors: (1) the density of the inflow, (2) the lake basin morphology, and (3) the volume of the lake basin. The density of the sediment-laden outburst discharges was significantly greater than that of the stilled lake water. As a result, inflows formed density currents that flowed along the lake bottom and deposited underflow fans (Kehew and Clayton, 1983). Underflows, because of their high density and low turbulence, may transport coarse sediment many kilometers past the inlet (Church and Gilbert, 1975). Underflow fans, because they do not prograde into lakes at the water level by fluvial action, tend to be very well sorted and gradually fine away from the inlet (Fenton and others, 1983; Kehew and Clayton, 1983). Underflow fans deposited by glacial-lake outbursts can be identified by their large areal extent and position at the mouth of a spillway eroded by a glacial-lake outburst. Outwash processes are ruled out for these deposits because they are not associated with valley trains, outwash plains, or moraines.

Lakes with irregular bottom topography developed local flow conditions that resulted in complex sediment deposition patterns. The capacity of the lake to contain the added volume of catastrophic inflows determined whether the recipient lake itself drained in a domino fashion. Where the influx triggered drainage of the recipient lakes, most of the fine-grained sediment--silt and clay, did not have sufficient time to settle out and was transported through the lake. In these cases, the lake acted similar to a very wide reach of a river with continuous throughflow rather than a lake in which the entire sediment load is dumped. Morphologic evidence of a

Figure 8. Exposure of glacial-lake outburst deposit in bar shown in Figure 7. Pick handle approximately 1 m long.

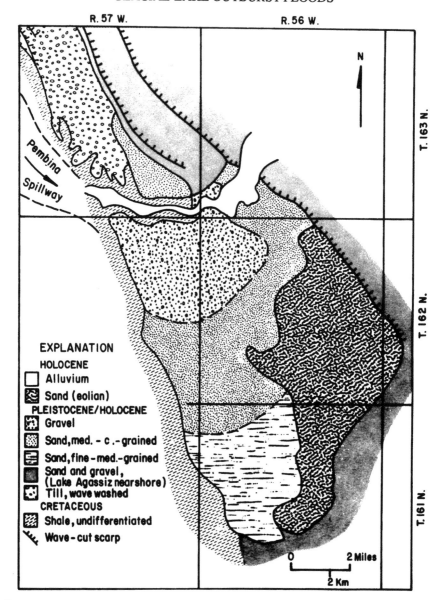

Figure 9. Generalized geologic map of Pembina underflow fan. Deposit is located where Pembina spillway terminates at Lake Agassiz basin.

domino-triggered lake drainage is incision into the fan by the spillway when the lake started to drain. Additional evidence for simultaneous outburst inflow and outflow from lake basins is summarized in Kehew and Clayton (1983).

Two examples illustrate the structure and composition of sediment fans deposited in glacial lakes by outburst floods. The Pembina fan was deposited in Lake

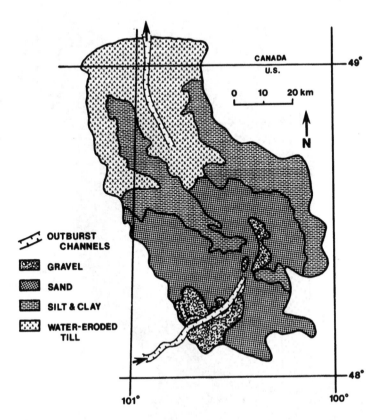

Figure 10. Generalized geologic map of Glacial Lake Souris basin. Modified from Bluemle (1982, 1985).

Agassiz by an outburst which originated from Lake Regina and travelled through the Lake Souris and Lake Hind basins (Fig. 9). This elongate body of sediment exhibits a gradual, systematic fining from gravel near the inlet to fine sand at the distal margin. Silt and clay were carried beyond the fan and were deposited over a large area of the lake bottom (Kehew and Clayton, 1983). The form of the underflow fan is relatively simple because it prograded into a deep, relatively flat-bottom lake basin which did not drain completely upon receipt of the outburst inflow. Partial incision of the upper portion of the fan may indicate that the inflow did trigger a drop in the level of Lake Agassiz. The presence of sand dunes on the fan, a consequence of the well sorted nature of the underflow fan sediments, is common to all underflow fan deposits in the western region. The Sheyenne and Assiniboine underflow fans are similar to the Pembina except that they formed in response to at least two outburst events (Brophy and Bluemle, 1983; Klassen, 1983).

The fan deposited in Lake Souris by the Lake Regina flood is an example of a complex fan. The basic pattern of an underflow fan is present, very well-sorted sediments fining away from the inlet (Fig. 10), but its development was complicated because Lake Souris was a generally shallow lake with a very irregular basin floor which drained upon arrival of the Lake Regina outburst. Lake Souris formed by the

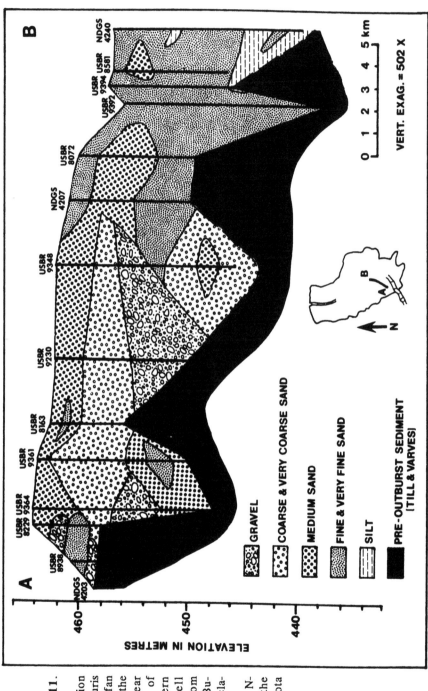

Figure 11. Geologic cross section of Lake Souris underflow fan from near the inlet to near the center of the southern basin. Well data is from the U.S. Bureau of Reclamation (USBR) and N-Files of the North Dakota Geological Survey (NDGS).

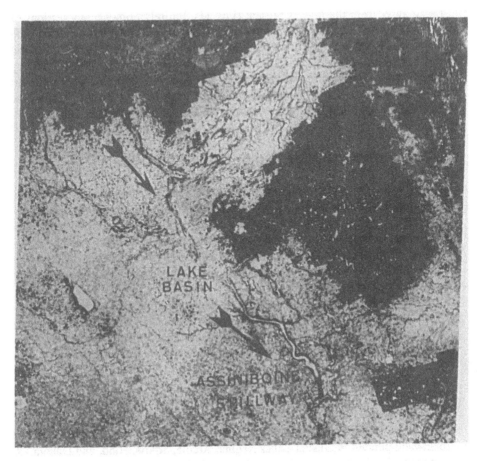

Figure 12. Landsat image showing beginning of Assiniboine spillway at a lake basin near the Manitoba-Saskatchewan border. A large spillway carried water into the lake basin from the north. North at top. Width approx. 200 km. Arrows indicate direction of flow in spillways.

coalescence of many isolated ice-walled or supraglacial lakes (Moran and Deal, 1970; Bluemle, 1985). When the Lake Regina outburst debouched into Lake Souris there was a deep basin (~30 m deep) in the southern portion of the lake and a moderately deep basin (~13 m deep) in the central portion separated by a linear, northwest-southeast trending till ridge. The basin geometry is reflected by the distribution of sediments; sand and gravel filled the basins whereas fine-grained sediment was deposited on the till high (Fig. 10). Two coarse-grained fans formed: one at the inlet and the other in the central basin where water was funneled through a gap in the till high.

A cross section from the inlet to the southern basin center reveals the textural complexity induced by irregular bottom topography; note that the basic pattern of fining away from the inlet is present (Fig. 11). Entirely homogenous sections of very well-sorted sands tens of meters thick are common in the basin centers. The outburst induced drainage of Lake Souris is evidenced by the scarcity of silt and clay in the fan sediments (Fig. 11) and by partial incision of the Souris spillway into the early deposited fan sediments (Fig. 10). Most of the silt and clay eroded to form the spillways was eventually deposited offshore in Lake Agassiz. Underflow fans like the Lake Souris fan were deposited in Glacial Lakes Hind and Dakota.

AREAL DISTRIBUTION OF OUTBURST FLOODS

Deposition of glacial sediment by the Laurentide Ice Sheet was greatest in the sedimentary lowlands bordering the Canadian Shield. Mesozoic and Cenozoic sedimentary rocks consist of poorly lithified marine and continental clastics that provided little resistance to glacial erosion; glacier movement was facilitated by deformation of these materials, particularly where the ice was frozen to its bed. The model of spillway erosion presented above applies only to areas of soft sediment of low resistance to fluvial erosion.

Two areas were selected for study of Laurentide outburst flood phenomena. These are the region west of Glacial Lake Agassiz, an area which contains several smaller glacial lakes and numerous spillways, and the Great Lakes region, from which floods travelled down the Mississippi Valley to the Gulf of Mexico.

Western Region

The western region (Fig. 2) includes parts of Saskatchewan, Manitoba, Minnesota, North Dakota, South Dakota, and Montana. West of Lake Agassiz, glacial meltwater ponded in Glacial Lakes Regina, Arcola, Souris, Hind, and Dakota. A network of spillways connects these lakes to Lake Agassiz, with the exception of Lake Dakota which drained into the Missouri River. Glacial lakes along the extreme southwestern margin of the Laurentide Ice Sheet also drained into the Missouri through a series of spillways in western Saskatchewan, Alberta, and Montana. These spillways were also produced by outburst floods, but will not be discussed further in this paper.

Lake Dakota (Fig. 2) was one of the first lakes in the western region to drain. Lake Souris also drained into Lake Dakota through the James spillway during an earlier period of Late Wisconsinan retreat. Later, after the ice readvanced across the Lake Souris basin, Lake Souris formed again along the ice margin and released an outburst down the Sheyenne spillway to Lake Agassiz (Kehew and Clayton, 1983). To the north and northwest, respectively, Lakes Hind and Regina were forming as melting progressed. These three lakes drained sequentially into Lake Agassiz, beginning with an outburst flood from Lake Regina. The arrival of this discharge at Lake Souris triggered an outburst through its northern outlet toward Lake Hind (Kehew and Clayton, 1983); subsequently, Lake Hind also drained as a result of the

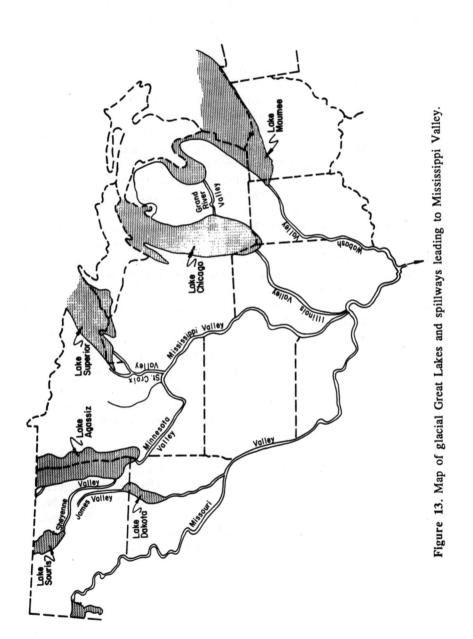

Figure 13. Map of glacial Great Lakes and spillways leading to Mississippi Valley.

huge influx of water. The drainage of Lake Hind incised the Pembina spillway, which terminated at Lake Agassiz. Lakes Regina, Souris, and Hind drained completely during these outburst events because of non-resistant sediment in the outlet areas and along the course of the spillways leading to the next glacial-lake basin.

As the ice retreated northward, new spillway systems developed. These included the Qu'Appelle and Assiniboine spillways in Saskatchewan and Manitoba (Fig. 2). The Qu'appelle spillway is a huge valley that conveyed meltwater from several lakes. In western Manitoba, the Qu'Appelle joins the Assiniboine spillway, which terminates at the Lake Agassiz basin at the Assiniboine "delta" near Brandon, Manitoba, a huge coarse-grained fan similar to those at the mouths of the Pembina and Sheyenne spillways. The Assiniboine spillway begins at a lake basin similar to Lakes Souris and Hind in that a large trench enters the basin from the north and a large spillway leaves the basin trending toward the south (Fig. 12). These relations suggest that the Assiniboine was cut by an outburst triggered by the inflow of an outburst at the northern end of the basin.

Lake Agassiz was probably much smaller than its maximum size (as shown on Fig. 2) when outbursts flowing down the Sheyenne, Pembina, and Assiniboine spillways entered the basin. The outlet during this period, prior to about 11,000 B.P. (Teller, 1985), was the Minnesota spillway, which begins at the southern end of the basin (Fig. 2). Three outbursts discharged down the Minnesota spillway from Lake Agassiz (Matsch, 1983). At Minneapolis, Minnesota, the Mississippi River joins the Minnesota spillway and flows in the trench cut by these floods. It is tempting to explain the outbursts leaving Lake Agassiz by floods entering the Lake from the Sheyenne, Pembina, and Assiniboine spillways. Unfortunately, these correlations cannot be proved or disproved with available evidence.

Downward incision of the Minnesota spillway was controlled by the presence of hard, Precambrian igneous and metamorphic rocks at a shallow depth beneath the glacial drift. These rocks prevented the lake from completely draining in the fashion of lakes to the northwest.

At approximately 11,000 B.P., the ice retreated sufficiently to drain Lake Agassiz through eastern outlets in Ontario (Teller, 1985; Teller and Thorleifson, 1983). The lake then dropped below the level of the southern outlet until the Marquette advance at 10,000 B.P. dammed the eastern outlets and raised Lake Agassiz back to the Campbell level, a level controlled by the southern outlet. Finally, beginning about 9500 B.P., the eastern outlets reopened, catastrophically draining Lake Agassiz through Lake Nipigon and into the Lake Superior basin (Teller, 1985).

Great Lakes Region

The Quaternary history of the Great Lakes is exceedingly complex. At times, however, lakes in the Lake Superior, Lake Michigan, and Lake Erie basins drained southward into spillways leading into the Mississippi Valley (Fig. 13). Reconnaissance investigation of the geomorphology of these spillways indicates that they are similar to spillways west of Lake Agassiz. Therefore, it is postulated that the im-

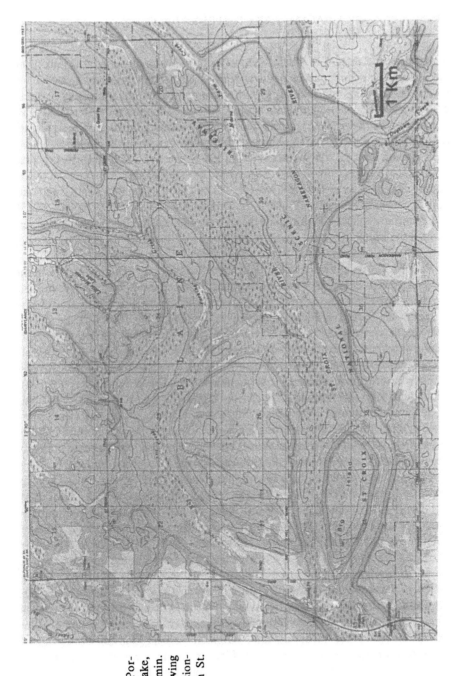

Figure 14. Portion of Webb Lake, Wisconsin 7.5 min. Quadrangle showing streamlined erosional topography in St. Croix spillway.

poundment and release of glacial meltwater in the form of outburst floods from these basins was similar to the occurrence of outbursts in the western region.

Ice advanced to moraines at the southwestern end of the Lake Superior basin at about 11,500 B.P. and then again during the Marquette advance at 10,000 B.P. (Farrand and Drexler, 1985). Lakes formed in the western Superior basin as the ice retreated from these positions. These early predecessors of Lake Superior, called Lake Duluth, drained via two spillways, the Portage and Brule spillways, which join to form the St. Croix Valley (Farrand and Drexler, 1985). Although the catastrophic eastern drainage of Lake Agassiz can be traced to the Superior basin, Lake Superior must have been well below the Duluth level at this time (Clayton, 1983) and therefore could not have drained through the Portage and Brule spillways. Even without the influx of Lake Agassiz water, however, floods from Lake Duluth must have been large, as evidenced by the topography of the St. Croix spillway.

The St. Croix spillway, which joins the Mississippi Valley downstream from the junction of the Mississippi Valley and the Minnesota spillway, is incised through glacial drift into bedrock along much of its course. For this reason, downcutting of the Portage and Brule outlets was limited and the level of Lake Duluth was stabilized near the elevation of the floor of the spillways. Although the morphology of spillways cut in bedrock differs from those incised entirely in unconsolidated glacial drift, the outburst origin of the St. Croix spillway is indicated by spectacular erosional topography along several reaches of the valley (Fig. 14). Streamlined erosional residuals formed from drift overlying bedrock rise as much as 30 m above the floor of the spillway (Big Island: Fig. 14), which provides a minimum estimate for water depth during the outburst floods. The width of the spillway in the area of Figure 14 is more than 5 km. These dimensions indicate that very large discharges passed through the spillway into the Mississippi Valley.

The southern outlet of the Lake Michigan basin, the Chicago outlet, consists of two spillways, the Des Plaines and Sag channels, which lead to the Illinois River valley (Fig. 13). The bedrock floor of these spillways provided a minimum elevation for Lake Chicago, the lake which drained through this outlet. Lake Chicago stabilized at three levels: the Glenwood, Calumet, and Toleston levels (Hansel et al., 1985). Traditionally, these levels have been interpreted as the result of episodic downcutting associated with rapid drops in lake-level elevation. This hypothesis is similar to the interpretation of the history of the southern outlet of Lake Agassiz. Alternatively, a new hypothesis based on radiocarbon dating of spillway sediment suggests that incision of the outlet to the bedrock elevation occurred during or before the Glenwood level (Hansel, et al., 1985; Hansel and Johnson, 1986). Such a scenario is consistent with the tendency of lake outlets to downcut during outbursts until the lake is drained or until a resistant horizon is reached. However, if the outlet was cut to bedrock (elev. 180 m) before the Glenwood Phase, this hypothesis requires an outlet water depth of 15 m during the entire Glenwood Phase (elev. 195 m) and 9 m during the Calumet Phase (elev. 189 m). Water depths of these magnitudes in an outlet spillway with a width of a kilometer or more would produce a huge discharge. Lake level could not remain stable during such an outlet discharge without a correspondingly large inflow to the basin. Although the Lake Michigan basin no doubt received outbursts from several other lakes (Hansel et al., 1985), it is difficult to imagine that these voluminous inflows and outflows could be associated with a stabilized lake level existing for hundreds of years.

Downstream from the Chicago outlet, the Illinois spillway conforms to the outburst-channel model described above (Fig. 15). It is a broad, steep-sided trench about 20 m in depth. Valley-side slopes and width are relatively constant. Streamlined erosional residuals within the inner channel (Buffalo Rock; Fig. 15) suggest deep, rapid flows of limited duration. Thus one or more outburst floods reached the Mississippi Valley from the Lake Michigan basin through the Illinois spillway.

Lake Maumee (Fig. 13) is the Lake Erie basin equivalent of Lake Chicago in the Lake Michigan basin and Lake Duluth in the Lake Superior basin -- an early, high-level lake impounded by lobate moraines of Late Wisconsinan glacial advanc-

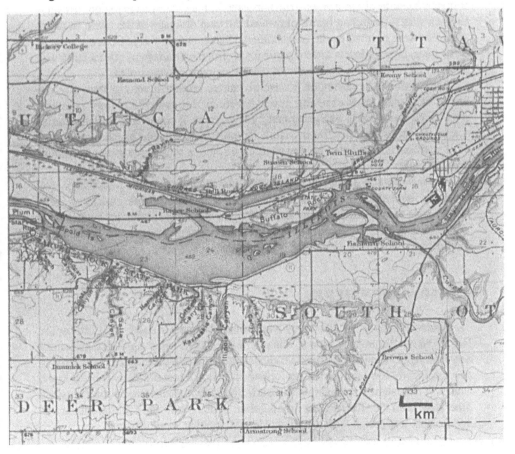

Figure 15. Portion of Ottawa, Ill. 15 min. Quadrangle showing Illinois spillway. Note streamlined hill in inner channel (Buffalo Rock).

es. Floods from Lake Maumee reached the Mississippi Valley via the Wabash Valley (Vaughn and Ash, 1983; Fraser and Bleuer, 1985). In addition, Lake Maumee and later lakes in the Erie and Huron basins drained through the Grand Valley (Fig. 13) into Lake Michigan Basin lakes at various times. The typical spillway morphology of the Grand Valley indicates that it was influenced by outburst floods.

CONCLUSIONS

Glacial lake outbursts played an integral part in deglaciation of the Laurentide Ice Sheet. Every major valley leading southward from the ice sheet, as well as numerous valleys trending in other directions, contains evidence of catastrophic glacial-lake outbursts. In the western region, complete drainage of smaller lakes was common because of the nonresistant substrate. In the Great Lakes region, resistant bedrock sills surrounding basins prevented the total drainage of the lakes. In many areas that were covered by the Laurentide Ice Sheet, outwash deposits are scarce. In these areas, impoundment and subsequent catastrophic drainage of glacial lakes was the primary mechanism for transfer of glacial meltwater to the ocean.

Outbursts can be attributed to several causes. Foremost among these was the retreat of glacial lobes exposing lower elevations through which drainage could pass. Once an outburst was generated by ice retreat, it could then initiate similar events in topographically lower basins. Domino-like drainage of lakes continued until stabilization occurred by downcutting of outlets to the elevation of resistant bedrock units.

The influence of glacial-lake outbursts upon the history of the Mississippi Valley is an interesting subject for further research. The probable impact of these floods upon sedimentation in the Mississippi fan has been pointed out by Kehew et al., 1986. Detailed studies of the depositional effects of these floods should shed further light on this important glaciofluvial processes.

ACKNOWLEDGEMENTS

Financial support for this study was provided to both Kehew and Lord by the North Dakota Geological Survey. The use of the NDGS auger rig is also gratefully acknowledged. The University of North Dakota supported the project in the form of several Faculty Research Grants to Kehew and two Doctoral Fellowships to Lord. The U.S. Bureau of Reclamation Missouri- Souris Project supplied boring logs and other data from the Glacial Lake Souris basin. Arden Mathison, Chief of the Drainage Branch, was particularly helpful and knowledgeable about this area.

REFERENCES

Baker, V. R., 1973, Paleohydrology and sedimentology of the Lake Missoula flooding in eastern Washington: Geological Society of America Special Paper 144, 79 p.

Bluemle, J. P., 1982, Geology of McHenry County, North Dakota: North Dakota Geological Survey Bulletin 74, Pt. I and North Dakota State Water Commission County Ground Water Studies 33, Pt. 1, 49 p.

Bluemle, J. P., 1985, Geology of Bottineau County, North Dakota: North Dakota Geological Survey Bulletin 78, Pt.I and North Dakota State Water Commission County Ground Water Studies 35, 57 p.

Boettger, W. M., 1986, Origin and stratigraphy of Holocene sediments, Souris and Des Lacs glacial-lake spillways, north central North Dakota: M.A. Thesis, Grand Forks, U. of North Dakota, 186 p.

Brophy, J. A. and Bluemle, J. P., 1983, The Sheyenne River; its geological history and effects on Lake Agassiz, in Teller, J. T. and Clayton, Lee, editors, Glacial Lake Agassiz: Geological Association of Canada Special Paper 26, p. 173-186.

Church, Michael and Gilbert, Robert, 1975, Proglacial fluvial and lacustrine environments, in Jopling, A.V. and McDonald, B.C., editors, Glaciofluvial and glaciolacustrine sedimentation: Society of Economic Paleontologists and Mineralogists Special Publication no. 23, p. 22-100.

Clayton, Lee, 1983, Chronology of Lake Agassiz drainage to Lake Superior, in Teller, J.T. and Clayton, Lee, editors, Glacial Lake Agassiz: Geological Association of Canada Special Paper 26, p. 290-307.

Ekblaw, G. E. and Athy, L. F., 1925, Glacial Kankakee Torrent in northeastern Illinois: Geological Society of America Bulletin, v. 36, p. 417-428.

Farrand, W. R. and Drexler, C. W., 1985, Late Wisconsinan and Holocene history of the Lake Superior basin, in Karrow, P.F. and Calkin, P.E., editors, Quaternary evolution of the Great Lakes: Geological Association of Canada Special Paper 30, p. 17-32.

Fenton, M. M., Moran, S. R., Teller, J. T., and Clayton, Lee, 1983, Quaternary stratigraphy and history in the southern part of the Lake Agassiz basin, in Teller, J. T. and Clayton, Lee, editors, Glacial Lake Agassiz: Geological Association of Canada Special Paper 26, p. 49-74.

Fraser, G. S. and Bleuer, N. K., 1985 Sedimentological consequences of two catastrophic flows in the Late Wisconsinan Wabash Valley: Geological Society of America Abstracts with Programs v. 17, no. 7, p. 586.

Hansel, A. K., Mickelson, D. M., Schneider, A. F., and Larsen, C. E., 1985, Late Wisconsinan and Holocene history of the Lake Michigan basin, in Karrow, P. F. and Calkin, P. E., editors, Quaternary evolution of the Great Lakes: Geological Association of Canada Special Paper 30, p. 39-54.

Hansel, A.K. and Johnson, W.H., 1986, New evidence for age of downcutting of the Chicago outlet: Geological Society of America Abstracts with Programs, v.18, no.4, p. 291.

Kehew, A. E., 1982, Catastrophic flood hypothesis for the origin of the Souris spillway, Saskatchewan and North Dakota: Geological Society of America Bulletin, v. 93, p. 1051-1058.

Kehew, A. E., 1983, Geology and geotechnical conditions of the Minot area, North Dakota: North Dakota Geological Survey Report of Investigations no. 73, 35 p.

Kehew, A. E. and Clayton, Lee, 1983, Late Wisconsinan floods and development of the Souris-Pembina spillway system in Saskatchewan, North Dakota, and Manitoba, in Teller, J. T. and Clayton, Lee, editors, Glacial Lake Agassiz: Geological Association of Canada Special Paper 26, p. 187-209.

Kehew, A. E. and Lord, M. L., 1986, Origin and large-scale erosional features of glacial-lake spillways in the northern Great Plains: Geological Society of America Bulletin, v. 97, p. 162-177.

Kehew, A. E., Lord, M. L., and Lowrie, Allen, 1986, Tracing catastrophic outbursts from the southern margin of the Laurentide Ice Sheet to deposition along the Louisiana offshore: Geological Society of America Abstracts with programs, v.18, no.6, p. 653.

Klassen, R. W., 1983, Assiniboine delta and the Assiniboine-Qu'Appelle valley system - implications concerning the history of Lake Agassiz in southwestern Manitoba, in Teller, J.T. and Clayton, Lee, editors, Glacial Lake Agassiz: Geological Association of Canada Special Paper 26, p. 211-230.

Komar, P. D., 1984, The lemniscate loop - Comparisons with the shapes of streamlined landforms: Journal of Geology, v. 92, no. 2 p. 133-146.

Lord, M. L., 1984, Paleohydraulics of Pleistocene drainage development of the Souris, Des Lacs, and Moose Mountain spillways, Saskatchewan and North Dakota: M.S. Thesis, Grand Forks, U. of North Dakota, 162 p.

Lord, M. L. and Kehew, A. E., in review, Paleohydrology and sedimentology of glacial-lake outbursts in southeastern Saskatchewan and North Dakota: Geological Society of America Bulletin.

Malde, H. E., 1968, The catastrophic late Pleistocene Bonneville flood in the Snake River Plain, Idaho: U.S. Geological Survey Professional Paper 595, 52 p.

Marcus, M. G., 1960, Periodic drainage of glacier-dammed Tulsequah Lake, British Columbia: Geographical Review, v. 50, p. 89-106.

Matsch, C. L., 1983, River Warren, the southern outlet to glacial Lake Agassiz, in Teller, J. T. and Clayton, Lee, editors, Glacial Lake Agassiz: Geological Association of Canada Special Paper 26, p. 231-244.

Moran, S. R. and Deal, D. E., 1970, Glacial Lake Souris - a study in the complexity of a glacial-lacustrine environment: Geological Society of America Abstracts with Programs v. 2, p. 398.

Teller, J. T., 1985, Glacial Lake Agassiz and its influence on the Great Lakes, in Karrow, P.F. and Calkin, P.E., editors, Quaternary evolution of the Great Lakes: Geological Association of Canada Special Paper 30, p. 1-16.

Teller, J. T. and Thorleifson, L. H., 1983, The Lake Agassiz - Lake Superior connection: in Teller, J.T. and Clayton, Lee, editors, Glacial Lake Agassiz: Geological Association of Canada Special Paper 26, p. 261-290.

Vaughn, D. and Ash, D.W., 1983, Paleohydrology and geomorphology of selected reaches of the upper Wabash River, Indiana: Geological Society of America Abstracts with Programs, v. 15, no. 6, p. 711.

6

Catastrophic flooding into the Great Lakes from Lake Agassiz

James T. Teller and L. Harvey Thorliefson

ABSTRACT

As the margin of the Laurentide Ice Sheet in Ontario retreated northward from the Lake Superior basin, the barrier that had prevented the vast Lake Agassiz watershed from overflowing into the Great Lakes was breached. Although this process must have occurred many times during the Quaternary, and at least twice during retreat of Late Wisconsinan ice, only the series of catastrophic flood bursts associated with retreat of the Marquette glacial advance from the Superior basin after 10,000 B.P. is easily documented.

While ice lay across the divide between the Agassiz and Superior basins, overflow from Lake Agassiz and its 2×10^6 km^2 watershed discharged south into the Mississippi River system. As ice retreated north along this divide, down the regional slope toward the Hudson Bay basin, a series of cols were uncovered, and water began to spill eastward into the Nipigon and Superior basins. The initial discharge through each outlet increased abruptly when the remaining glacial dam across the channel or channel complex failed. The resulting catastrophic burst of water stripped away glacial sediment and plucked giant slabs of jointed Proterozic diabese, eroding deep channels as it made its way down the steep gradient to the Nipigon basin and then to the Superior basin east of Thunder Bay, Ontario. As the level of Lake Agassiz declined, so did the discharge, reaching an equilibrium flow of about 30,000 m^{3}s^{-1}. Successive flood bursts occurred as lower cols in the divide were uncovered during northward retreat of the Laurentide ice.

Maximum flows of 200,000 m^{3}s^{-1} were estimated using the Manning equation and shear stress calculations based on the size of bedload in these channels. The largest catastrophic bursts discharged 3000-4000 km^3 of water from Lake Agassiz into Lake Superior in only a year or two. In turn, this water overflowed east into the Huron basin and then through the North Bay outlet to the Ottawa River Valley and the St. Lawrence Lowland. During these catastrophic floods, the level of Lake Superior may have fluctuated dramatically and flow through the St. Lawrence Valley to the Atlantic Ocean increased by up to four times. Many of the numerous beaches and cut strandlines in the Lake Agassiz basin are related to periods of equilibrium flow between these floods.

INTRODUCTION

During part of the late Quaternary, Lake Agassiz was the largest lake in North America. Its drainage basin included most of central Canada, from the Rockies to the Great Lakes. Lake Agassiz filled the topographically low and isostatically depressed region south of the retreating Laurentide Ice Sheet in Manitoba, North Dakota, Minnesota, Saskatchewan, and northwestern Ontario, overflowing either south to the Mississippi River or east into the Great Lakes-St. Lawrence Valley system. Although water would have been impounded in this region each time advancing or retreating Laurentide ice impeded flow of the northward-draining rivers of Hudson Bay basin, the record of this proglacial lake and its overflow is largely related to the last deglacial period.

The areal extent and volume of Lake Agassiz varied substantially through the late glacial period, depending on the position of the ice margin, isostatic rebound, and on which outlet was carrying the overflow. The lake came into existence when ice of the Red River Valley Lobe retreated north of the Mississsippi-Hudson Bay (Lake Agassiz) divide for the last time about 11,700 B.P. The earliest lacustrine record is found in the southern part of the basin of North Dakota and Minnesota, and consists of interbedded tills and fine-grained sediment that reflects the seesaw battle between advance and retreat of the Red River-Des Moines Lobe after it reached its maximum in Iowa at 14,000 B.P.(e.g. Clayton and Moran, 1982; Fenton et al., 1983). Although ice-marginal fluctuations continued through the life of Lake Agassiz, waters in the basin gradually expanded northward into Manitoba and northwestern Ontario. Until about 10,800 B.P. overflow was south through the

Figure 1. Maximum areas covered by Lake Agassiz, Nipigon, Superior, and Nakina in the eastern outlet region during the Nipigon Phase of Lake Agassiz (stippled). Names of groups of channels (systems) that linked Lake Agassiz with Lake Nipigon are shown: individual outlets, numbered 1 to 17, are named in Table 1. Numbers 18 and 19, respectively, refer to the Dog-Kaministikwia and Kashabowie-Seine channels which carried water from Lake Agassiz directly to Lake Superior during the Moorhead Phase. Overflow into Lake Agassiz (heavy arrow) occurred during the Emerson Phase when Superior Lobe ice at the Marks-Dog Lake moraine formed Lake Kaministikwia (coarse stipples). Channels numbered 20 to 26 are named Wolf, Wolfpup, Shillabeer, Black Sturgeon, Nipigon, Cash and Pijitiwabik, respectively, and carried overflow from Lake Nipigon to Lake Superior mainly when they were at their highest levels; later extensions of these channels to lower levels are shown by dashed lines and arrows (reprinted from Teller and Thorleifson, 1983, Fig. 7, with permission of the Geological Association of Canada.).

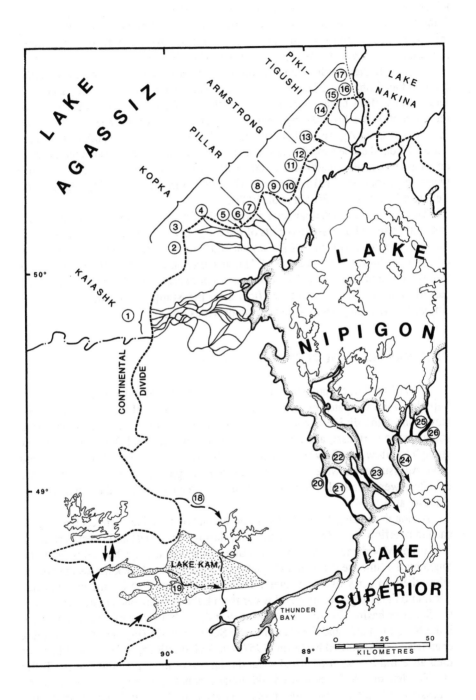

LAKE AGASSIZ

PIKI-TIGUSHI

ARMSTRONG

PILLAR

KOPKA

LAKE NAKINA

⑰
⑯
⑮
⑭
⑬
⑫
⑪
⑧ ⑨ ⑩
④ ⑤ ⑥ ⑦
③
②

KAIASHK

50°

①

CONTINENTAL DIVIDE

LAKE NIPIGON

49°

⑱

LAKE KAM.

⑲

THUNDER BAY

㉕
㉖
㉔
㉒
㉓
⑳ ㉑

LAKE SUPERIOR

90°   89°

0   25   50
KILOMETRES

River Warren outlet to the Minnesota River valley and the Mississippi River (Matsch, 1983). Lake levels declined from the highest (Herman) strandlines during this period in response to isostasy and erosion of the southern outlet channel (Upham, 1895; Leverett, 1932; Elson, 1967; Matsch, 1983).

As ice retreated north of the Thunder Bay, Ontario area, exposing a lower outlet through the Kashabowie-Seine valleys to the Kaministikwia River valley and Lake Superior, overflow began to spill eastward (Fig. 1). Continued northward retreat of the Laurentide ice down the regional slope, made still lower avenues available and the level of Lake Agassiz fell substantially. Most of the region south of the International Boundary became dry, as did a large area to the north, and an unconformity developed between 10,800 and 10,000 B.P. during this, the Moorhead low water phase (e.g. Johnston, 1916; Elson, 1957; 1967; Harris et al., 1974; Arndt, 1977; Teller and Last, 1981; Fenton et al.,1983).

The Marquette glacial advance invaded the Superior basin about 10,000 B.P., and again damned the eastern outlets of Lake Agassiz, causing overflow to return to the southern outlet (Elson, 1967; Clayton and Moran, 1982). Until this ice retreated from the Superior basin north of Thunder Bay about 9500 B.P., overflow was through the southern outlet. The main record of catastrophic discharge through the eastern outlets is the result of overflow after the Marquette advance, although a similar scenario may have existed during previous times.

THE EASTERN OUTLET CHANNELS

Two of the largest proglacial lakes in North America, Lake Agassiz and Lake Superior, were separated by less than 200 km during much of late glacial time. The elevation of the divide between these two basins declined then, as today, from southwest to northeast, and no cols south of the International Boundary were low enough to allow overflow from Lake Agassiz to Lake Superior. Thus, even when Lake Agassiz levels were highest during its early phase, no avenues of eastward overflow were available until the ice margin retreated north to the Thunder Bay, Ontario, area.

Although the Dog-Kaministikwia and Kashabowie-Seine channels (see Fig. 1) carried overflow directly to Lake Superior during the earliest and latest part of the low-water Moorhead Phase, the elevation of these spillways had rebounded too much by the time ice from the Marquette advance retreated back across the area to allow them to carry overflow again. Therefore, not until the ice margin retreated north into the Nipigon basin could water from Lake Agassiz once more overflow to the Superior basin. This time the routing was through a series of channels into the Lake Nipigon basin and then south to the Superior basin (Fig. 1). Many of these channels were identified by Elson (1957) and Zoltai (1965a, b; 1967), and they (or their ancestral forms) also carried water during one or more previous stages. Teller and Thorleifson (1983) recognized additional channels, grouping those between the Agassiz and Nipigon basins into five complexes, as shown in Figure 1, each con-

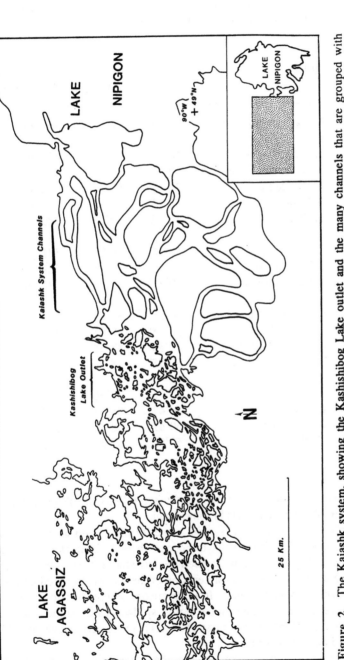

Figure 2. The Kaiashk system, showing the Kashishibog Lake outlet and the many channels that are grouped with it. Only the southernmost group of channels were deglaciated when outflow began. Lake Agassiz was at the lower Campbell level when this outlet opened and is here based on the 457 m (1500') contour line. Lake Nipigon was slightly above 305 m (1000') (after Thorleifson, 1983).

Table 1. The Nipigon Phase outlets of Lake Agassiz. Locations of systems and numbered outlets are shown in Figure 1 (Thorliefson, 1983).

System	Outlet	Location		Elevation of
		Latitude	Longitude	Water Level (m)
Kaiashk	1. Kashishibog Lake	49°47'	90°06'	451-468
Kopka	2. Aldridge Lake	50°06'	89°49'	439-465
	3. Gastmeier Lake	50°12'	89°52'	434-444
	4. Beagle Lake	50°15'	89°44'	415-442
	5. Loop Lake	50°13'	89°33'	396-423
	6. Boulder Lake	50°12'	89°26'	389-399
Pillar	7. Chief Lake	50°16'	89°20'	381-395
	8. Track Lake	50°17'	89°15'	381-395
	9. Badwater Lake	50°18'	89°08'	381-395
Armstrong	10. Little Caribou Lake	50°25'	89°04'	378-392
	11. Big Lake	50°28'	89°01'	381-392
	12. Michell Lake	50°30'	88°58'	375-386
	13. Linklater Lake	50°35'	88°53'	375-386
Pikitigushi	14. Arnston Lake	50°46'	88°50'	370-386
	15. Pickett Lake	50°47'	88°44'	350-377
	16. Clearbed Lake	50°47'	88°40'	350-377
	17. Whiteclay Lake	50°52'	88°39'	328-355

sisting of a number of outlets (Table 1). Thorleifson (1983) and Teller and Thorleifson(1983) describe these channels in some detail.

All channels are cut into bedrock, with some forming steep-walled canyons more than 100 m deep. The deepest canyons occur where the bedrock is Proterozoic diabase, because the jointing in this layered intrusive has allowed relatively easy erosion. In several places, the headward erosion of the diabase has produced deep plunge basins within these channels. In comparison, where overflow occurred across the massive Archean granites of the region, spillways are shallower and less steep-sided. Gradients along most of these channels average between about 0.0025 and 0.004 (or 2.5 and 4 m per km).

The southernmost outlet to Lake Nipigon, the Kaiashk system, is a complex of anastomosing channels that extends eastward from Kashishibog Lake (Fig. 2). Overflow from Lake Agassiz reached the continental divide and the Kaiashk system through a 15-km-wide corridor dominated by islands. To the north, outlets are progressively lower in elevation (Table 1), and also form an anastomosing network. Many of these channels converge downslope, so that the 17 identified outlets (Fig. 1 and Table 1) entered Lake Nipigon at only about 7 places (Figs. 3 and 4).

Large rounded boulders and cobbles are common in the channels (Fig. 5). Many of these exceed a half meter in diameter, and occasionally are more than 3 m in diameter. In the Kaiashk channel system there are areas where the entire floor is paved with cobbles and boulders that lack any finer matrix. As the gradient on the channel floor decreases toward the east, the coarse lag is replaced by gravel and sand. Where elevations fall below the former level of Lake Nipigon, surface deposits are even finer grained, being composed mainly of fine sand, silt, and clay, much of which is attributed to density underflows into deep water.

A number of large and deep canyons lie south of the Nipigon basin (Fig. 1), where temporarily ponded Lake Agassiz overflow and glacial meltwater from the retreating Laurentide Ice Sheet eroded mainly Proterozoic diabase and metasedimentary and metavolcanic bedrock en route to Lake Superior. In general, the elevations of these channels in the west are greater than those in the east. Like many spillways that lie west of Lake Nipigon, those to the south vary in width from less than a kilometer to several kilometers, and are bounded by vertical walls that today have a talus apron of angular diabase blocks along their base.

As the levels of both Lake Nipigon and Lake Superior declined through the late glacial period from their maximum, shown in Figure 1, there was a concomitant constriction of their boundaries, causing overflow channels to lengthen their course across the newly exposed relatively flat lake floor and to decrease their overall gradient. Entrenchment of many spillways followed as a result of crustal rebound and a continued decline in the level of these lakes.

PALEOHYDROLOGY
OF THE EASTERN OUTLETS

On numerous occasions, the level of Lake Agassiz remained constant long enough to allow beaches and cut scarps to develop along its margin. These strandlines were mapped by Upham (1895), Leverett (1932), and Johnston (1934, 1946), as well as by more recent workers (e.g. Prest, 1963; Elson, 1959; Clayton et al., 1980; Klassen, 1983). Figure 6 shows the major shorelines indentified in the Lake Agassiz basin and their relationship to the southern (Herman to Campbell beaches) and eastern (lower Campbell to Gimli beaches) outlets. Each strandline represents a period of time when inflow to the lake roughly equalled the outflow, and during which the lake level remained relatively constant. Given the number of beaches that formed during the last few thousand years of the lake's history, and the estimated rapid rate of crustal rebound during the early stages of deglaciation (e.g. Peltier and Andrews, 1983), the length of time available for each beach to form (i.e. its equilibrium time) was probably less than 50 years.

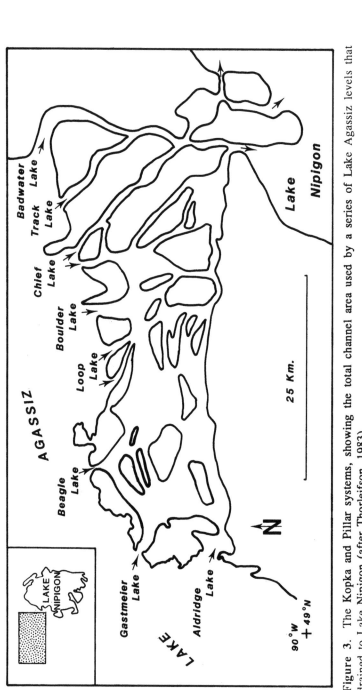

Figure 3. The Kopka and Pillar systems, showing the total channel area used by a series of Lake Agassiz levels that drained to Lake Nipigon (after Thorleifson, 1983).

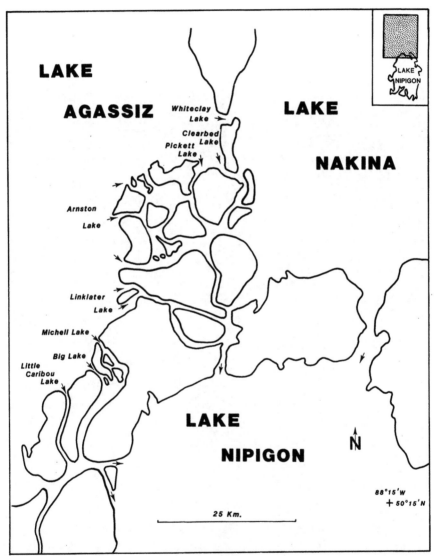

Figure 4. The Armstrong and Pikitigushi systems, showing the total channel area used by a series of Lake Agassiz levels that drained to Lake Nipigon. At the southwestern corner of the map is the Badwater outlet of the Pillar system (Fig. 3), which received some outflow from the Armstrong system (after Thorleifson, 1983).

Figure 5. Aerial view of boulder lag in the Kaiashk system.

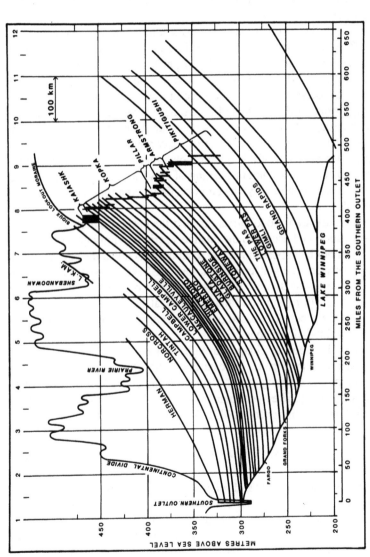

Figure 6. A southwest to northeast Lake Agassiz strandline diagram, showing former water planes that now rise in elevation toward the northeast because of differential crustal rebound. The eastern outlets to the Nipigon basin, as well as the southern outlet, are shown. All features projected onto a plane normal to regional isobases of crustal rebound (Teller and Thorleifson, 1983, Fig. 2, reprinted with permission of the Geological Association of Canada. Based on data from Upham, 1895; Johnston, 1946; Elson, 1967; Zoltai, 1967; Dredge, 1983; and Klassen, 1983).

Equilibrium overflow has been calculated for the southern (Matsch, 1983) and eastern (Teller, in press) outlets of Lake Agassiz. For the eastern outlets, this baseline flow ranges from 26,640 m^3s^{-1} to 32,660 m^3s^{-1}, which is determined by summing an estimated 400 km^3yr^{-1} from runoff due to precipitation and 440 to 630 km^3yr^{-1} of meltwater generated from the Laurentide ice as it retreated north across the Lake Agassiz watershed.

Discharge through these outlets was not constant, however, and was estimated by Matsch (1983) to range from 17,000 to 66,000 m^3s^{-1} for the southern outlet and by Catto et al., (1982) to range from 13,500 m^3s^{-1} to 70,700 m^3s^{-1} for the eastern outlets. Teller (1981) and Teller and Thorleifson (1983) recognized that outflow through the eastern spillways took place in a series of "steps", rather than uniformly through time, as the northward-retreating glacial dam failed at successively lower cols through the divide between the Lake Agassiz and Lake Nipigon basins (Table 1). Thus the maximum value of eastward discharge calculated by Catto et al. (1982) on the basis of precipitation plus meltwater alone, must be increased to take into account the abrupt opening of new and lower channels along the divide where, at least temporarily, the cross-sectional area for outflow was dramatically increased.

Estimating the maximum flood bursts through the eastern outlets is difficult. Post-flood modification and access to the channels seriously limit data collection. Furthermore, it has not been possible yet to accurately establish the chronological interrrelationship between the glacier margin, level of Lake Agassiz, and isostatically-controlled slope on the divide. It is not even clear how much erosion of the channels relates to the last (Nipigon Phase) floods at 9500-8500 B.P., and discharge calculations are necessarily based on modern channel geometry, rather than estimates of pre-last flood dimensions.

Calculations of the flood discharge through the southernmost of the eastern outlets, the Kaiashk system (Fig. 1), indicate that initial flow may have exceeded 200,000 m^3s^{-1} during the opening of this channel, if the entire 15-km-wide corridor through the Kashisibog Lake hydrologic divide (Fig. 2) was undammed at about the same time (Teller and Thorleifson, 1983). The Manning equation,

$$Q= (1/n)\, AR^{2/3}S^{1/2} \tag{1}$$

was used for this estimate, where Q is discharge in m^3s^{-1}, n is the Manning roughness coefficient (estimated to be about 0.03 to 0.05 for the outlet channels), A is the channel cross-sectional area, R is the hydraulic radius (area/wetted perimeter), and S is the dimensionless slope. An estimate of the depth of Lake Agassiz at the head of the Kaiashk channel was based on extrapolation of water planes (strandlines) to the drainage divide, because no clear high-water mark is present. Teller and Thorleifson (1983) also use the size of the bedload in the channels to provide an independent estimate for the hydraulic radius of the channel (and thus a check on the water depth estimate) by first relating critical mean shear stress (τ) for initiating particle movement to the maximum transported particle size (D) as follows (Baker and Ritter, 1975):

$$D = 65 \, \tau^{0.65} \qquad (2)$$

The hydraulic radius (R) can then be determined using the DuBoys equation,

$$\tau = \gamma R S \qquad (3)$$

where γ is the specific weight of the transporting fluid and S is the slope. Because particle diameter commonly exceeds 0.5 m in the Kaiashk outlet, as well as in channels to the north, the solution for discharge indicates that flows exceeded 100,000 $m^3 s^{-1}$ (Teller and Thorleifson, 1983).

After the initial catastrophic burst of water that followed the undamming of each of the eastern outlets, flow would have declined until equilibrium conditions were reached. The first flood through the Kaiashk channel occurred about 9500 B.P., when Lake Agassiz covered an area of about 350,000 km^2 and stood at the lower Campbell level. By the time the lake had fallen 12 m to the next equilibrium level, the McCauleyville beach (Fig. 6), about 4000 km of water had been lost from Lake Agassiz storage; this is about nine times the volume of present-day Lake Erie and two times the flood volume that produced the Channeled Scabland of Washington and Idaho (Pardee, 1942). If the average discharge during this catastrophic flood was 100,000 $m^3 s^{-1}$, then it would have taken only 15 months for the 4000 km^3 of water stored between the lower Campbell and McCauleyville beaches to flow out through the Kaiashk outlet to Lake Nipigon and into Lake Superior.

HISTORY OF LAKE AGASSIZ
WATERS "DOWNSTREAM"

Initial overflow from Lake Agassiz was southward to the Mississippi River Valley and then to the Gulf of Mexico. In combination with the isotopically light meltwater generated from the glacial margin to the east, these water strongly influenced the $\delta^{18}O$ record in the Gulf (e.g. Leventer et al., 1982; Kennett and Shackleton, 1975; Teller, in press). By about 12,000 B.P. most meltwater from the eastern region had been re-directed through the St. Lawrence Valley to the North Atlantic Ocean, and the dominant isotopic modification on Gulf of Mexico waters was by overflow from Lake Agassiz. Similarly, glacially-related deposition and erosion within the Mississippi River Valley and delta (e.g. Flock, 1983; Knox, this volume) continued until both the Great Lakes and Hudson Bay (Lake Agassiz) drainage basins could overflow into the St. Lawrence Valley. This occurred about 10,800 B.P. , but was interupted by a return to overflow to the Mississippi between 10,000 and 9500 B.P. when Marquette glacial ice readvanced into and across the Superior basin (Drexler et al., 1983).

Catastrophic flooding into the Great Lakes from Lake Agassiz is known to have occurred after retreat of the Marquette advance from the eastern outlets about 9500 B.P. Similar floods probably occurred between 10,800 and 10,000 (the Moorhead low water phase), although that record apppears to have been destroyed by the Marquette advance and ensuing floods.

The abrupt addition of thousands of cubic kilometers of Agassiz waters into the Superior basin would have briefly raised the level of that lake (Farrand and Drexler, 1985; Teller, 1985). Farrand and Drexler (1985) suggest that the large drift barrier that constricted outflow from Lake Superior to Lake Huron after retreat of the Marquette advance was progressively eroded by these Agassiz floods. They point out that there may be a cause-effect relationship between the five major outlet systems from Lake Agassiz (Fig. 1) and the five or six mapped shorelines below the highest Minong level that formed during the first thousand years after retreat of Marquette ice from the Superior basin.

Outflow of Lake Agassiz waters from the Superior basin probably was largely, if not entirely, into North Channel of the Huron basin after 9500 B.P., rather than into the main Lake Huron basin (Teller, 1985). By this time the North Bay outlet from the Huron basin to the Ottawa River and St. Lawrence valleys had opened (Fig. 7).

The sedimentary record in the Great Lakes and St. Lawrence Valley must have been affected not only by the catastrophic bursts that introduced up to 4000 km^3 of Lake Agassiz water to the Superior basin within a few years, but also by the baseline overflow of about 30,000 m^3s^{-1} from Lake Agassiz. When this baseline flow was added to that estimated by Catto et al. (1982) for discharge through the Ottawa River Valley from the eastern regions alone, the total flow in the Ottawa and St. Lawrence valleys more than doubled during the late glacial period when the western (Lake Agassiz) drainage basin was spilling its waters into the Great Lakes. During the brief periods of catastrophic overflow from Lake Agassiz to the Nipigon and Superior basins, discharge into the North Atlantic through the Ottawa-St. Lawrence system may have increased by another four times.

The possible affect of Lake Agassiz waters on the St. Lawrence valley has been noted by Teller (1985), Catto et al. (1982), and Graham and Teller (1984), although no serious attempt has yet been made to relate the history of the eastern region to outflow from Lake Agassiz. It is possible that the widespread red clay in the St. Lawrence Valley described by Gadd (1986 may, depending on its age, be related to catastrophic pulses of water emanating from the Superior basin, which, unlike other basins downstream, is underlain in many areas by red glacial sediment and bedrock. Others have aempted to relate the isotopic record of the North Atlantic Ocean to the influx of meltwater through the St. Lawrence during late glacial time (Fillon and Williams, 1984; Teller, in press).

There is little doubt that the integration of the 2,000,000 km^2 Lake Agassiz drainage basin into the Great Lakes-St. Lawrence system, which produced a drain-

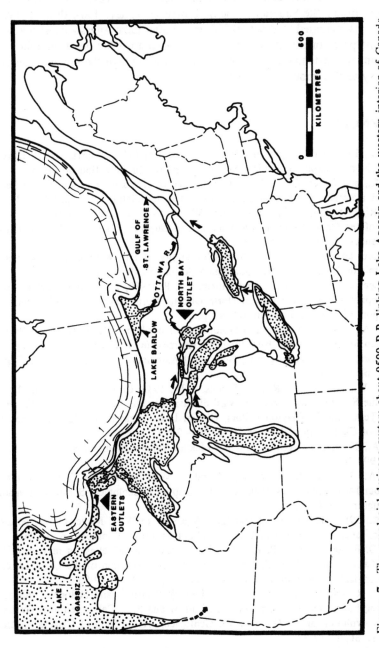

Figure 7. The proglacial drainage system about 9500 B.P. linking Lake Agassiz and the western interior of Canada through the Great Lakes and Ottawa-St. Lawrence valleys to the Atlantic Ocean (Teller, in press, Fig. 20, partly from data in Eschman and Karrow, 1985; Vincent and Hardy, 1979; Prest, 1970; and Elson, 1967).

age network that spanned the continent from the Rockies to the Atlantic during late glacial time, had a substantial influence on sedimentation and erosion in the downstream areas from Lake Superior to the St. Lawrence Valley. Flow through these eastern regions increased to about 50,000 m^3s^{-1} after Lake Agassiz overflow was added. Superposed on this flow were a number of catastrophic floods into the Superior basin, which temporarily increased the flow through the Great Lakes-St. Lawrence system by another three or four times.

REFERENCES

Arndt, B.M., 1977, Stratigraphy of offshore sediment of Lake Agassiz, North Dakota: North Dakota Geological Survey Report of Investigation 60, 58 p.

Baker, V.R., and Ritter, D. F., 1975, Competence of rivers to transport coarse bedload material: Geological Society of America Bulletin, v. 86, p. 975-978.

Catto, N. R., Patterson, R. J., and Gorman, W. A., 1982, The late Quaternary geology of the Chalk River region, Ontario and Quebec: Canadian Journal Earth Sciences, v. 19, p. 1218-1231.

Clayton, Lee, and Moran, S. R., 1982, Chronology of late Wisconsinan glaciation in middle North America: Quaternary Science Reviews, v.1, p. 55-82.

Clayton, Lee, and Moran, S. R., and Bluemle, J. P., 1980, Explanatory text to accompany the geologic map of North Dakota: North Dakota Geological Survey Report of Investigation No. 69, 93 p.

Dredge, L. A., 1983, Character and development of northern Lake Agassiz and its relation to Keewatin and Hudsonian ice regimes: in J. T. Teller and L. Clayton, editors, Glacial Lake Agassiz, Geological Association of Canada Special Paper 26, p. 117-131.

Drexler, C. W., Farrand, W. R. and Hughes, J. D., 1983, Correlation of glacial lakes in the Superior basin with eastward discharge events from Lake Agassiz: in J. T. Teller and L. Clayton editors, Glacial Lake Agassiz, Geological Associaton of Canada Special Paper 26, p. 309- 329.

Elson, J. A., 1957, Lake Agassiz and the Makato-Valders problem: Science, v. 126, p. 999-1002.

Elson, J. A., 1959, Surficial geology Brandon, Map 1067A: in E. C. Halstead, Ground-water resources of the Brandon map-area, Geological Survey of Canada Memoir 300, 67 p.

Elson, J. A., 1967, Geology of glacial Lake Agassiz: in W. J. Mayer-Oakes, editor, Life, land, and water, University of Manitoba Press, Winnipeg, p. 37-95.

Eschman, D. F., and Karrow, P. F., 1985, Huron basin glacial lakes: a review: in P. F. Karrow and P. E. Calkin editors, Quaternary evolution of the Great Lakes, Geological Association of Canada Special Paper 30, p. 79-93.

Farrand, W. R., and Drexler, C. W., 1985, Late Wisconsinan and Holocene history of the Lake Superior basin: in P. F. Karrow and P. E. Calkin, editors, Quaternary evolution of the Great Lakes, Geological Association of Canada Special Paper 30, p. 17-32.

Fenton, M. M., Moran, S. R., Teller, J. T., and Clayton, L., 1983, Quaternary stratigraphy and history in the southern part of the Lake Agassiz basin: in J. T.

Teller and L. Clayton editors, Glacial Lake Agassiz, Geological Association of Canada Special Paper 26, p. 49-74.

Fillon, R. H., and Williams, D. F., 1984, Dynamics of meltwater discharge from Northern Hemisphere ice sheets during the last deglaciation: Nature, v. 310, no. 5979, p. 674-677.

Flock, M., 1983, The late Wisconsinan savanna terrace in tributaries to the Upper Mississippi River: Quaternary Research, v. 20, p. 165-176.

Gadd, N. R., 1986, Lithofacies of Leda Clay in the Ottawa Basin of the Champlain Sea: Geological Survey of Canada Paper 85-21, 44 p.

Harris, K. L., Moran, S. R., and Clayton, L., 1974, Late Quaternary stratigraphic nomenclature, Red River Valley, North Dakota and Minnesota: North Dakota Geological Survey Miscellaneous Series 52, 47 p.

Johnston, W. A., 1916, The genesis of Lake Agassiz: a confirmation: Journal of Geology, v. 24, p. 625-638.

Johnston, W. A., 1934, Surface deposits and groundwater supply of Winnipeg map-area, Manitoba: Geological Survey of Canada Memoir 174, 110 p.

Johnston, W. A., 1946, Glacial Lake Agassiz with special reference to the mode of deformation of the beaches: Geological Survey of Canada Bulletin 7, 20 p.

Kennett, J. P., and Shackleton, N. J., 1975, Laurentide Ice Sheet meltwater recorded in Gulf of Mexico deep-sea cores: Science, v. 188, p. 147-150.

Klassen, R. W., 1983, Lake Agassiz and the late glacial history of northern Manitoba: in J. T. Teller and L. Clayton, editors, Glacial Lake Agassiz, Geological Association of Canada Special Paper 26, p. 97- 115.

Knox, J. C., this volume, Stratigraphic evidence of large floods in the upper Mississippi Valey: in L. Mayer and D. B. Nash, editors, Catastrophic flooding, 18th Annual Geomorphology Symposium, Miami University, Ohio.

Leventer, A., Williams, D. F., and Kennett, J. P., 1982, Dynamics of the Laurentide Ice Sheet during the last degalaciation: evidence from the Gulf of Mexico: Earth and Planetary Science Letters, v. 59, p. 11-17.

Leverett, F., 1932, Quaternary geology of Minnesota and parts of adjacent States: U. S. Geological Survey Professional Paper 161, 149 p.

Matsch, C. L., 1983, River Warren, the southern outlet of glacial Lake Agassiz:in J. T. Teller and L. Clayton, editors, Glacial Lake Agassiz, Geological Association of Canada Special Paper 26, p. 231-244.

Pardee, J. T., 1942 Unusual currents in glacial Lake Missoula, Montana: Geological Society of America Bulletin, v. 53, p. 1570-1599.

Peltier, W. R., and Andrews, J. T., 1983, Glacial geology and glacial isostasy of the Hudson Bay region: in D. E. Smith and A. G. Dawson, editors, Shorelines and isostasy, Institute of British Geographers Special Publ. 16, p. 285-319.

Prest, V. K., 1963, Red Lake-Lansdown House area, northwestern Ontario: surficial geology: Geological Survey of Canada Paper 63-6, 23 p.

Prest, V. K., 1970, Quaternary geology: in R. J. W. Douglas, editor, Geology and economic minerals of Canada, Geological Survey of Canada Economic Geology Report No. 1, p. 675-764.

Teller, J. T., 1981, The catastrophic drainage of glacial Lake Agassiz: Geological Society of America, Abstracts with Programs, v. 13, no. 7, p. 565.

Teller, J. T., 1985, Glacial Lake Agassiz and its influence on the Great Lakes: in P. F. Karrow and P. E. Calkin, editors, Quaternary evolution of the Great Lakes,

Geological Association of Canada Special Paper 30, p. 1-16.

Teller, J. T., in press, Proglacial lakes and the southern margin of the Laurentide Ice Sheet: *in* W. F. Ruddiman and H. E. Wright editors, North America and adjacent oceans during the last deglaciation, Geological Society of America, The Decade of North American Geology, v. K-3.

Teller, J. T., and Last, W. M., 1981, Late Quaternary history of Lake Manitoba, Canada: Quaternary Research, v. 16, p. 97-116.

Teller, J. T., and Thorleifson, L. H., 1983, The Lake Agassiz-Lake Superior connection: *in* J. T. Teller and L. Clayton, editors, Glacial Lake Agassiz, Geological Association of Canada Special Paper 26, p. 261-290.

Thorleifson, L. H., 1983, The eastern outlets of Lake Agassiz: M. Sc. Thesis, University of Manitoba, Winnipeg, 87 p.

Upham, W., 1895, The glacial Lake Agassiz: U. S. Geological Survey Monograph 25, 658 p.

Vincent, J-S., and Hardy, L., 1979, The evolution of glacial Lakes Barlow and Ojibway, Quebec and Ontario: Geological Survey of Canada Bulletin 316, 18 p.

Zoltai, S. C., 1965a, Glcail features of the Quetico-Nipigon area, Ontario: Candaian Journal Earth Sciences, v. 2, p. 247-269..

Zoltai, S. C., 1965b, Thunder Bay surficial geology: Ontario Department Lands and Forests Map S265, scale 1:506:880.

Zoltai, S. C., 1967, Eastern outlets of Lake Agassiz: *in* W. Mayer-Oakes (ed.), Life, land, and water, University of Manitoba Press, Winnipeg, p. 107-120.

7

Drainage of Lake Wisconsin near the end of the Wisconsin Glaciation

Lee Clayton and John Attig

ABSTRACT

Lake Wisconsin existed in late Wisconsin time whenever the Green Bay Lobe of the Laurentide Ice Sheet overrode the east end of the Baraboo Hills, damming the drainage from central and northern Wisconsin; water then drained from the northwest end of the lake, down the East Fork Black River to the Mississippi River. When the ice margin retreated from the Baraboo Hills, the lake drained southward down the Wisconsin River. The geometry of the area of the ice dam, the sandstone gorges at Wisconsin Dells, a large area around the gorges that has been scoured free of lake sediment, and fans of coarse sediment downstream from the outlets indicate that the southward drainage of the main basin and the Lewiston basin was catastrophic.

INTRODUCTION

Proglacial lakes are generally unstable and subject to rapid drainage. When a lake rises and spills over a low point in the basin's drainage divide, a flood may occur as water rapidly cuts through easily eroded sediment at the divide. Larger floods can occur when the lake water quickly cuts across, through, or under an ice dam. Floods should be considered a possibility at most glacial-lake outlets. Since Baker (1971) confirmed the catastrophic flooding of Lake Missoula, first recognized by Bretz (1923), an argument for the catastrophic drainage of several large glacial lakes in the midcontinent region has been advanced (for example, Teller and Thorleifson, 1983; Kehew and Lord, 1986). In this paper we discuss the drainage of Lake Wisconsin near the end of the Wisconsin Glaciation. The geometry of the ice dam, the distribution of the sandstone gorges at Wisconsin Dells, the scouring of lake sediment, and fans of coarse sediment downstream from lake outlets indicate that Lake Wisconsin drained rapidly, probably catastrophically.

Lake Wisconsin has been recognized for many years. Alden (1918) mapped the southern part of the lake basin. Harloff (1942) described the distribution of the silt and clay deposits in the main basin. Bretz (1950) described the Alloa delta (discussed below) at its southeastern outlet. He remarked that, except for a delta at the head of a fiord in Greenland, he had never seen a more bouldery delta. He referred to the Alloa delta as a "torrential delta," and recognized that it resulted from the drainage of Lake Wisconsin through the Lewiston basin to Lake Merrimac, but he said (p. 135), without giving reasons, that "no debacle occurred during this failure of the ice barrier at the east end of the [Baraboo] Ranges." Alden (1918, p. 288), who later became a critic of Bretz's Lake Missoula flood hypothesis (Baker, 1978, p. 8-9), noted that "the discharge of such a volume of water must have been an event of considerable moment."

More recently, Socha (1984) mapped the glacial geology in the southern part of the lake basin. Brownell (1986) augered a series of testholes, constructed several stratigraphic sections across the main basin, and described the lake sediment. Clayton (in press, a and b; in preparation, a and b) has mapped most of the lake plain; the southern and extreme western parts will be mapped in the near future, and more detailed work is being done by Attig and Clayton in the Devils Lake and Wisconsin Dells areas.

DESCRIPTION OF LAKE WISCONSIN

Lake Wisconsin was dammed any time the Green Bay Lobe of the Laurentide Ice Sheet overrode the east end of the Baraboo Hills (Fig. 1), blocking the south-flowing drainage from central and northern Wisconsin. The water ponded north of the ice dam until it rose to the level of the northwest outlet into the East Fork Black River, which drained to the Mississippi River. The lake was then about 100 km long. This outlet functioned whenever the ice front was at or between the position of the Johnstown moraine and one of the Elderon moraines (Figure 1), during much of the Wisconsin Glaciation after 25,000 B.P. When the ice margin retreated east of that Elderon moraine, the contact between the glacier and the east end of the Baraboo Hills was below the elevation of the outlet to the East Fork Black River,

Figure 1. Lake Wisconsin, shown just as the water started to flow from the Alloa outlet (in the southeast); water was still flowing down the East Fork Black River, probably by way of the west Babcock outlet (in the northwest). Flow between the various basins is indicated by arrows. Immediately prior to the opening of the Alloa outlet, water flowed from the Lewiston basin to the Hulburt and main basins, opposite to the direction shown here. The contours indicate the elevation (in metres) of the water plane just before water began to flow through the Alloa outlet; elevations are relative to present sea level and crustal position. Water in all the Lake Wisconsin basins was at nearly the same level, but water in Lake Merrimac (in the southeast) was about 30 m lower.

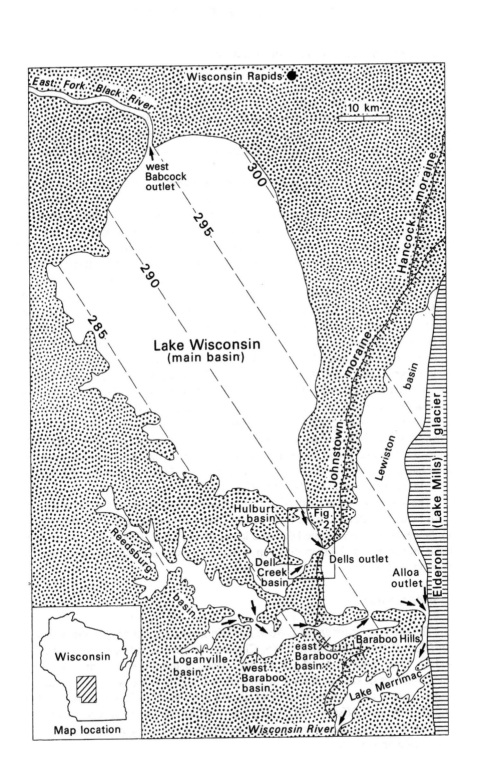

Wisconsin Rapids

East Fork Black River

10 km

west Babcock outlet

300

295

290

285

Lake Wisconsin (main basin)

Hancock moraine

Johnstown moraine

Lewiston basin

Elderon (Lake Mills) glacier

Reedsburg basin

Hulburt basin

Fig 2

Dells outlet

Dell Creek basin

Alloa outlet

east Baraboo basin

Baraboo Hills

Loganville basin

west Baraboo basin

Lake Merrimac

Wisconsin River

Wisconsin

Map location

and water from the lake could spill southward.

The water occupied several basins. The main basin (Fig. 1) northwest of Wisconsin Dells is about 25 m deep and 80 km long. (The names "Dells" and "Wisconsin Dells" refer both to the city of Wisconsin Dells and to the dells or dalles--which are sandstone gorges--along the Wisconsin River near the city.) The main basin drained out the northwest outlet until the ice margin retreated from the east end of the Baraboo Hills; then the main basin drained southeastward into the Lewiston basin.

Southwest of the Dells is Dell Creek basin (Fig. 1), which is about 15 km long. Throughout its existence, this part of Lake Wisconsin flowed northeastward into the main basin through a narrow channel (Figs. 1 and 2). To the southwest is the Reedsburg basin. Part of the time the water in this basin drained into the Dell Creek basin by way of a channel, now clogged with shore sediment, 6 km east of Reedsburg. At other times it drained into the west Baraboo basin through Rock Springs gorge (also called Ablemans Gorge or the Upper Narrows), indicated by an arrow in Fig. 1). The Loganville basin occupies a small area in Narrows Creek valley. The water in it drained either into the Reedsburg basin or the west Baraboo basin. The west Baraboo basin occupies part of the Baraboo valley west of the city of Baraboo. Water in it drained westward into the Reedsburg basin through Rock Springs Gorge whenever the ice margin was west of the Lower Narrows of the Baraboo River. (The Lower Narrows is indicated by an arrow on the northeast side of the east Baraboo basin in Fig. 1.) It drained into the east Baraboo basin after the ice margin retreated from the Lower Narrows. Water in the east Baraboo basin drained into the west Baraboo basin until the Lower Narrows became free of ice. Then it drained into the Lewiston basin, which is 50 m deep and 60 km long. The water drained into the main basin through a narrow gap in the Johnstown moraine until the ice melted from the east end of the Baraboo Hills. It then drained through the Alloa outlet, into Lake Merrimac (Fig. 1).

The water in the the basins described above was all at nearly the same level; it is all here considered to have been part of Lake Wisconsin. Lake Merrimac (Fig. 1), however, was a separate lake, at a lower level (Bretz, 1950). The water from Lake Merrimac drained through a gap in the Johnstown moraine and then down the Wisconsin River.

Offshore sediment

The offshore sediment of Lake Wisconsin is well known only in the main basin (Harloff, 1942; Brownell, 1986; Clayton, in press, a and b; in preparation, a and b). Over most of the lake plain, the surface material is eolian sand overlying several metres of sand deposited by underflow currents in the lake. This in turn is underlain by a few metres of offshore silt and clay (the New Rome Bed), which is underlain by more offshore sand. One or more additional clay beds occur at depth in some areas. In the central part of the basin the offshore sediment is typically several tens of metres thick.

Shorelines

Before the history of drainage of Lake Wisconsin could be evaluated, it was necessary to locate the shorelines so that the water depth and the location of the outlets could be determined. This had not been previously done; therefore, a detailed discussion of the shorelines is given here.

Permafrost existed until after Lake Wisconsin drained (Attig and Clayton, 1986). As a result, solifluction has greatly modified any shore features. The beaches of proglacial lakes in the Superior and Michigan basins formed later, after permafrost had melted, and are nearly continuous, easily recognizable features. In contrast, beaches or shore terraces formed around Lake Wisconsin are, at best, obscure. Remnants are difficult to recognize because their form has been modified by solifluction and slopewash when permafrost was present as well as by later slope processes and wind erosion or deposition. Few have exposures big enough for sedimentologic analysis. As a result, criteria other than geomorphic form and sedimentologic characteristics had to be used for identification of shorelines. These criteria include the highest occurrences of offshore silt and clay, the lowest occurrence of braided-meltwater-stream surfaces west of the Johnstown moraine, the highest occurrences of rafted erratics, the occurrence of shore-ice-collapse trenches (Clayton and Attig, 1986 and in press), and the elevation of outlets and deltas.

Only shoreline features associated with the outlet to the East Fork Black River have been recognized, perhaps because drainage of the lake was so fast after this outlet was abandoned. There is little evidence that the outlet was downcut significantly during its existence, probably because it was floored by resistant Precambrian rock. This shoreline, therefore, should occur as a single feature at the outlet and also at a point on the opposite shore to the southeast. To the northeast and southwest of a line through the outlet and that point, this shoreline should be multiple, with an apparent pivot along that line, as a result of crustal depression as the glacier margin advanced and crustal rebound as it retreated. On either side of the pivot line a number of shorelines can be recognized, but only the highest shoreline features in an area can be correlated with each other with much confidence. Below that are scattered shoreline remnants that are difficult to correlate.

In the northeast half of the main basin, northeast of the outlet, the highest shoreline rises N. 60^0 E. at a gradient of 0.6×10^{-3}. Because this is the steepest shoreline, it is interpreted to have formed at the time of greatest crustal depression, at the time of maximum glacier-margin advance, when the Johnstown and Hancock moraines (Fig. 1) formed, roughly 15,000 B.P. (Attig and others, 1985). The East Fork Black River has long been known to have carried the outflow of Lake Wisconsin (Martin, 1932, p. 340), but the exact position of the outlet across the modern drainage divide has received little attention because it occurs in an area of flat swampy land with no obvious channel form and because the exact level of the water plain and the amount and direction of crustal rebound have been unknown until now. Slightly different outlets probably functioned at different times. These outlet channels have apparently been nearly filled with postglacial river sediment, but two faint linear depressions of an appropriate width occur west of Babcock; probing by power auger and ground-penetration radar suggest the presence of channel forms below these depressions. Because this was the time of greatest crustal depression, the eastern of these two outlets, 2 km west of Babcock, at a present-day elevation of about 298 m, probably functioned at this time.

In the Lewiston basin (Fig. 1), the only recognized shoreline rises about N. 55° E. at a gradient of roughly 0.3×10^{-3} Because the basin was inundated only after the ice margin retreated from the Johnstown moraine, this Lewiston shoreline is interpreted to have formed after the Johnstown shoreline. It is less steep than the Johnstown shoreline because it formed after some crustal rebound had occurred. It formed no later than the time the Elderon moraine formed, around 13,000 B.P. (Attig and others, 1985). Because some crustal rebound had occurred since Johnstown time, the western, and fresher looking, of the two outlets near Babcock probably functioned at this time. It is 5 km west of Babcock, at a present-day elevation of about 297 m.

In the southwestern half of the main basin, the highest shoreline rises N. 45° E. at a gradient of about 0.1×10^{-3}. It had to have formed at a time of minimum crustal tilt, at either the very beginning or the very end of the existence of Lake Wisconsin. This shoreline seems too well preserved to have formed in pre-Wisconsin time. So it must have formed either just as the ice margin covered the east end of the Baraboo Hills, perhaps around 23,000 B.P., or just before the east end of the hills were uncovered, about 13,000 B.P. (Attig and others, 1985). Because the late shoreline has already been identified in the Lewiston basin (previous paragraph), it is interpreted to be the early one, formed around 23,000 B.P.

Furthermore, the early shoreline should be less steep than the late one because the crustal response lags behind the glacial fluctuation; the southwestern shoreline is considerably less steep than the one in the Lewiston basin. Because it is less steep, the outlet may have been west of the west Babcock outlet, discussed in the previous paragraph. The next low point on the present drainage divide to the west is 3 km southeast of City Point, at about the same elevation as the west Babcock outlet, but no channel form is present there.

RAPID DRAINAGE

Rapid drainage probably occurred when Lake Wisconsin spilled across low points of divides, but here we discuss only the flood that was triggered when the ice dam at the east end of the Baraboo Hills failed, with probable catastrophic flow through the Alloa and Dells outlets (Figs. 1, 2, and 3).

Alloa outlet

The west side of the Alloa outlet (Fig. 1) consists of Precambrian quartzite, rising about 40 m in a horizontal distance of 300 m. At the start of discharge, water stood at an elevation of about 294 m (relative to modern sea level and crustal position; Fig. 3). Down to an elevation of about 275m, the channel is today one sided; during the height of the flood the east side consisted of ice. Downslope from that elevation the channel is about 1 km wide, and the east bank consists of thin Pleistocene sediment over Cambrian sandstone. The channel, when the water stood at about 275 m and the Alloa delta (described below) to the south was built, was only about 2 km long. When the water dropped below that level, it began to cut through the Alloa delta; the channel through the delta is about 0.4 km wide. The high point of the channel talweg is about 255 m. The thickness and character of Pleistocene fill below the bottom of the channel is unknown.

The Alloa outlet is similar to the outlet of Russell Lake in Alaska (see photos, Eliot, 1987). Russell Lake formed when the Hubbard Glacier advanced to Gilbert Point, damming Russell Fiord. The water level in Russell Lake rose until October 8, 1986, when the ice dam failed and the lake catastrophically drained into the ocean. Similarly, the west edge of the Green Bay Lobe retreated until only a small amount of ice prevented the lake from draining southward at the east end of the Baraboo Hills (Fig. 1). When the ice dam failed, water impounded in Lake Wisconsin flowed southward into Lake Merrimac and then into the Wisconsin River (Fig. 1).

Until the drainage of Lake Wisconsin, Lake Merrimac was dammed by the Johnstown moraine near Prairie du Sac (south edge of Fig. 1) and filled the basin between the Johnstown moraine and the Alloa outlet (Bretz, 1950). The dam has been eroded but the lake elevation was at about 255 m if an abandoned outlet 4 km north of Prairie du Sac was in use then. A large delta was deposited in Lake Merrimac at the lower end of the Alloa outlet. The delta is at an elevation of about 264 m or about 9 m higher than the elevation of the lake at the outlet north of Prairie du Sac. This difference is in agreement with the 9 m of differential rebound that would be expected based on the differential rebound of the main Lake Wisconsin shoreline.

The Alloa delta is the main evidence for a flood through the Alloa outlet. The delta and the channel cut into it have been described by Bretz (1950) and Socha (1984). Foreset beds exposed in gravel pits consist of bouldery cobble gravel with foreset beds more than 12 m high. Bretz (1950) noted that the Alloa delta has abundant boulders more than 1 m in diameter.

The flood from the Lewiston basin presumably deepened and widened the outlet of Lake Merrimac, causing it to drain. The main high terrace of the Wisconsin River at Prairie du Sac is the outwash plain of the glacier when it stood at the Johnstown moraine (Fig. 1); except for the part of the terrace next to the moraine, it does not contain many boulders of Baraboo quartzite. The next lower terrace, occupied by Sauk City, contains a conspicuous number of quartzite boulders; this surface was probably formed by the combined floods of Lake Wisconsin and Lake Merrimac. The boulders were probably carried by talus-laden ice from the east end of the Baraboo Hills that was washed through the Alloa outlet as icebergs that floated down the Wisconsin River on the flood.

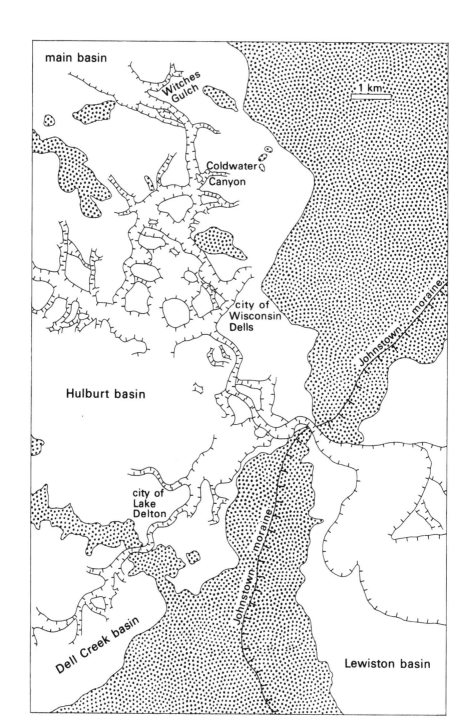

main basin

Witches
Gulch

1 km

Coldwater
Canyon

city of
Wisconsin
Dells

Johnstown moraine

Hulburt basin

city of
Lake
Delton

Dell Creek basin

Johnstown moraine

Lewiston basin

Dells outlet

The configuration of the Lake Wisconsin basins at the start of the Alloa flood is shown in Figure 1. Just before the Alloa outlet opened, water flowed from the Lewiston basin to the Hulburt basin to the main basin, the reverse of the flow indicated by arrows in Figure 1. The Dells outlet at that time was only big enough to carry the discharge of the Lewiston basin, which included meltwater from the glacier margin northward from the Baraboo Hills for about 350 km.

As soon as the water level in the Lewiston basin began to drop, flow reversed in the Dells outlet. If the Lewiston basin did drain catastrophically, as suggested above, the Dells outlet might have downcut rapidly, causing the Hulburt basin, the Dell Creek basin, and the main basin to catastrophically drain through the Dells area (Fig. 2).

The Dells outlet would have had to downcut through no more than about 15 m of easily eroded sandy till of the Johnstown moraine, which was here only about 0.5 km wide; the moraine would have been quickly breached. When the moraine was removed, the water would have been flowing across about 10 km of easily eroded sand and poorly cemented sandstone of the Cambrian Mt. Simon Formation. Judging from the effects of catastrophic floods from other glacial lakes, a network of gorges should have resulted. Here, in fact, are a set of gorges, including the Dells of the Wisconsin River, which we consider to be evidence for a Lake Wisconsin flood (Figs. 2 and 3).

The Dells have long been a favorite Midwest tourist site because of the spectacular scenery in the sandstone gorges (Martin, 1932). The gorge of the Wisconsin River, which is the central gorge shown in Figure 2, is typically about 0.2 km wide but it narrows to 15 m at water level in the Narrows. Through most of the Dells the sandstone cliffs rise about 25 m above water level. At the Narrows the river is 10 to 20 m deep. Side gorges such as Witches Gulch and Coldwater Canyon (north end of Fig. 2) are as deep as several tens of metres but are only 1 m wide in some places. Gorges that lack modern streams are less conspicuous because the sandstone cliffs have begun to erode. The gorges form a complex network (Fig. 2), with smaller channels crossing local divides.

The Dells outlet is strikingly different from the Alloa outlet because the Baraboo quartzite at the Alloa outlet was nearly unerodible, whereas the Mt. Simon sandstone at the Dells was highly erodible. The sandstone of the cliff walls of the gorges give the appearance of considerable durability, but once the case-hardened surface has been removed, it can be dug with a shovel.

Figure 2. Wisconsin Dells area. Lines with tick marks are banks of gorges, which were at least partly cut by a flood or floods from the Dell Creek basin, the Hulburt basin, and the main basin of Lake Wisconsin. All gorges upstream (west) from the Johnstown moraine have sandstone walls; those downstream (east) from the moraine are partly sandstone and partly Pleistocene sediment. Other sandstone gorges occur up to 9 km north of the north edge of the map. The areas with a dotted pattern were above shoreline just before the flood. The location of the map is shown in Figure 1.

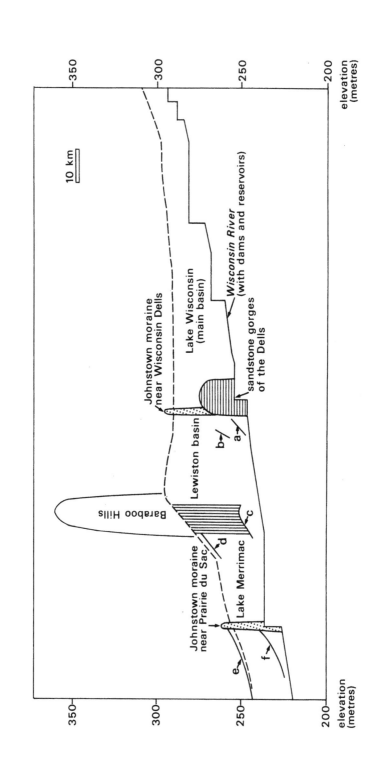

elevation (metres)

350 — 300 — 250 — 200

10 km

Johnstown moraine near Wisconsin Dells

Lake Wisconsin (main basin)

Wisconsin River (with dams and reservoirs)

sandstone gorges of the Dells

b
a

Baraboo Hills

Lewiston basin

c

Johnstown moraine near Prairie du Sac.

d

Lake Merrimac

e
f

elevation (metres)

Other origins for the gorges at Wisconsin Dells have been suggested. Martin (1932, p. 349) thought some were cut by meltwater from the glacier to the east. However, this does not explain those coming from the west, such as the one from Dell Creek basin (Fig. 2), where there was no meltwater source. In any case, meltwater flowed westward from the glacier in this area only when Lake Wisconsin was in existence; the lake drained when the ice stood at one of the Elderon moraines (Fig. 1), when meltwater could reach this area only from northern Wisconsin by way of the Wisconsin River.

Martin (1932, p. 351) also suggested that the major gorges were cut by the modern streams. The gorges of the Wisconsin River and Dell Creek certainly have been modified by postflood erosion. Witches Gulch and Coldwater Canyon (northern part of Figure 2) might also have been deepened by the streams now occupying them, but these extremely narrow gorges seem to consist of a series of interconnected potholes (Thwaites and others, 1935, p. 135), probably formed during a flood from the main basin of Lake Wisconsin. Many of the gorges have no modern stream in them, and many of them cross modern divides, indicating that they were cut by premodern streams.

In addition, it can be shown that all the gorges were cut down to an elevation of at least 263 m (Fig. 3) before about 13,000 B.P. because tundra polygons occur on postflood fluvial sediment at that elevation 3.5 km north of the north edge of Figure 2. This sediment was deposited by streams that eroded gullies back into the flood-cut scarp at the foreset face of a proglacial delta into Lake Wisconsin. The polygons formed before 13,000 B.P. because the permafrost in the region melted at about that time (Attig and Clayton, 1986).

Other evidence for a flood through the Dells area includes the washed upland between the gorges. The main basin and the other smaller basins are largely floored with lake sediment. The main exception is the Dells area. In the main basin, the lake sediment tends to thicken into the deeper part of the basin. In the deepest southern part of the basin, however, where the lake sediment would be expected to be thick, the Cambrian sandstone has been scoured free of lake sediment. The scoured area includes the area upstream from the Johnstown moraine and below the shoreline shown in Figure 2 plus an area several kilometres wide extending several kilometres north of Figure 2. The lake sediment has been scoured down to an ele-

Figure 3. Profile of Lake Wisconsin when the Lewiston basin started to drain into Lake Merrimac. Profile was drawn along the Wisconsin River from just south of Wisconsin Rapids (right end of profile; north end of Figure 1) to near Sauk City (left end of profile; south end of Figure 1). The dashed line is the present-day elevation of the water surface; Lake Wisconsin is curved because the profile is some places perpendicular to and some places oblique to crustal rebound. The vertically ruled area is the Alloa outlet. Surfaces a and b are fans with their apexes at the Dells outlet. Surface c is the present bottom of the Alloa channel. Surface d is the top of the Alloa delta. Surface e is the outwash plain formed when the ice was at the Johnstown moraine north of Prairie du Sac. Surface f is the fluvial plain (terrace) at Prairie du Sac and Sauk City, formed during the flood from Lake Wisconsin.

vation of at least 260 m (the land below that elevation is largely covered with later fluvial sediment). That is, most of the Dells gorges, as seen today and shown in Figure 2, were cut by the flood down to an elevation of at least 260 m (Fig. 3). Below that elevation, the deepest parts of the deepest gorges might have been cut either by the flood or by postflood erosion.

Where did the sediment go that was washed from the Dells area? It might contain some evidence of the flood. The bottoms of the gorges have been filled with postflood sediment, so no channel lag has been observed. In any case, the Cambrian sandstone was probably too soft to form gravel lag but instead disintegrated to sand. Downstream from the Johnstown moraine (Fig. 2), two fan surfaces can be recognized. The higher surface (Fig. 3, surface b) resulted from the erosion of the moraine and offshore sand on the sandstone upstream from the moraine. The lower one (Fig. 3, surface a) resulted from the cutting of at least the deepest part of the sandstone gorges, perhaps after the flood. The material of these fans is poorly exposed and has not yet been studied. No flood bedforms have been observed on the fan surfaces, but because of postlacustrine solifluction, any originally present are not likely to be preserved.

SIZE OF FLOOD

It seems probable that Lake Wisconsin drained catastrophically. There is considerable evidence for a flood at the places where it would be expected, but the size of the flood is harder to reconstruct.

The flood from the Alloa outlet is not easily reconstructed because the channel is not known with adequate precision. Early in the flood, the east side of the channel was ice and the width of the channel is therefore unknown. In addition, the thickness of postflood sediment in the bottom of the channel is unknown, and the total channel depth is therefore unknown. At the start of the flood, water in the Lewiston basin stood at about 295 m and water in Lake Merrimac stood at about 265 m (Fig. 3), but early in the flood the Johnstown moraine dam of Lake Merrimac was probably washed away. Therefore, a single Alloa flood might have lowered lake level 45 m or more to the bottom of the Alloa channel.

The part of the Alloa outlet cut through the Alloa delta during the dying stages of the flood may be an alluvial channel--not a rock-walled channel. It is about 0.4 km wide. On the basis of typical modern rivers, this channel had a discharge of roughly 3,000 m^3s^{-1} (Leopold and others, 1964, Fig. 7-21). The largest recorded flood on the Wisconsin River is 2,000 m^3s^{-1}. This channel was cut after the delta formed--after the main outburst; at its peak, the flood must have been considerably bigger.

The size of the flood from the Dells outlet cannot be determined with the information now available, but the size would have depended on the rate at which the water level in the Lewiston basin dropped and the rate at which the Johnstown moraine and the sandstone of the gorges could be removed. The moraine was probably removed in an extremely short time, causing a sudden drop of about 10 m in the water level in the main basin; the high fan downstream from the Dells outlet (surface b in Fig. 3) may have formed then. A 10 m drop in the main basin would result in a flood with a volume of roughly 10 km^3.

Bretz (1950, p. 132) suggested that the lake discharge would cut slowly through the sandstone in the Dells area because the sandstone is "far more durable" than the Pleistocene material. However, it seems likely that the sandstone would be more easily eroded than the boulder lag that would develop as the moraine was eroded. Alden (1918, p. 288) noted that "with this fall [of 30 m as the river cut through the sandstone of the Dells] and so friable a sandstone to work upon after the morainal drift dam south of [the city of Wisconsin Dells] had been cut through, the newly established Wisconsin River cut down very rapidly with no widening of its channel."

NUMBER OF FLOODS

In the above discussion, we have generally assumed that Lake Wisconsin drained only once at the end of the Wisconsin Glaciation. However, it probably drained and refilled many times. It seems likely that the water would have at first just trickled across the ice at the Alloa outlet, but it would have rapidly increased in flow as the ice was eroded and the hydraulic head increased. When the channel had cut down to solid rock or some other barrier, downcutting would decrease or stop, and lake lowering would catch up to outlet erosion, causing hydraulic head to decrease, reducing the discharge. The glacier, which was serving as the east bank of the outlet might then be able to advance into the outlet, causing the lake to partly refill. The rejuvenated lake could then recut the outlet and flood again, and the cycle could be repeated several times, as is expected to happen at Russell Fiord in Alaska.

Furthermore, the Green Bay Lobe fluctuated several times during the Elderon Phase (Attig and others, 1985). Even if the number of floods postulated in the previous paragraph did not occur, it seems likely that the Elderon glacial fluctuations would result in more than one Lake Wisconsin flood. Therefore, the network of sandstone gorges at the Dells is probably a complex feature, formed by the combined effects of several floods. In addition, the area was also glaciated before late Wisconsin time, and it seems likely that the present network is in part the result of exhumation of earlier flood gorges.

ACKNOWLEDGMENT

We thank Dave Mickelson for reviewing the paper.

REFERENCES

Alden, W. C., 1918, Quaternary geology of southeastern Wisconsin: U.S. Geological Survey Professional Paper 106, 356 p.

Attig, J. W., and Clayton, Lee, 1986, History of late Wisconsin permafrost in northern Wisconsin: American Quaternary Association Program and Abstracts of the Ninth Biennial Meeting, p. 115.

Attig, J. W., Clayton, Lee, and Mickelson, D.M., 1985, Chronology of late Wisconsin glacial phases in the western Great Lakes area: Geological Society of America Bulletin, v. 96, p. 1585-1593.

Baker, V. R., 1971, Paleohydrology of catastrophic Pleistocene flooding in eastern Washington: Geological Society of America Abstracts with Programs, v. 3, p. 497.

Baker, V. R., 1978, The Spokane Flood controversy in Baker, V.R., and Nummedal, Dag, The Channeled Scablands: Washington, D.C., National Aeronautics and Space Administration, p. 3-15.

Bretz, J. H., 1923, The Channeled Scablands of the Columbia Plateau: Journal of Geology, v. 31, p. 617-649.

Bretz, J. H., 1950, Glacial Lake Merrimac: Illinois Academy of Science Transactions, v. 43, p. 132-136.

Brownell, J. R., 1986, Stratigraphy of unlithified deposits in the central sand plain of Wisconsin [M.S. thesis]: Madison, Wisconsin, University of Wisconsin, 140 p.

Clayton, Lee, in press, a, Pleistocene geology of Portage County, Wisconsin: Wisconsin Geological and Natural History Survey Information Circular 56.

Clayton, Lee, in press, b, Pleistocene geology of Adams County, Wisconsin: Wisconsin Geological and Natural History Survey Information Circular 59.

Clayton, Lee, in preparation, a, Pleistocene geology of Wood County, Wisconsin: Wisconsin Geological and Natural History Survey Information Circular.

Clayton, Lee, in preparation, b, Pleistocene geology of Juneau County, Wisconsin: Wisconsin Geological and Natural History Survey Information Circular.

Clayton, Lee, and Attig, J.W., 1986, Catastrophic-flood origin of the Wisconsin Dells: American Quaternary Association Program and Abstracts of the Ninth Biennial Meeting, p. 123.

Clayton, Lee, and Attig, J. W., in press, Lake-ice collapse trenches in Wisconsin, U.S.A.: Earth Surface Processes and Landforms.

Eliot, J. L., 1987, Glaciers on the move: National Geographic, v. 171, no. 1, p. 104-119.

Harloff, N. C., 1942, Lacustrine clays of Glacial Lake Wisconsin as determined by fire protection well records [Ph.B. thesis]: Madison, Wisconsin, University of Wisconsin, 116 p.

Kehew, A. E., and Lord, M.L., 1986, Origin and large-scale features of glacial-lake spillways in the northern Great Plains: Geological Society of America Bulletin, v. 97, p. 162-177.

Leopold, L. B., Wolman, M. G., and Miller, J. P., 1964, Fluvial processes in geomorphology: San Francisco, W.H. Freeman and Company, 522 p.

Martin, Lawrence, 1932, The physical geography of Wisconsin: University of Wisconsin Press, 608 p.

Socha, B. J., 1984, The glacial geology of the Baraboo area, Wisconsin, and application of remote sensing to mapping surficial geology [M.S. thesis]: Madison, Wisconsin, University of Wisconsin, 154 p.

Teller, J. T., and Thorleifson, L. H., 1983, The Lake Agassiz-Lake Superior connection *in* Teller, J. T., and Clayton, Lee, Glacial Lake Agassiz: Geological Association of Canada Special Paper 26, p. 261-290.

Thwaites, F. T., Thwaites, A. M., and Bays, C. A., 1935, Road log *in* Ninth Annual Field Conference: Kansas Geological Society Guide Book, p. 129-144.

8

Stratigraphic evidence of large floods in the upper Mississippi Valley

James C. Knox

ABSTRACT

The most common overbank stratigraphic evidence of former large floods in meandering river systems of the Upper Mississippi Valley is textural reversals in fining-upward alluvial sequences. The overbank deposits of large floods can be grouped into coarse-grained (gravel) and fine-grained (sand, silt, and clay) deposits for purposes of paleoflood analyses. Graphical moments statistics have very limited usefulness for estimating relict flood magnitudes, but they are useful for differentiating whether overbank gravels were deposited from suspended load or bedload. The distinction is important because competent flow depths can be determined from the coarsest overbank particles transported as bedload. Relative changes in magnitudes of graphical moments statistics recorded across fine-grained overbank sediments at vertical increments of 1-3 cm show that deposits of large floods are more poorly sorted, less right skewed, and more platykurtic than adjacent deposits from small floods. A typical fine-grained depositional sequence for a large flood begins with high silt and low sand percentages in the basal section, progressively changes to high sand and low silt percentages in middle and upper middle sections, and culminates with a return to high silt and low sand percentages at the top of the sequence. Repetition of this sequential pattern across several geomorphic surfaces of different elevation is evidence of a very large flood, whereas variance between surfaces implies deposition from a small or moderate flood. Radiocarbon dated alluvial deposits indicate non-random recurrences for large floods during time scales spanning several millennia.

INTRODUCTION

One of the most common problems in flood frequency analysis is estimating the magnitudes and frequencies of rare and large floods that might occur in a watershed. Stream gaging records in most regions are too short to provide the desired level of reliability in formulating such estimates. Because large floods usually are associated with high magnitudes of erosion and sedimentation, evidence of these large events often remains preserved in the stratigraphic record of alluvial deposits.

This paper examines the relationships between textural discontinuities in floodplain alluvium and occurrences of large floods for a humid environment. Holocene and historical (since c. 1820 A.D.) alluvial sediments in Upper Mississippi Valley tributaries in southwestern Wisconsin and northwestern Illinois are evaluated to establish a sedimentological link with large floods. Identification of the sedimentological criteria that denote overbank alluvial deposits resulting from large floods provides a basis for detecting rare and extreme flood events that may not be represented in short-term records of historical observation. These criteria also provide a basis for evaluating the evolutionary behavior of floodplain vertical accretion as differentially influenced by high frequency and low frequency flood events.

A universally acceptable threshold that defines a flood as "catastrophic" does not exist. Therefore, I will discuss stratigraphic evidence of "large floods." Because I am concerned with stratigraphic evidence of large floods in a humid region typified by meandering rivers, I will define large floods as those which produce stratigraphic phenomena that are associated with textural discontinuities that reverse the "fining-upward" sequence common in alluvial sediments. For convenience of discussion, I recognize two principal categories of grain size distributions common of overbank flood deposits. The first category, which involves sediment coarser than 2 mm diameter, not only indicates the past occurrence of an extreme flood, but in many instances can be used to estimate the depth of the relict flood based on competency requirements. The second category, which involves overbank deposition of sediment finer than 2 mm diameter, normally leaves a relatively subtle imprint in the stratigraphic record. However, these fine-grained sediments have the advantage of being relatively ubiquitously distributed across alluvial surfaces. Thus, they can be extremely useful for identifying occurrences of extreme floods and in many cases tracing the flood's lateral and vertical limits of deposition.

BACKGROUND

Several types of stratigraphic evidence exist that suggest former occurrences of large floods. Stratigraphic phenomena that indicate former magnitudes and frequencies of large floods can be generalized into three categories, including: (1) textural discontinuities (Jahns, 1947; Wolman and Eiler, 1958; Baker 1973; Costa, 1974a, 1974b; Ritter, 1975; Knox, 1979; Patton, Baker, and Kochel, 1979), (2) inclusion of anomalous sediment clasts in floodplain alluvium, usually involving clay balls, peat blocks, and other debris eroded from channel margins by large floods (Jahns, 1947; Costa, 1974a), and (3) primary bedform structural features that imply flow regime or water depth relationships (Harms and Fahnestock, 1965; McKee, Crosby, and Berryhill, 1967). The following discussion will be restricted to the application of textural discontinuities in floodplain alluvium as a method for evaluating occurrences of past floods.

Textural discontinuities in floodplain and terrace sediments probably are one of the most common indicators of a former large flood or floods. The alluvial deposits of meandering channels, which are especially prominent in humid climate regions such as represented by the Upper Mississippi Valley during Holocene and historical times, typically show a fining-upward tendency (Wolman and Leopold, 1957). Floodplain alluvium in most places consists of lateral accretion coarsegrained (sand and gravel) point bar deposits overlain by fine-grained (silt-dominated) deposits formed by overbank flooding. The high energy conditions of large floods often are associated with intense erosion, transportation, and deposition of sediments. The occurrence of a large flood normally transports sediment of a given grain size to higher surfaces than is possible by the more frequent floods of smaller magnitude, causing a reversal in the fining-upward trend in sediment texture. Deposition may occur on most alluvial surfaces inundated by flood waters, but it is especially favored in slack water areas such as upstream of valley constrictions or in mouths of tributaries adjacent to major constricted channelways (Patton, Baker, and Kochel, 1979; Ely and Baker, 1985). Jahns (1947, p. 116-120) provided one of the first detailed descriptions of textural discontinuities in alluvium deposited by large floods when he examined erosion and deposition caused by the floods of 1936 and 1938 in the Connecticut Valley. He noted (p. 129) that the presence of coarse textured stratigraphic units in fine-grained alluvium on higher surfaces was especially indicative of a former large flood. Jahns reported (p. 116) that "The thickness and texture of each member in a given flood sequence can be correlated with the height and duration of a definite period of the flood..." Jahns also noted that the coarser textured flood deposits often have angular unconformities with the more flat-lying sediments below. A similar observation was made by Costa (1974a) who indicated that a very large flood in 1972 on floodplains in Maryland deposited well-laminated and cross-bedded coarse quartz and feldspar sand on pre-existing fine sand and silt.

Coarse units in overbank deposits resulting from extremely large floods often include gravel. Wolman and Eiler (1958) found that the extreme flood of August 1955 in Connecticut deposited gravel on a terrace surface approximately 2.1 m above the bed of West Branch Salmon Brook in Connecticut. However, the use of overbank gravels to reconstruct characteristics of relict floods requires knowledge of the depositional mode. Costa (1978) concluded that primary sedimentary structures and texture alone do not adequately differentiate channel and point bar gravels from overbank gravels deposited by large floods. Ritter (1975) also observed that an overbank flood gravel on Sexton Creek in southern Illinois was indistinguishable from the texture of adjacent channel and point bar gravels, but that the convex-shaped lens of overbank gravel had an almost flat lower surface where it was in contact with overbank silt and clay sediments. He found that the morphologic shape of the buried overbank gravel was very similar to the morphologic shapes of surficial overbank flood gravels deposited by a recent flood, implying that overbank flood gravels might commonly be deposited as lenses of coarse sediment that have convex upper surfaces but relatively flat bases where they come into contact with underlying overbank silt and clay sediments.

The location of flood gravels that can definitely be connected to former large floods is highly selective. I have found that the most easily deciphered and relatively complete record of large floods represented by deposits of coarse gravels over silt and clay sediments typically occurs on the upper surfaces of point bars, whereas higher surfaces represented by floodplain and low terraces have relatively few deposits of flood gravels (Knox, 1979). Ritter (1975) has suggested two modes by which gravel reaches higher overbank alluvial surfaces to interrupt the normal fining-upward sequence. He suggested that in some situations the gravel is rolled or saltated up the surface of enchannel bars which grow vertically to the level of the higher surfaces and thereby serve as a ramp for the moving gravel. He concluded that a second mechanism involved gravel being transported in the upper layers of the flood waters as particulate matter in momentary suspension. In Wisconsin, I also observed that elevated enchannel bars were frequently the mode for transporting gravels to low floodplain surfaces, but overbank gravels on high floodplain surfaces and low terraces appear to have been transported in suspension. In either situation, coarse gravels overlying fine-textured sediments are strongly indicative of the former occurrence of a large flood.

In summary, the most common stratigraphic record of former large floods in humid regions typified by meandering channels is a reversal of the normal fining-upward sequence in overbank alluvium. Despite the widely recognized significance of reversals in fining-upward sequences, meager quantitative information exists that links the physical properties of these deposits with flood events. This paper provides detailed particle size analyses of both coarse gravels and fine-grained overbank flood deposits. The general objective of the paper is to show that specific sedimentological properties, including particle size and unique patterns of variation in graphical moments statistics, denote occurrences of large floods.

SEDIMENTOLOGICAL EVIDENCE OF LARGE FLOODS IN THE UPPER MISSISSIPPI VALLEY

The study area for this investigation is the southern part of the Driftless Area, an unglaciated region of southwestern Wisconsin and northwestern Illinois (Fig. 1). Bedrock is exposed in most small headwater tributaries, but larger tributaries and trunk streams rest on sandy gravels. Bedrock includes sandstone, dolomite, and limestone which produce bedload textures that range from sand through large boulders. The wide range in grain size for coarse sediments helps delineate for large

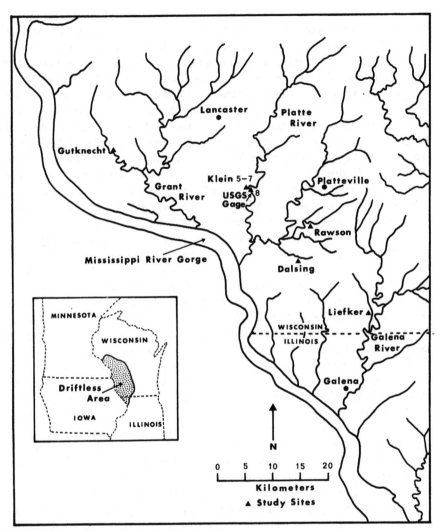

Figure 1. Location of key study sites within the southern part of the Upper Mississippi Valley Driftless Area.

floods the competent depths required to move the coarsest sediment fraction. The bedrock is thinly covered with red-clay residuum that in turn is overlain by loess. Loess thickness varies from negligible on steep hillslopes to several meters on specific interfluves. Loess deposits are the principal source of suspended sediments in stream flows, causing the grain size distributions of most floodplain vertical accretion sediments to resemble loess in a textural sense. The most frequent, and smaller floods of the region result from snowmelt, while the largest floods of record have resulted from intense, large rainfalls associated with summer frontal systems. Because the landscape is highly dissected by stream channels, and local relief averages between 50 and 120 m per km^2, floods are quite responsive to climatic events.

In the following, I discuss sedimentological properties of overbank deposits resulting from large floods in relation to sedimentological properties of other nearby or adjacent alluvial units that were deposited by small magnitude, high frequency floods. The analysis is presented in two parts that include (1) overbank coarse-grained stratigraphic units, and (2) overbank fine-grained stratigraphic units. Although the concepts and principles that I present are based on data collected from many site throughout southwestern Wisconsin and northwestern Illinois, I will focus much of the discussion on the stratigraphic record of an extreme flood that occurred on southwestern Wisconsin's Platte River in July 1950. A site on the Platte River 1.4 km upstream of the U.S. Geological Survey stream gage near Rockville, Wisconsin has several physical characteristics that are especially suitable for illustrating the stratigraphic record of large floods. The site: (1) includes overbank deposits of boulder gravels that can be used to estimate the competent depth of flow during the 1950 flood, (2) is in close proximity to the U.S. Geological Survey stream gage where recorded flood depths of recent large floods, including the 1950 extreme flood, can be used to evaluate estimates of flood depths at the study site, and (3) includes several levels of floodplain and terrace surfaces that are ideally suited to examine the influence that relative height of a depositional surface has on the nature of the fine-grained stratigraphic record of overbank large floods.

The extreme flood on southwestern Wisconsin's Platte River in July 1950 resulted from an excess rainfall in which the official weather bureau recording station on the northwest edge of the watershed observed 155 mm of rainfall within a five-hour timespan. A rainfall of this magnitude and duration has an expected return period of less than once per hundred years (Hershfield, 1961). The U.S. Geological Survey estimated the peak runoff at the downstream gage near Rockville, Wisconsin to be about 1230 cms, a discharge magnitude that corresponds with an exceedence probability of about 0.004 when the recurrence frequency is determined by the Log-Pearson Type-III probability distribution (U.S. Water Resources Council, 1981). A local newspaper (Grant County News, 1950) reported that the flood was the largest since 1876. However, since the stratigraphic record suggests the 1950 flood was significantly larger than the 1876 flood, which was the largest known previous flood, the 1950 flood probably has been the largest flood observed on the Platte River since European settlement began in the 1820s.

Coarse-Grained Overbank Sedimentation by Large Floods

Deep, turbulent flows during large floods often have sufficient energy to transport coarse sediments from the channel bed onto the adjacent floodplain or other alluvial surfaces such as low terraces (Baker, 1984). The overbank deposition of sediments, which are normally found on the channel bed and adjacent lateral bars, produces a striking discontinuity in the alluvial stratigraphy. While the presence of a major textural discontinuity in overbank alluvium suggests that a high-energy flood previously occurred, characteristics of the grain size distribution in overbank flood gravels are influenced by their mode of transport from the channel margins to higher overbank positions. Knowledge of the probable mode of transport is important because use of the coarsest grainsize fractions to estimate competent flow depths presumes rather specific conditions of sediment transport (Bradley and Mears, 1980; Costa, 1983; Williams, 1983a).

The most useful information for reconstructing aspects of the associated flood is contained within the coarse tail of the sediment distribution where the competent flow depth may be reflected by the dimensions of the largest transported particles. Calculations of competent flow depths, are based on the DuBoys equation for boundary shear:

$$\tau = \gamma RS \qquad (1)$$

where τ is the critical mean shear stress in Nm^{-2} for initiating particle transport, γ is the specific weight of the water and sediment mixture in Nm^{-3}, R is hydraulic radius in m, and S is the dimensionless energy slope (in m/m). The method of calculating flood depths is similar to that determined by Baker and Ritter (1975) and Williams (1983b), except that the present regression equation was established based on particle diameters ranging between 50-3300 mm (Knox, 1987). The equation is:

$$D = 0.0001A^{1.21}S^{-0.57} \qquad (2)$$

where D is the competent flood depth in meters, A is the intermediate axis in mm, representing the coarsest particles moved (in this case either the mean of the coarsest 5 or coarsest 10 particles), and S is an approximation of the energy slope in m/m. The equation assumes a constant of 9800 Nm^{-3} for the specific weight of the water and sediment load. The standard error of estimate is 0.34 $\log_{10}$ units and the equation is statistically significant at the 0.01 probability level with 74% of the variance in competent flow depth being explained by the two independent variables.

Further discussion of the basic data underlying theeEquation is given in Knox (1987). Two critical requirements associated with the determination of flood depths from competency relationships are: (1) the maximum particle size in the flood deposit should be smaller than the maximum size of particles that were available for fluvial transportation, and (2) the particles of maximum size must have been fluvially transported and not have been locally introduced by mass wasting, ice rafting, or some other external process. These requirements and other assumptions associated with the use of the competency method for paleoflood reconstructions have been reviewed by Baker and Ritter (1975).

The floodplain and low terrace surfaces at the Platte River site located 1.4 km upstream of the U. S. Geological Survey stream gage contain numerous large boul-

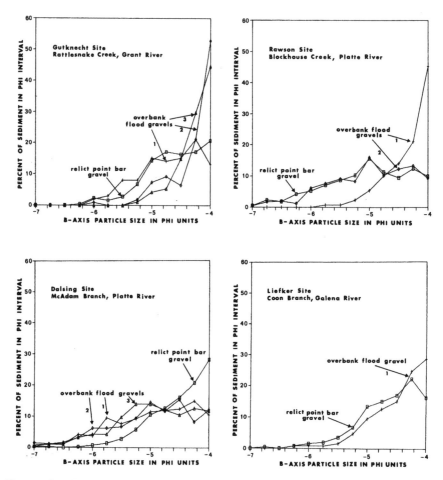

Figure 2. Grain size distribution for selected Holocene point bar and overbank flood gravels. Overbank gravels with grain size distributions similar to underlying point bars appear to have been elevated to overbank positions during large floods by rolling or saltating up the former surfaces of enchannel bars. Overbank gravels with grain size distributions distinctly finer than underlying point bars appear to have been transported to their present positions as part of the suspended load of former large floods.

ders that were deposited by the deep, high velocity flood waters of July 1950. Hereafter the location of the boulder deposits will be referenced as Klein Site 8. A local newspaper (Grant County News, 1950) reported that deposition of boulders on pastures and fields of the Platte River floodplain occurred in many places. The mean

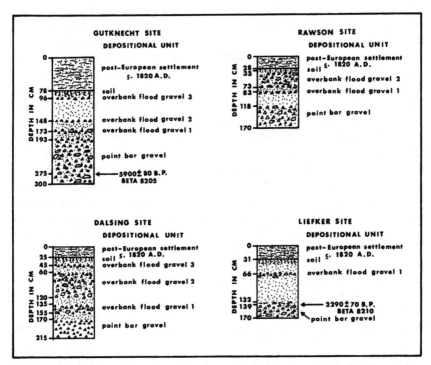

Figure 3. Stratigraphic settings for the point bar and overbank flood gravels shown on Figure 2.

dimension of the intermediate axis for 18 boulders deposited on the floodplain at Klein Site 8 was 309 mm and the corresponding standard deviation was 85 mm. The range in intermediate axis dimensions was from 175 mm to 440 mm. The boulders rested on sandy gravel that was deposited during the 1950 flood. All of the boulders are now partially buried by post-1950 floodplain sedimentation that ranges from about 12 cm deep on higher floodplain surfaces to about 20 cm deep on lower floodplain surfaces.

Application of Equation (2), using the mean of the coarsest 10 boulders and the mean of the coarsest 5 boulders on the Platte River floodplain, indicates that the July 1950 flood was 2.7 to 3.1 m deep over the floodplain at Klein Site 8. These estimates agree closely with heights of flood scars on valley floor trees and with the overbank flood depth of about 3 m that was estimated at the gage site 1.4 km downstream. The estimates also indicate that thalweg depths of flood waters at and near Klein Site 8 ranged from a little over 5 m deep in narrower valley reaches to about 4 m deep in the wider reaches. Flood waters extended laterally about 180 m across the entire valley floor to cover the floodplain and low terraces.

The boulders on the floodplain and low terraces at Klein Site 8 appear to have been transported to their present positions by rolling and saltating up the ramp-like surfaces of enchannel bars. Because they were moved as part of the bedload, Equation (2) provides a reasonable approximation of the 1950 flood depth at Klein Site 8. The application of Equation (2) may be less suitable when applied to overbank flood gravels that were not transported as part of the bedload. Overbank flood gravels that have grain size distributions that are significantly finer textured than the grain size distributions in the point bar and channel bed source sediments probably have been transported as part of the suspended load during large floods. It may be important to differentiate the two types of overbank flood gravels in paleoflood analyses because the DuBoys equation is an expression of boundary shear and thus has been closely associated with the movement of bedload sediment. Therefore the competency method, using Equations (1) and (2), probably is not strictly appropriate for application to all overbank flood gravels because the relationship with suspended load sediment transport is unclear. Estimates of flood depths from overbank suspended load flood gravels probably provide gross approximations of true depths because, as an expression of the depth-slope product, the DuBoys equation incorporates several internal forces of flow (Bogardi, 1978, p. 83), but these estimates probably error on the conservative side.

Differentiating overbank flood gravels into those transported as bedload versus those transported as suspended load can be approximated by examination of the grain size distribution and other morphologic properties of the deposits. Figure 2, for example, shows the distributions of grain sizes between negative four phi (16 mm) and negative seven phi (128 mm) for relict flood gravels representing point bar and overbank deposits in four small watersheds of the study area. The four sites are representative of 70 relict point bar and overbank gravel deposits that were examined in the region. Note that grain size distributions for some overbank floods are nearly identical to those of point bars that represent the source sediment, while other overbank flood gravels have a tendency to be finer textured with a much greater percentage of the sediment being concentrated near the modal range, and a much smaller percentage of the sediment being found in the coarse tail of the distribution.

The most likely explanation for the two categories of grain size distributions in overbank sediments involves the mechanism by which coarse sediments are transported to higher surfaces. Overbank flood deposits that roll and saltate up the ramps of enchannel bars appear to produce floodplain deposits that have grain size distributions that are nearly identical to those of point bars within the margins of the active channel. In the study region, these types of deposits tend to be sandy gravels with bed thicknesses that range from a few centimeters to a few tens of centimeters. Overbank flood deposits that were formerly transported as part of the suspended load usually are of finer texture and sometimes involve only widely distributed coarse clasts. Where lenses of gravel have been deposited, the total thickness of the flood deposit usually is considerably less than those which originated by rolling and saltating up enchannel bars.

Examples of overbank flood gravels involving both bedload and suspended load transport processes are shown in Figures 2 and 3. Note, for example, that textural properties for two of the overbank gravels at the Gutknecht Site 5 on Rattlesnake Creek have grain size distributions that are quite different from those of the underlying point bar (Fig. 2 and Table 1). At the Rattlesnake Creek site, the finer-textured flood gravels thought to have been deposited from suspended load are located higher in the stratigraphic section than another overbank flood gravel that appears to have been deposited by rolling and saltating up an enchannel bar (Figs. 2 and 3). However, overbank flood gravels deposited from suspended load may also occur below overbank flood gravels deposited as part of the bedload that was transported to the

Table 1. Summary statistics for relict point bar and overbank fluvial gravels in southwestern Wisconsin drainage Basins*

Site Location	Depositional Process	Sorting Coefficient	Skewness Coefficient	Kurtosis Coefficient
Liefker 1	relict point bar	0.50	0.25	1.01
Liefker 1	ramp transported flood gravel	0.46	0.28	1.80
Dalsing 2	relict point bar	0.48	0.30	0.96
Dalsing 2-1	ramp transported flood gravel	0.76	0.26	0.81
Dalsing 2-2	" "	0.78	0.25	1.00
Dalsing 2-3	" "	0.69	0.08	1.00
Gutknecht 5	relict point bar	0.48	0.22	1.02
Gutknecht 5-1	ramp transported flood gravel	0.52	0.20	0.95
Gutknecht 5-2	suspended load flood gravel	0.43	0.49	0.95
Gutknecht 5-3	" "	0.32	0.21	1.04
Rawson	relict point bar	0.78	0.15	0.86
Rawson 1-1	suspended load flood gravel	0.38	0.17	0.93
Rawson 1-2	ramp transported flood gravel	0.73	0.27	0.94

*Graphical moments statistics for skewness and kurtosis based on methods of Folk and Ward (1957); graphical moments method for sorting based on method of Inman (1952).

higher surface via an enchannel ramp-like bar. An example of the latter is illustrated by Rawson Site 1 on Blockhouse Creek where a relatively fine-textured flood gravel, perceived to be of suspended load origin, underlies an overbank flood gravel having a grain size distribution that is nearly identical to the point bar gravel underlying both units (Figs. 2 and 3). The close textural similarity between the upper flood gravel and the point bar gravel suggests that the upper flood gravel probably was transported to its present position by rolling and saltating up a ramp-like enchannel bar.

In most cases, the grain size distribution of point bar sediments is either equal to or coarser than the grain size distribution of overbank flood deposits as is illustrated by the Gutknecht, Rawson, and Liefker Sites on Figure 2. However, when major environmental and hydrologic changes have occurred within the period of stratigraphic record, relict point bar sediments may be finer textured than the overbank flood gravels that overlie them. An example is Dalsing Site 2 on McAdam Branch where three very coarse textured late Holocene overbank flood gravels overlie relatively fine textured early Holocene point bar gravels (Figs. 2 and 3).

The explanation for the seemingly anomalous relationship between overbank and point bar gravels at Dalsing Site 2 is explained by hydrologic response to climatic change. Point bar sediments reflect textural characteristics of the channel bed which is a function of a hydrologic regime that varies with climate. The early Holocene environment at Dalsing Site 2 was relatively warm and dry, and flood magnitudes were relatively small, but climatic conditions of the late Holocene became cooler and wetter resulting in an increase in the magnitude and frequency of floods (Knox, 1985). Elsewhere, I have estimated long-term magnitude variations in overbank Holocene floods from the competent depth of flow required to transport the coarsest particles in overbank flood gravels (Knox, 1987). This analysis indicated that recurrence probabilities of extreme floods have not been stationary during the Holocene. During much of the early Holocene when the region's climate was somewhat warmer and drier than present, large overbank floods were very rare. The available stratigraphic history of overbank floods from about 6000 yrs B.P. indicates that overbank floods with large depths occurred from 6000-4500 yrs B.P., between 3000-1800 yrs B.P. and between about 1000 and 500 yrs B.P. In addition to the early Holocene, depths of overbank floods were prominently small between about 4500-3000 yrs B.P. The analysis showed that overbank flood gravels can serve as an important source of proxy information about long-term fluctuations in the magnitudes of large floods.

Overbank flood gravels, therefore, provide strong evidence for the former occurrence of a large flood. Graphical moments standard descriptive statistics, such as the mean, mode, median, sorting, skewness, and kurtosis, when applied to sediments equal to or coarser than 16 mm, provide little specific information that is directly related to flood magnitudes, but these statistics are useful for indicating the mode of sediment transport (Table 1). Because the descriptive statistics in Table 1 represent grain size distributions in which the fine tail was truncated by the sampling method employed, the calculated statistics can only be interpreted relative to each other. Therefore, standard nominal classifications associated with interval ranges for the various statistics are not appropriate. However, given these limitations, overbank flood gravels that have been classified as originating from the suspended load tend to be better sorted, less coarse skewed, and finer textured than overbank flood gravels that have been classified as originating from rolling and saltating up ramp-like enchannel bars. If graphical moments statistics indicate overbank flood gravels were transported as bedload, then particles in the coarsest fraction can be used to calculate the competent depth of flow required for their movement. Where it is possible to assemble a large number of competent flow depths for overbank flood gravels in radiocarbon dated alluvial sequences, a proxy record of long-term variations in the magnitude and frequency of large floods can be established.

Fine-Grained Overbank Sedimentation by Large Floods

Whereas the occurrence of fluvial gravels in overbank vertical accretion sediments is an easily recognizable stratigraphic indicator denoting the former occurrence of a large flood, the limited spatial occurrence of these deposits restricts their usefulness as a proxy indicator of former large floods in a given watershed. Therefore, other forms of evidence are required. The overbank alluvium of most floodplains in humid regions is dominated by various fractions of sand, silt, and clay. Although these fine-grained sediments have limited usefulness for quantitatively reconstructing flood depths over the floodplain, variations in the grain size distributions and thickness of depositional sequences provide much information about the history of floods at a site.

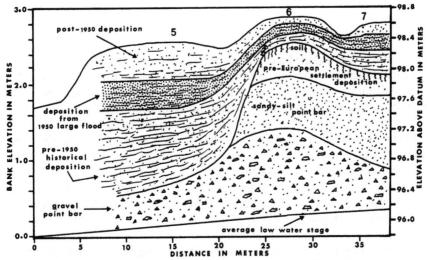

Figure 4. Stratigraphic sequence at Klein Sites 5,6, and 7, Platte River, Grant County, Wisconsin (see Fig. 1 for location).

Table 2. Summary statistics for overbank fine-grained sediments at Klein Sites 5, 6, and 7, Platte River, near Rockville, Wisconsin*

Site Location	5	6	7	5	6	7	5	6	7
Sediment Type	(mean sand %)			(mean silt %)			(mean clay %)		
Post-1950	19.8	19.1	16.3	65.3	65.3	67.8	15.0	15.6	15.9
1950 Flood	20.2	18.6	26.2	62.6	67.8	60.8	17.2	13.6	13.0
1820-1950	18.2	7.8	3.4	64.5	74.4	76.8	17.3	17.8	19.8
Graphical Moments Statistic	(mean sorting)			(mean skewness)			(mean kurtosis)		
Post -1950	2.61	2.82	2.80	0.36	0.33	0.34	1.49	1.72	1.71
1950 Flood	2.98	2.31	2.87	0.37	0.33	0.39	1.56	1.83	1.45
1820-1950	2.80	2.54	2.65	0.39	0.42	0.53	1.38	1.35	1.41

*Graphical moments statistics for skewness and kurtosis based on methods of Folk and Ward (1957); graphic moments statistics for sorting based on method of Inman (1952).

Relatively few detailed quantitative analyses of sedimentological properties of overbank floodplain sediments exist where the objective has involved assessment of flood magnitudes and frequencies. Relatively little is known about how the various fractions of sand, silt, and clay reflect magnitudes of overbank floods, or conversely, how the various fractions of sand, silt, and clay are influenced by the relative height of the depositional surface above the channel bed of a flood of a given magnitude. I will use the overbank stratigraphic record in vertical accretion alluvial sediments of southwestern Wisconsin's Platte River flood of 1950 to illustrate the fine-grained sedimentological imprint of a large flood. As noted above, the 1950 flood is the most extreme flood known to have occurred in the watershed since settlement c. 1820, and application of the Log-Pearson Type-III method indicates an exceedence probability of about 0.004.

Because I am concerned with the nature of the fine-grained stratigraphic record as it might be variably preserved on surfaces with different relative heights above the channel thalweg, detailed sedimentological analyses were undertaken along a

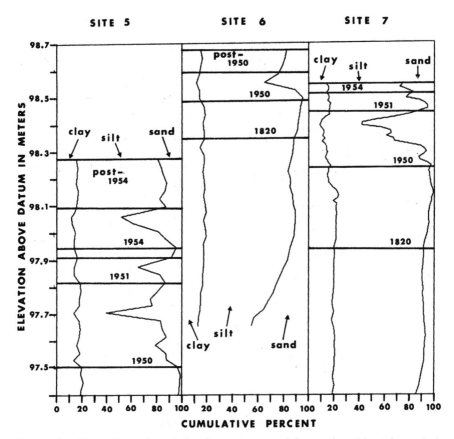

Figure 5. Clay, silt, and sand fractions represented in stratigraphic units underlying Klein Sites 5, 6, 7, (see Fig. 4). Note that the sand fraction is the most sensitive recorder of relict large floods that occurred in 1950, 1951, and 1954, while the clay fraction is relatively insensitive for recording variations in flood magnitudes.

transect that paralleled the right cut-bank of the channel and included three geomorphic surfaces. The orientation of the channel bank is approximately perpendicular to the median axis of the downvalley flow direction for the 1950 flood. A continuous core, 10 cm in diameter, was extracted hydraulically with a truck-mounted Giddings probe at each of the three stratigraphic surfaces approximately 2-3 meters away from the right stream bank. These corings are referenced as Klein Sites 5, 6, and 7 and they are located along the Platte River directly opposite Klein Site 8 (Fig. 1). No evidence of levee development was detected at the coring sites, probably because the bank has eroded laterally several meters since the 1950 flood. The relative positions of the core sites and their associated stratigraphy are shown on Figure 4.

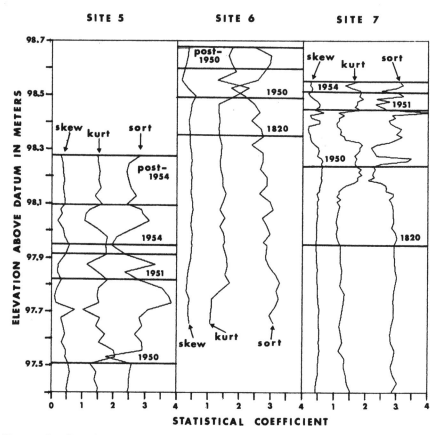

Figure 6. Graphical moment statistics show that grain size distributions of overbank fine-grained sediments deposited by large floods tend to be more poorly sorted, more platykutic, and less positively skewed in comparison to other overbank fine-grained sediments associated with floods of smaller magnitude.

The cores were briefly described in the field to verify that they represented the continuous stratigraphic units that were exposed in the right channel bank. The cores were wrapped in aluminum foil and sealed for transportation to the laboratory where they were described and sampled in detail. The average width of sampling horizons for sediment analyses ranged from 1-3 cm, but care was taken to ensure that all stratigraphic units were sampled. To evaluate the possible effects of smoothing sedimentological variations by sampling at too coarse an interval, all of the historical (post-1820 A.D.) sediments for Klein Site 7 were sampled at intervals not exceeding 1 cm thickness. Samples for which at least 30 g dry weight were available were analyzed by the hydrometer method according to the specifications outlined in designations D421 and D422 of the American Society for Testing and Materials (1977, p. 68-80). Samples for which less than 30 g dry weight were available were analyzed by the pipette method following standard methods reviewed by Hallberg (1978). After completion of the hydrometer and pipette analyses, the sand fraction was separated from other sediments by wet sieving, then dried and sieved at one phi intervals. Following completion of the laboratory procedures, down-core variations in the percentage of sediment fractions in each one phi interval from coarse sand (0-1 phi) to coarse clay (8-9 phi) were plotted to determine the relative importance of respective sediment fractions preserved on the three different topographic surfaces. Graphical moments statistics, following the procedures outlined by Folk and Ward (1957) and Inman (1952), were also determined for each of the sediment samples to help differentiate an overbank deposit resulting from a rare large flood from an overbank deposit resulting from more frequent small floods.

The 1950 flood produced a prominent stratigraphic sequence on all three geomorphic surfaces, but the thickness and pattern of grain size variation changed systematically across the levels. The wide variation in characteristics of the grain size distributions is noteworthy because only modest differences in relative heights exist between the three topographic surfaces. For example, the height of the alluvial surface that immediately pre-dated the 1950 flood was about 1.6 m above the average low water channel bed at Klein Site 5, 2.4 m at Klein Site 6, and 2.1 m at Klein Site 7. Major differences in the pattern of grain size variation with modest variation in the height of the depositional surface above the channel bed elevation illustrate how overbank depositional units become increasingly filtered of high frequency and low magnitude flood events as the relative height of the depositional surface increases.

The percentage of sediment within a given range of grain size in a sedimentary sequence deposited during the passage of a large flood tends to display systematic patterns of variation (e.g., Fig. 5). Widely ranging magnitudes of the various sediment fractions in overbank vertical accretion deposits often cause sequence-wide mean values of sediment fractions to be relatively poor indicators of differences between stratigraphic sequences deposited by large floods and stratigraphic sequences deposited by small floods. The data of Table 2, which present means for grouped observations of grain size categories and for graphical moments statistics, show that sediments associated with the extreme flood of 1950 tend to be slightly coarser and more poorly sorted than pre- and post-1950 overbank flood deposits, but that the degree of difference is often small due to the grouping effect. Closer inspections of excursions from mean values show that sorting, skewness, and kurtosis each undergo a systematic pattern of variation during overbank sedimentation from the large 1950 flood. For example, Figure 6 indicates that all three levels at the Klein Sites show a tendency for sediments to become progressively more poorly sorted throughout the depositional sequence until near the final stages of deposition when sediments become slightly better sorted.   As might be anticipated from the trends in sorting, grain size distributions for sediments deposited during the early phase of the 1950 flood are more leptokurtic than those deposited during the lat-

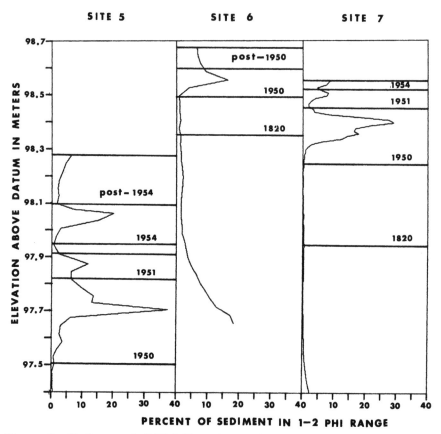

Figure 7. Medium sand (0.25-0.50 mm) proved to be one of the best sediment fractions for delineating the top and bottom stratigraphic boundaries of units representing individual large floods. Finer textures were not sufficiently sensitive to subtle changes in flood stage while coarser textures often were not represented on the relatively higher alluvial surfaces. Note that even relatively small differecnes in elevation among Sites 5, 6, and 7, account for major differences in the stratigraphic detail of former floods.

ter phase of the flood. Skewness appears to be the least responsive graphical moments statistic for signalling a former occurrence of an extreme flood, probably because grain size distributions for most overbank flood deposits tend to be positive to very positively skewed. Nevertheless, grain size distributions became less positively skewed as overbank deposition progressed from the early to late stages during the 1950 flood.

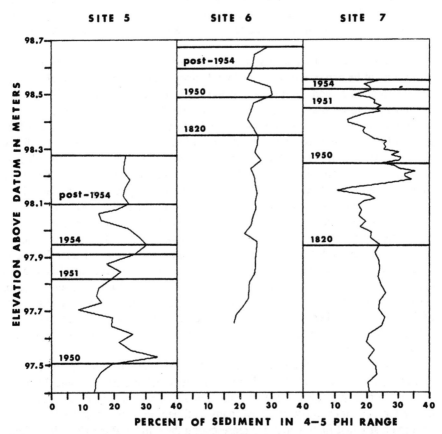

Figure 8. The percent of the grain size distribution that is coarse silt (0.031-0.063 mm) was useful for identifying flood cycles in study area overbank alluvium. The typical sequence begins with a very high percentage of coarse silt that is progressively reduced upward in the sequence until near the upper limit of the flood unit where a reversal toward higher percentages occurs.

Not surprisingly, the coarser sediment fractions (sand) prove to be better recorders of large floods than finer sediment fractions (silt and clay). Plots showing cumulative frequencies of clay, silt, and sand in the total sediment distribution indicate that the percentage of clay sediments varies conservatively between about 15-20%, whereas sand ranges from about 5-50% (Fig. 5). The relatively large variations in silt, ranging from about 25-30% to about 65-75%, primarily reflect the large fluctuations in sand. The insensitivity of clay sediments to record flood events probably reflects the tendency for clay to be easily transported throughout a water column once it is in suspension. On the other hand, the concentrations of coarse silt and sand fractions in flood waters tend to decrease rapidly with vertical distance above the channel bed and toward the water surface (Allen, 1984, p. 98-102).

While it is generally true that magnitude variations of sand in overbank sediments tend to be better indicators of flood history than are magnitude variations in silt, information on vertical fluctuations in each is useful for accurate delineation of the upper and lower boundaries of flood depositional units. In the present analyses, full grain size distributions were subdivided into one phi intervals for detailed assessment of sediment fractions. Although all fractions were examined, I will use variations in medium sand (1-2 phi; 0.25-0.50 mm diameter) and coarse silt (4-5 phi; 0.031-0.063 mm diameter) to illustrate the stratigraphic imprint of the 1950 flood in relation to other overbank flood depositional sequences at the Klein Sites 5, 6, and 7 on southwestern Wisconsin's Platte River (Figs. 7 and 8).

Medium sand proved to be the best overall sediment fraction for defining boundaries of flood units. It was sufficiently coarse to be sensitive to changes in flood energy levels during rising and falling stages, and it was sufficiently fine to be present on all three depositional surfaces. Coarser sand fractions often did not appear in the depositional sequence for the 1950 flood until the higher stages of flow were achieved, and they were poorly represented on the higher alluvial surfaces of deposition. Another factor that contributed to medium sand being the most useful paleoflood indicator was its greater abundance in source sediments relative to other sand fractions.

The 1950 flood is prominently represented by variations in the amount of medium sand deposited on all three alluvial surfaces at Klein Sites 5, 6, and 7 (Figs. 4 and 7). Note that at each of the sites the amount of medium sand increased from negligible quantities in pre-1950 overbank sediment to as much as 15-35% of the total grain size distribution in strata representing the middle of the 1950 flood depositional sequence. The percentage of sediment deposited as medium sand on each alluvial surface decreased systematically as the relative height of the original depositional surface increased. In similar fashion, the thickness of the 1950 flood unit was 31 cm on the lowest surface (Site 5), 20 cm on the intermediate surface (Site 7), and 12 cm on the highest surface (Site 6).

The very large flood of 1950 is clearly represented on Klein Sites 5, 6, and 7 whose alluvial surfaces were respectively 1.6 m, 2.4 m, and 2.1 m above the low water channel bed of the adjacent Platte River. Since 1950, two additional moderate magnitude overbank floods occurred in 1951 and 1954, each overtopping the three geomorphic surfaces. In comparison to the exceedence probability of 0.004 for the 1950 flood, these floods have exceedence probabilities of only about 0.05 even though their stage heights were only 1.2-1.4 m lower than the 1950 stage height at the U.S. Geological Survey gaging station 1.4 km downstream. The much lower energy levels associated with the 1951 and 1954 floods is clearly evident in the stratigraphic record of medium sand at Klein Sites 5, 6, and 7 (Fig. 7). On the lowest alluvial surface at Site 5, both the 1951 and 1954 floods have relatively thick deposits. Note that at Site 5, insufficient time had elapsed since the 1950 flood to allow for deposition of silty sediment from low energy floods of high frequency. Consequently, the sandy alluvium of the 1951 flood directly over-lies the 1950 deposit. In contrast, moderate overbank flooding between 1951 and 1954 resulted in deposition of finer textured sediments to clearly separate these major flood events (Fig. 7).

The stratigraphic imprint of the 1951 and 1954 floods in the medium sand fraction is less clear at Klein Sites 6 and 7. In fact, on the highest surface at Site 6, it is not possible to differentiate the two flood events although they probably are responsible for the relatively high magnitude of medium sand in all of the approximately 8.5 cm of overbank sediments that overlie the 1950 flood unit. At Site 7, on the alluvial surface of intermediate height, two very modest rises in the amount of medium sand are indicated in the 10.5 cm of alluvial sediments deposited since the 1950 event. As at Site 6, it is apparent that overbank deposition since the 1954 flood has been negligible.

Coarse silt also was especially useful for identification of flood cycles within the alluvial sequences. A typical depositional sequence resulting from the passage of a large flood involved progressive decreases in the amount of coarse silt until near the end of the event when a reversal occurred (Fig. 8). Initial high magnitudes for coarse silt probably represent the tendency for flood waters during the rising stage to be highly charged with silty bank sediments derived from bank slumps that had fallen since the previous flood. Following the rising stage, the concentration of silt in the flood waters apparently is progressively diluted until near the end of the flood passage when lower energy levels cause the transportation of sand to decrease. Because sediment fractions are expressed as relative percentages of the total sediment grain size distribution, a decrease in sand will contribute to the increased importance of silt. Therefore, a typical depositional sequence associated with a large flood begins with high silt percentages, which become progressively reduced until near the end of the depositional sequence when silt percentages increase as other coarse fractions decrease in importance during the final falling stages of flow. Hence, a plotted curve representing silt percentages produces a pattern of variation that is typically asymmetric (Site 5, Fig. 8). If subsequent floods are smaller, overbank sedimentation will continue to be low in sand and high in silt composition, and silt percentages will tend to return to higher magnitude values that are more typical of sediments deposited by floods of low to moderate magnitude and high frequency of occurrence as is illustrated at Klein Site 5 following the 1954 flood (Fig. 8). On the other hand, if another relatively large flood immediately follows a large flood, as occurred in 1950 and 1951 at Klein Site 5, then any tendency for continued fining upward is interrupted (Fig. 8).

I previously described for medium sand sediment how the relative height of a depositional surface above the bed of the flood channel influences the thickness and composition of depositional units. Note that most of the generalizations presented for medium sand also apply to coarse silt (Figs. 7 and 8). The extreme flood of 1950 was sufficiently deep that even the highest alluvial surface at the Klein Sites was covered by approximately 2.5 m of water, but the floods of 1951 and 1954 appear to have overtopped the 1950 flood deposit on the highest alluvial surface by only 20-30 cm depth. Consequently, on the highest depositional surface at Klein Site 6, the 8.5 cm of post-1950 overbank deposition is rather blurred by a mixture of silt and clay sediments, whereas coarse silt in sediments on the slightly lower depositional surface at Klein Site 7 contains a distinct record of both the 1951 and 1954 floods (Fig. 8). Note, however, that the thicknesses of the individual flood deposits are much smaller at the intermediate level surface of Site 7 compared to those on the lower level surface of Site 5, and that the range of variability in coarse silt decreases with increases in height of the depositional surface.

The examples from the Klein sites show that even relatively modest differences in the heights of low terraces and floodplains are responsible for considerable differences in grain size distributions and thicknesses of alluvial sequences deposited by individual overbank floods. On the other hand, a truly exceptional flood of great overbank depth usually deposits relatively similar sequences of fine-grained sediment on low terraces and floodplain surfaces, although the thicknesses of individual units decrease as the height of the depositional surface increases. Therefore, for comparable distances from the flood channel, if the tracing of overbank fine-grained flood deposits across alluvial surfaces of different height indicates rather similar successional sequences in grain size variation, the deposit probably is the result of a very large flood. If the tracing shows that the successional pattern of grain size in the sequence varies from one geomorphic surface to another, then the deposit probably resulted from overbank deposition by a moderate to small flood.

The rapid decrease in thickness of depositional sequences resulting from overbank floods, as the relative height of a depositional surface increases in relation to the channel bed elevation, slightly contradicts the hypothetical model of floodplain vertical accretion presented by Wolman and Leopold (1957, p. 100). They assumed that each time a stream overflows a given level it deposited a specific thickness of sediment, resulting initially in a very rapid increase in the elevation of the floodplain but eventually shifting to a very slow increase in elevation as low frequency, high magnitude floods were necessary to deposit sediment on the higher surface elevations. The differential between the amount of sediment deposited by overbank floods on alluvial surfaces of different elevation probably decreases with increases in flood size, but differences in the amount of deposition were quite substantial even for the large flood of 1950 at Klein Sites 5, 6, and 7 where the lowest surface received 31 cm of sediment and the highest surface received only 12 cm of sediment (Fig. 4). Note also, that on the highest surface at Site 6, only 13 cm of historical overbank sedimentation had occurred prior to the large flood of 1950. Since then, approximately 21 cm of additional overbank sedimentation has occurred, and nearly all of it was deposited by moderate to large floods of 1950, 1951, and 1954.

Therefore, as relative heights of alluvial surfaces increase with deposition over time, the position is soon passed whereby further overbank sedimentation becomes increasingly related to floods of low frequency and high magnitude even though grain size may be decreasing upward.

SUMMARY AND CONCLUSIONS

Textural discontinuities represented by reversals in the normal fining-upward sequence in overbank flood deposits are one of the most common types of evidence for the former occurrence of large floods on rivers in humid climate regions. Deep turbulent flows of large floods normally transport sediment of a given grain size to higher surfaces than is possible by the more frequent floods of smaller magnitude. When the elevation of the flood channel is known in relation to overbank flood deposits it is possible to use textural discontinuities in overbank alluvium as a proxy flood record of long-term variations in magnitudes and frequencies of large floods. It is usually convenient for paleohydrologic studies to partition overbank deposits from large floods into those that are coarse-grained (gravel) versus those that are fine-grained (sand, silt, and clay). Graphical moments statistics for overbank gravels provide little specific information that is directly indicative of flood magnitudes because grain size distributions of most fluvial deposits tend to be poorly sorted, right skewed, and platykurtic. However, graphical moments statistics are useful for differentiating overbank gravels that were deposited from suspended load versus those deposited from bedload.

The distinction is important because overbank deposits that were carried as bedload are suitable for further analyses to determine the competent flow depth necessary to transport the coarsest particles in the deposit. Reconstructed long-term variations in large overbank floods controlled by radiocarbon dated alluvial deposits indicate that large floods were not randomly distributed in time during the Holocene because of climatic changes. As in the case of overbank gravels, it also is not possible to associate specific values of graphical moments statistics for fine-grained overbank flood deposits with specific magnitudes of floods. However, graphical moments statistics indicate that fine-grained overbank deposits from large floods have grain size distributions that are more poorly sorted, less right skewed, and more platykurtic than adjacent overbank flood deposits associated with smaller floods. As might be anticipated, fine-grained sediments reflect magnitude differences in floods most strongly in the sand fractions, while variations in clay fractions show very little relationship with flood size.

A typical overbank depositional sequence in fine-grained sediment representing a large flood begins with high silt and low sand percentages in the basal section, shifts progressively to high sand and low silt percentages in the middle section, and culminates with low sand and very high silt percentages at the top. When this kind of depositional sequence can be traced across several alluvial surfaces of different relative heights, but of approximately comparable distances from the flood channel, the deposit usually is the result of a very large flood. When the successional pattern of grain size in the depositional sequence varies from surface to surface, the deposit usually has resulted from overbank deposition by a moderate to small flood. The thicknesses of depositional sequences resulting from overbank floods decrease rapidly as the relative height of a depositional surface increases in relation to the channel bed elevation. The tendency for upward decreasing thickness of units and the tendency for overbank deposition to become increasingly related to floods of low frequency and high magnitude as the alluvial surface of deposition increases in height explains the well-known principle that rates of overbank vertical accretion on floodplains tend to decrease with time.

ACKNOWLEDGEMENTS

The research related to the use of overbank flood flood gravels to reconstruct magnitude of Holocene floods was supported by National Science Foundation Grant EAR-8306171. The research related to stratigraphic evidence of extreme floods in fine-grained overbank alluviumwas support by National Science Foundation Grant EAR-8511280. I thank David Leigh, David Richardson, Steven Luecke, Richard Dunning, and Frank Magilligan who assisted with field and laboratory investigations.

REFERENCES

Allen, J. R. L., 1984, Sedimentary structures, their character and   physical basis: New York, Elsevier Science Publishers, 663 p.

American Society for Testing and Materials, 1977, Standard method for particle size analysis of soils, designation D422-463, *in* 1977 annual book of ASTM standards: Philadelphia, American Society for Testing and Materials.

Baker, V. R., 1973, Paleohydrology and sedimentology of Lake Missoula flooding in eastern Washington: Geological Society of America, Special Paper 144, 79 p.

Baker, V. R., 1984, Flood sedimentation in bedrock fluvial systems, *in* Koster, E. H., and Steel, R. J., editors, Sedimentology of gravels and conglomerates: Canadian Society of Petroleum Geologists, Memoir 10, p. 87-98.

Baker, V. R., and Ritter, D. F., 1975, Competence of rivers to transport coarse bedload material: Geological Society of America Bulletin, v. 86, p. 975-978.

Bogárdi, J., 1978, Sediment transport in alluvial streams: Budapest, Hungarian Academy of Sciences Publishing House, 826 p.

Bradley, W. C., and Mears, A. I., 1980, Calculations of flows needed to transport coarse fraction of Boulder Creek alluvium at Boulder, Colorado: Geological Society of America Bulletin, v. 91, Part II, p. 1057-1090.

Costa, J. E., 1974a, Stratigraphic, morphologic, and pedologic evidence of large floods in humid environments: Geology, v. 2, p. 301-303.

Costa, J. E., 1974b, Response and recovery of a piedmont watershed from tropical storm Agnes, June 1972: Water Resources Research, v. 10, p. 106-112.

Costa, J. E., 1978, Holocene stratigraphy in flood frequency analysis: Water Resources Research, v. 14, p. 626-632.

Costa, J. E., 1983, Paleohydraulic reconstruction of flash-flood peaks from boulder deposits in the Colorado Front Range: Geological Society of America Bulletin, v. 94, p. 986-1004.

Ely, L. L., and Baker, V. R., 1985, Reconstructing paleoflood hydrology with slackwater deposits: Verde River, Arizona: Physical Geography, v. 6, p. 103-126.

Folk, R. L., and Ward, W. C., 1957, Brazos River bar: a study in the significance of grain size parameters: Journal of Sedimentary Petrology, v. 27, p. 3-26.

Grant County News, 1950, v. 66, No. 1, Platteville, Wisconsin.

Hallberg, G. R., 1978, Standard procedures for evaluation of Quaternary materials in Iowa: Iowa Geological Survey, Technical Information Series No. 8, 109p.

Harms, J. C., and Fahnestock, R. K., 1965, Stratification, bed forms, and flow phenomena (with an example from the Rio Grande): Society of Economic Paleontologists and Mineralogists, Special Publication 12, p. 84-115.

Hershfield, D. M., 1961, Rainfall frequency atlas of the United States: U. S. Weather Bureau Technical Paper 40.

Inman, D. L., 1952, Measures for describing the size distribution of sediments: Journal of Sedimentary Petrology, v. 22, p. 125-145.

Jackson, R. G., 1978, Preliminary evaluation of lithofacies models for meandering alluvial streams, in Miall, A. D., editor, Fluvial Sedimentology: Canadian Society of Petroleum Geologists, Memoir 5, p. 543-576.

Jahns, R. H., 1947, Geologic features of the Connecticut Valley, Massachusetts, as related to recent floods: U. S. Geological Survey Water Supply Paper 996, 158 p.

Knox, J. C., 1979, Geomorphic evidence of frequent and extreme floods, in Henry, W. P., editor, Improved hydrologic forecasting, why and how: American Society of Civil Engineers, New York, p. 220-238.

Knox, J. C., 1985, Responses of floods to Holocene climatic change in the Upper Mississippi Valley: Quaternary Research, v. 23, p. 287- 300.

Knox, J. C., 1987, Climatic influence on Upper Mississippi Valley floods, in Baker, V. R., Kochel, R. C., and Patton, P. C., editors, Flood geomorphology: New York, John Wiley & Sons (in press).

McKee, E. D., Crosby, E. J., and Berryhill, H. L., 1967, Flood deposits, Bijou Creek, Colorado, June 1965: Journal of Sedimentary Petrology, v. 37, p. 829-851.

Patton, P. C., Baker, V. R., and Kochel, R. C., 1979, Slack-water deposits: a geomorphic technique for the interpretation of fluvial paleohydrology, in Rhodes, D. D., and Williams, G. P., Adjustment of the fluvial system: Dubuque, Kendall-Hunt Publishing Co., p. 225-253.

Ritter, D. F., 1975, Stratigraphic implications of coarse-grained gravel deposited as overbank sediment, southern Illinois: Journal of Geology, v. 83, p. 645-650.

U. S. Water Resources Council, Hydrology Committee, 1981, Guidelines for determining flood flow frequency: Bulletin 17B, Washington D. C.

Williams, G. P., 1983a, Paleohydrological methods and some examples from Swedish fluvial environments: I - cobble and boulder deposits: Geografiska Annaler, v. 65, p. 227-243.

Williams, G. P., 1983b, Improper use of regression Equations in earth sciences: Geology, v. 11, p. 195-197.

Wolman, M. G., and Eiler, J. P., 1958, Reconnaissance study of erosion and deposition produced by the flood of August 1955 in Connecticut: American

    Geophysical Union Transactions, v. 39, p. 1-14.
Wolman, M. G., and Leopold, L. B., 1957, River flood plains: some observations
    on their formation: U. S. Geological Survey Professional Paper 282C, p. 87-
    109.

9

Reconstruction of a flood resulting from a moraine-dam failure using geomorphological evidence and dam-break modeling

P. A. Carling and M. S. Glaister

ABSTRACT

On October 29th 1927 a small moraine-dam on the eastern flank of the Helvellyn range in the English Lake District failed after heavy rainfall. The moraine had impounded a small mine-reservoir at Keppel Cove. The discharge from the breach, totalling some 1.24×10^5 m^3, swept down the course of the Glenridding Beck devastating the village of Glenridding 2.7 km to the east.

In excess of 13000 m^3 of material was eroded from the dam-breach and some of this debris exacerbated property damage in Glenridding. However most of the moraine debris was deposited as a series of boulder lobes and berms in expanding reaches throughout the stream course.

The event at the time was poorly documented, there being, for example, no estimate of flood velocity or peak discharge. However limited published information coupled with detailed field survey and a number of unpublished contemporary photographs have provided sufficient data to establish maximum water depths at sections throughout the valley. These data form the limiting conditions in an hydraulic reconstruction of the flood-wave dynamics using a published dam-break model.

The propagation of flood waves from breached dams clearly is of interest to the engineer and hydrologist. In addition, the activity of the flood wave is manifest today chiefly in the form and location of the berm structures. The location of these in respect of channel slope, cross-sectional geometry and flood dynamics is discussed. The value of the dam-break to the geomorphologist is that the unfortunate event represents a "controlled experiment" whereby a known volume of water is routed through a catchment. As such, it forms a prescribed model of catastrophic natural flooding which in certain circumstances may be related to a local magnitude-frequency model of flood occurrence and geomorphic effectiveness.

INTRODUCTION

River gauging stations for upland streams in the U.K. are relatively few and records are usually short compared with those available for lowland rivers (N.E.R.C., 1975). Even when supplemented by historical records of extreme flood events, the magnitude-frequency relationship of rare large floods to the gauged record is uncertain. As the number of gauging stations increases and the length of records is extended, the recorded maximum specific discharge per unit catchment area also has increased (Werrity and Acreman, 1985). Yet these discharge values, representative of catchment areas ranging over four orders of magnitude, are often less than the theoretical maximum possible flood.

Factors controlling run-off are complex and the representativeness of records is complicated by possible climatic changes over the historic gauging period (Lamb, 1977), so that design criteria presently based on extant records may not be appropriate if applied to periods of more than a decade or two. In many cases, particularly in North America and Australasia, gauging records have been extended using the stratigraphical record of Holocene floods to provide a more substantial time-base for assessing flood frequency and magnitude (Costa and Baker, 1981). Such an approach is in its infancy in the U.K. but geomorphological and stratigraphical evidence of large ungauged floods can be found in our upland valleys. As our understanding of the significance of this record grows, interpretive analysis of geomorphological evidence may be useful to constrain hydraulic reconstructions of past flood episodes.

Catastrophic flooding in upland channels is associated with unsteady flow and steep slopes. The physical effects are often reminiscent of an upland dam-break event. Unlike the natural ungauged flood, in the latter case the total volume of flow in a dam-break flood is usually known from reservoir capacity records, so that these unfortunate events may have value as 'semi-controlled experiments' to assess the impact of rare natural events.

The specific aims of the present study are three-fold; firstly, to ascertain whether geomorphological evidence, supplemented by meager historical records, is adequate input to a dam-break simulation of a small dam-burst some 60 years after the event occurred (if not, it may be concluded that either the geomorphological control is inadequate or that the model is inappropriate for such reconstructions); secondly, to evaluate the event as a model of natural catastrophic flooding in the particular locality and finally, there is the intrinsic interest in ascertaining the nature of an unpleasant episode which is still firmly implanted in the memory of the local populace.

BACKGROUND

Keppel Cove Tarn was a small natural cirque lake (2.55ha) occupying a moraine-dammed depression on the eastern flank of the Helvellyn range in the English Lake District at an altitude of about 530 m a.s.l. (Fig. 1 and 3b). The bedrock consists of the Borrowdale series of Ordovician volcanic ashes, tuffs and breccias interstratified with lava flows and disrupted by numerous igneous intrusions (Eastwood, 1921; Gough, 1965). The rocks are heavily mineralized with lead veins predominating especially on the southern flank of Greenside mountain (749 m a.s.l).

The small stream draining the tarn is joined by a similar feeder from Brown Cove (Cove being a local name for a cirque) to the south-west. These two streams together with Swart Beck, Red Tarn Beck and Mires Beck are the chief tributaries feeding Glenridding Beck, a true mountain stream which flows generally eastwards, passing through the village of Glenridding (164 m) to discharge into a large lake (Ullswater - 8.94 km^2) through a delta.

In its course, Glenridding Beck, draining an area of 10.45 km^2, falls some 402 m in 4.35 km giving an average gradient of 0.092. In its upper course, gradient varies between 0.059 and 0.236 only being reduced to 0.024 in its lower reaches. In its upper sections the stream is cut through thick drift deposits often exposing the bedrock beneath (Fig. 2a). Few of the reaches here may be described as alluvial, instead the stream is constrained in a V-shaped defile with a bare rock or boulder strewn bed. The few alluvial reaches exhibit poorly developed alluvial plains, 200-300 m wide, consisting of coarse gravels and little fine sediment (Fig. 2b). There are few sites in these upper reaches which are actively supplying sediment to the stream from the stream banks or slopes. Throughout its course the profile is interrupted by a number of small falls and a major narrow gorge at Greenside Mine which includes a 10 m near vertical fall. Below the gorge the stream course is cut in thick drift deposits (Hay, 1934) with steep active exposures providing considerable coarse bedload (Fig. 2c). Here the streams' course is locally alluvial, braided and unstable, changing its position frequently. In earlier times much of the bed material below the gorge must have consisted of fine tailings from the lead mine at Greenside. Although this source is not important today, the area of the Glenridding delta in Ullswater was considerably augmented in the late nineteenth century by fine mineral waste from the mine (Mill, 1895); a process which continued in the twentieth century (Hay, 1937).

The Greenside Mining Company was formed about 1822 and in total produced about 1/4 million short tons of lead concentrate until closure in 1962 (Shaw, 1983). The significance of the mining operations lies in the requirement for water, chiefly used for hydro-electric power. Water for this purpose was supplied from small dams constructed of peat, earth and stone on the Swart Beck and at Red Tarn. Keppel Cove Tarn was also used for supplying water to a power-house through a leat (channel) cut across the slope below Catstycam to Red Tarn Beck (Fig. 1). To augment flow into Keppel Cove, which naturally receives drainage only from the screes below the cirque headwall, a leat was cut from the stream in Brown Cove to

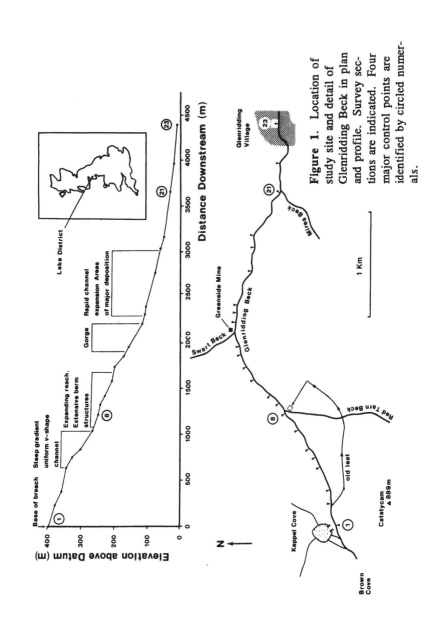

Figure 1. Location of study site and detail of Glenridding Beck in plan and profile. Survey sections are indicated. Four major control points are identified by circled numerals.

feed the tarn. The suggestion (Hay, 1928) that the moraine impounding the tarn may have been raised artificially to increase the reservoir volume, would seem incorrect (Anon., 1927). There is no evidence for this in the vicinity of the breach and the controlled outflow was originally over a spillway to the west at the natural point of efflux (Anon., 1927; Challis, 1979).

At sometime in its history, an attempt was made to increase the discharge from the tarn into the power-house leat, by drilling through the base of the impounding moraine and installing a six- to nine-inch pipe and valve assembly (Anon., 1927). Presumably flow from the pipe was reduced, if not completely shut off, when the mines were not working, as each day a workman walked up to the Cove to open the valve (Westmorland Gazette - 5 Nov. 1927). This of course meant that, should the reservoir be full, the only outlet was over the controlled spillway channel. Measuring only 60 cm by 60 cm in section, built of rough stone-work and meeting the dam crest at an acute angle the spillway must have offered a greater than minimal resistance to high flows over the crest. A simple undershot gate controlled flow to the spillway and this would overtop with a depth greater than 1.25 m measured from the base of the gate.

It is evident in the field today, that should a storm occur, the spillway might easily have proved inadequate especially if the pipe-valve was partially or completely closed.

METEOROLOGY AND HYDROLOGY

The last week in October 1927 was notably mild with only light rainfall over the Lake District. However from the 27th-29th October an unusually deep depression tracked rapidly S.W. to N.E. over the British Isles. The centre of the depression rapidly occluded with a warm wet sector to the south as the main cold front moved quickly N.E. Barometric pressure rose in the rear of the depression giving intense gales and exceptionally heavy rainfall. As the centre passed over Southern Scotland and the Lake District between about 2300 and 2400 hr on 28th October, winds of up to 129 km h^{-1} were recorded on the Lancashire coast (Crichton, 1927). In the 1920's the rain gauge network in the Lake District was sparse. However gauges in the valleys (in Ambleside and Grasmere) recorded 44 mm and 69.9 mm respectively in the 24 hr preceding 0900 hr on the 29th of October. Rainfall would have been considerably higher over the mountains themselves.

Glenridding Beck has no permanent gauging station but has a history of rapid rises, responding to heavy rainfall in less than an hour (Hay, 1937). Spot gaugings by the local water authority at Glenridding demonstrate that low flows may be less than 1 m^3 s^{-1}. Although no high flows have ever been monitored, moderate storm flows might be expected in the range of 10-15 m^3s^{-1} although larger floods have occurred. The mean annual flood was estimated using catchment characteristics (N.E.R.C., 1975) as equal to 25 m^3s^{-1}, whilst the magnitude of the maximum probably flood was estimated as about 260 to 280 m^3 s^{-1} using the method of Far-

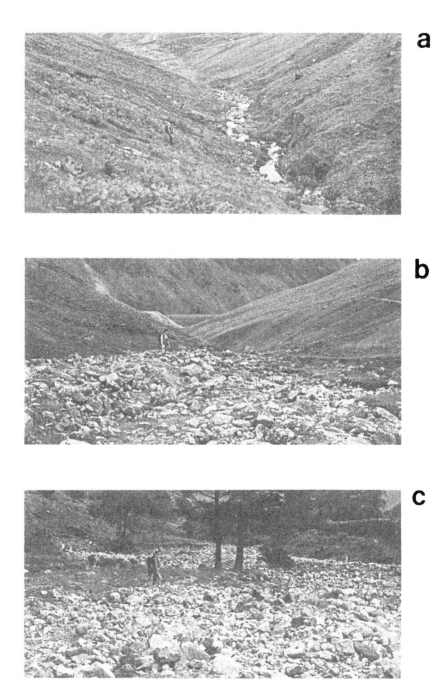

Figure 2. Glenridding Beck (a) View of channel and catchment at section 5, (b) typical valley-section expansion at section 9, (c) view of downstream reach close to section 18.

Figure 3. (a) Photograph of the site of Rattlebeck Bridge taken immediately after the 1927 flood. Note the trashline across the road and the truncated stone-wall. (b) View of the moraine-breach looking north toward the tarn basin.

quharson et al., (Sutcliffe, 1978). As a result of the storm rainfall, water levels in Glenridding Beck would have been rising rapidly by about 0100 GMT when the dam failed.

THE DAM-BURST

Eye-witness accounts indicated that the flood wave from the breached dam swept through Glenridding village about 0130 to 0140 GMT on the 29th October 1927. The main force of the crest of the flood was spent in 15 to 30 minutes and maximum depth was estimated to be in the region of 2 m in the vicinity of the village.

Rattlebeck Bridge, a lightly built structure upstream of the village (Figs. 1 and 3a), failed shortly before the crest of the flood passed the site and was probably lifted from its footings. Although housing and other property were extensively damaged and livestock lost, fortunately no one was killed although there were several close escapes (Westmorland Gazette, 5 Nov. 1927; Hay, 1928; Cooper, 1960; 1966).

The moraine had been breached to the east of the spillway in the vicinity of the pipe-work. Although the reservoir was believed not to have been "overfull" as the sluice was open the previous afternoon (Westmorland Gazette - 5 Nov. 1927; Challis, 1979), nevertheless failure at the time was attributed to "the cloudburst". Given the time of year and the nature of operation, the reservoir must have been close to capacity. It is probable that the moraine failed primarily by overtopping, owing to heavy rainfall possibly exacerbated by wind-wave action (Anon., 1927; Challis, 1979). However, the rapidity of the failure as reflected in the catastrophic flood wave so soon after the storm centre passed over the Lake District suggests a sapping process (piping) may have already been present along the length of the pipe-work which had been packed around with peat and clay. This possibility was strongly contested at the time (Anon., 1927) although the opinions expressed may have been partisan. The sapping process would be intensified by an excessive head of water in the reservoir. The result would be partial collapse of the moraine, increased spillage over the slumped crest and consequent rapid erosion as the moraine contains little cohesive material (< 0.2% silt and clay).

The result of the breach however is unequivocable. The moraine had a base thickness of some 80 to 90 m and a height of about 10 m. The discharge through the moraine resulted in a V-shaped breach some 35 m wide at the top and 8 to 12 m deep (Fig. 3b). The volume of material swept out of the breach totalled 13 to 14,000 m^3 (about 25,000 metric tons) and the tarn, of a volume 124,000 m^3, was completely emptied.

Immediately below the breach, at the confluence with the Brown Cove stream, Hay (1928) noted that the bed level was scoured down approximately 1m whilst turf was stripped from the valley side-wall to a height of about 6.1 m. Throughout the upper course of the stream, vegetation and turf were stripped off the bed rock and dry-stone walls were demolished. Additional debris must have been entrained from deposits in Glenridding Beck but by far the greater portion of material available for deposition consisted of moraine from the breach.

Upstream of the gorge there are two major deposits with boulders up to 1.5 m^3 in volume. These gravels form a complex of fans and berms extending over 0.8 km. Immediately below the gorge two similar deposits occurred (Hay, 1928). Although these latter deposits are now deeply dissected, they contain some of the largest boulders (< 3 m^3) transported during the event. Throughout the lower course, the stream-bed was raised by 1 m with finer gravels, and material was deposited in the fields to the north of the village as well as on the delta. Frequent overbank discharges persisted for some years after the event until the bed level was restored (Hay, 1928; 1937).

Other floods, although not on the scale of the dam-burst, have eroded channel sediments and deposited flood gravel bars. Further floods occurred for example in 1931 and 1932 (Hay, 1937; Cooper, 1960) although without serious consequence. These floods flushed out finer sediments from the lower course of the beck (Hay, 1937) and may have dissected the 1927 deposits and constructed smaller berms

That the coarsest flood gravels were deposited by one unique event was evident in that all the major boulder deposits support an 'even-aged' population of the saxicolous lichen Rhizocarpon geographicum Agg. with individual thalli of a similar and limited size range. Fresh flood deposits have few if any lichens present and it has not been possible to identify other deposits with mature lichen communities which might be of considerable age. The average of the five largest lichens present (53 mm in 1986) indicates a date of 1927 on a Rhizocarpon growth curve for Eastern Cumbria (Harvey et al, 1984) and the largest lichen (57 mm) suggests the deposits are about 62 years old. Given the limitations of the lichenometric method (Innes, 1985) this result is consistent with the dam-burst date of 1927.

DAM-BREAK MODEL

The dam-break model selected was developed originally by Fread (1977) for the U.S. National Weather Service, though the version implemented (DAMBRK) has been described more recently by Fread (1984) Although the model simulates a dambreak flood in one computational exercise, it first routes an incoming flood through the storage basin of the reservoir. Breaching is simulated when the reservoir waterlevel reaches a predetermined value. Instantaneous flow through the breach is calculated using broad-crested weir equations for a variety of optional breach geometries. Breach dimensions develop over a given time period to limiting values defining the basic final configuration. The outflow hydrograph is estimated using a hydrological storage-continuity method or alternatively, a dynamic routing procedure. The second part of the simulation routes the resultant flood wave through the downstream channel using the Saint-Venant flow equations and non-linear finite-difference method.

Full details of the model are given in Fread (1984) and Land (1981) describes a simpler version which has been used recently to model the 1982 Larimer County, Colorado dam failures (Jarrett and Costa, 1985).

DATA INPUT AND CALIBRATION

In order to use the model, initial and limiting conditions need to be specified as well as the physical geometry of the reservoir and channel reaches.

The tarn basin was surveyed in detail to establish both the original reservoir volume and the maximum top-water altitude and surface area. The breach was surveyed to determine its configuration and to calculate the quantity of eroded material.

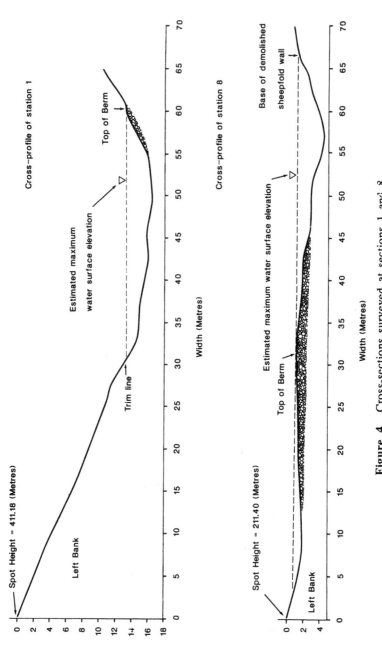

Figure 4. Cross-sections surveyed at sections 1 and 8.

The channel downstream was described by twenty-three surveyed cross-sections. This number was required to maximize the models capabilities in such a high-gradient stream typified by significant variation in channel form throughout its length. Sections were selected to represent the range of channel morpho-types i.e., narrow and broad sections and the gorge and, in addition, selected critical sections were surveyed at locations where historic records or geomorphological control indicated maximum or minimum flood water levels. All sections were used as computational nodes but certain of these were control points as described below.

Constraints on the simulated hydrograph consisted of a range of data. As the storm occurred around midnight only about one and one-half hours were available for the storm flow to concentrate, the dam to breach, and the flood wave to travel the length of the stream. Further the main flood was described as passing in 15 to 30 mins. Important observations at the time included the rough estimate of 2 m as a typical water depth in the channel at Glenridding and a water depth on a building, measured immediately after the flood, of 2.38 m (Westmorland Gazette - Nov. 5 1927). Further, at the location of section 1 (Fig. 1), Hay's record of 6.1 m of turf stripped from the right-bank possibly indicates a maximum flood trimline immediately below the dam. On inspection of the site at Section 1, Hay's trimline could not be discerned. However an inactive coarse-gravel berm was evident on the right-bank at 3.2 m vertically above the bed. This length can be taken as one side of a right-angled triangle the hypotenuse of which has a length of about 6 m and is equivalent to the valley-side wall length inundated up to the bar-top (Fig. 4). On the left-valley side-wall, which consists of the unbreached moraine, an apparent trimline in the moraine was also at a vertical height of 3.2 m. These data provided an important control, delimiting an expected value for the water-surface elevation at Section 1.

At downstream sections boulder-berm heights were used to indicate minimum water depths. Costa (1984) suggests that berm heights often closely match maximum water depths although some examples exist in the literature where they may be below the flood maximum height. Costa's observation was found to be applicable to a flood berm in the Pennines (Carling, in press) and, at Section 8, the base of a dry-stone wall demolished by the flood was at the same elevation as the crest of a large berm within the section (Fig. 4).

Finally a series of contemporary photographs confirmed overbank flow occurred through downstream reaches. At Rattlebeck Bridge one photograph taken after the bridge had been destroyed shows a truncated stone-wall and trash-line across the cart-track (Fig. 3a). The height of this trash-line was surveyed.

Altogether four major control points were identified. These, together with four less important and less well-verified minor control points, gave a total of eight sections with estimated water levels (Table 1).

Final input data consisted of estimates of Manning's roughness coefficient and expansion and contraction coefficients, as defined by Fread (1984). The further option of designating active and inactive cross-section flow areas was not used as very little opportunity existed in the valley for significant flood storage. Manning's n

Table 1. (1) Expected Water Depths, (2) Observed Water Depths and (3) Peak Discharges at Main Control Points.

Section	(1) Min	(1) Max	(2) (m)	(3) $(m^3 s^{-1})$	Geomorphological Controls*
1	3.20	3.20	3.21	107.8	Berm & trim-line
2	1.22	-	2.06	107.7	Small berm
3	-	3.80	3.94	107.9	Trimline
8	3.50	-	3.65	102.6	Berm & wall
16	2.68	-	2.65	101.1	Berm
18	1.28	-	1.74	100.9	Berm
21	2.04	2.04	2.08	99.9	Wall & trashline
23	2.38	2.38	2.55	99.8	Water mark

* See text for details

was established for each of the eight control points using Jarrett's (1984) regime equation for high gradient streams;

$$n = 0.39\ S^{0.38}\ R^{-0.16} \tag{1}$$

wherein the hydraulic radius R = A/P was defined using the estimated depth to delimit the critical wetted perimeter (P) and cross-sectional area (A). The calculated values were in accord with values expected from the authors experience of flow resistance in U.K. mountain streams. Values of n for the remaining 14 reaches were estimated from a series of photographs and field observations of the 22 reaches interpolating consistent with those obtained from equation (1). Initial estimates used in the model ranged between 0.087 and 0.149, averaging 0.133. Expansion and contraction coefficients were estimated from photographs and field observation and ranged between -0.9 and 0.19.

SIMULATION

The time-step of the computation was allowed to be calculated automatically and varied between 9 seconds immediately after the breach and 96 seconds in the downstream reaches near Glenridding village. Initial conditions in the reservoir were set with the water-surface elevation equal to the dam-crest; this was to simulate an over-topping failure (owing to wind-wave action) aggrevated by piping. The importance of the latter is speculative but the rapidity of the breach formation, in the little time available between initial rainfall and the flood wave arriving in Glenridding, would seem to indicate catastrophic failure within 15 to 30 minutes.

Initial trial runs used a baseflow here and throughout in Glenridding of 10-15 $m^3 s^{-1}$ and 2-3 $m^3 s^{-1}$ immediately below the dam. The small base-flow at the dam-site caused computational difficulties which seem typical when depths are shallow (Cunge et al, 1980; Jarrett and Costa, 1985). This was alleviated by incrementally increasing the upstream base-flow to a value of 11 $m^3 s^{-1}$ at which minimal value the model performed satisfactorily. Baseflow was maintained by having both an inflow to the reservoir and a pipe outflow equal to this value. The baseflow is not excessive and is small compared to the flood wave peak.

Given a typical downcutting rate for earth dams of about 1 m min^{-1} (Fread, 1984) the breach time was initially set by default as 10 minutes. Initial water-levels associated with the peak of the flood wave at section 1 were slightly too high. For this section, model calibration values available for fine-tuning consisted of Manning's roughness coefficient, the expansion-contraction coefficient and breach time. Experience throughout the routing procedure demonstrated little effect on water-levels was obtained by adjusting the second parameter whilst the roughness coefficient realistically could not be reduced below its estimated value. Instead the required water depth at Section 1 was achieved by increasing the breach-time to 24 minutes.

Having obtained acceptable initial outflow conditions, the simulation proceeded by routing the flow reach by reach down the Glenridding valley utilizing the continuity method. At first, model runs would not converge, especially in upstream reaches, owing usually to excessive supercritical flow being predicted through steep reaches or highly variable flow conditions existing from one reach to the next. The problem was overcome by iteratively obtaining the minimum Manning's necessary to allow convergence to be attained for that particular reach. Generally this required slightly higher values of n than were calculated using Jarrett's model. These data ranged between 0.050 and 0.220, averaging 0.145. The procedure was repeated reach by reach until the model would route the flow successfully through all 23 reaches; flow remaining supercritical in the steepest channel segments.

The first successful complete run gave estimated water depths typically not more than ± 30 cm of expected water depths at the 7 control points downstream of section 1. The simulation was completed by fine-tuning the results at the seven control points to match the expected values by slightly altering the Manning's roughness coefficients as required. Final simulated water depths and discharges are given in Table 1 and Figure 5.

It should be noted that, although there is no independent evidence for the accuracy of the peak discharge estimate, the results are consistent reach by reach showing slight and steady flood attenuation and diminuation of the hydrograph amplitude downstream. Altering the breach time and the time-base for the complete routing within realistic limits did not significantly affect the result. The conclusion re-

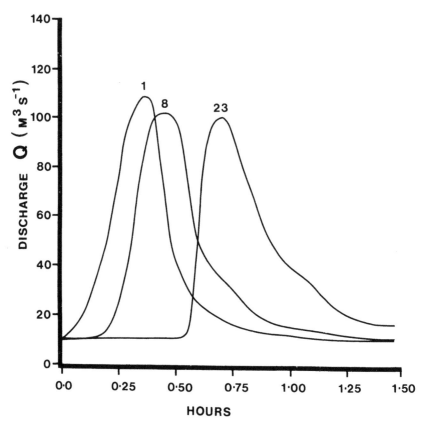

Figure 5. Examples of the predicted dam-breach hydrographs at sections 1, 8 and 23. Abcissa scale is time since breach-formation.

mains that after subtracting a baseflow of about 10 $m^3 s^{-1}$, a flood wave of 90 to 100 $m^3 s^{-1}$, routed over a time period of 1.5 hrs, passed through Glenridding village in some 35 minutes.

The discrepancy between the simulated water depth at section 23, 2.55 m and the measured depth of 2.38 m on the building needs some comment. Considerable flow through the village was not within the stream channel but through buildings and gardens behind the houses which lined the channel bank. The frictional effect of the buildings and flow losses to the gardens could not be accommodated in the model so that one would expect the simulated depth to be slightly in excess of the recorded flood level.

As a final check on the reliability of the simulation, outflow characterisations vis-a-vis aspects of breach size and development time were found to compare favourably with observed characteristics of well documented earth-dam failures com-

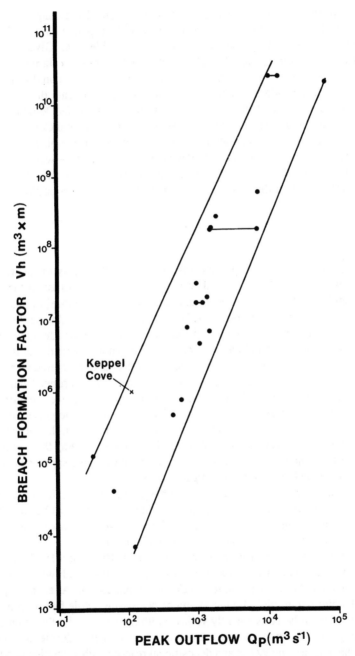

Figure 6. Relationship between peak outflow rate and the breach formation factor for Keppel Cove compared with data and envelope curves extracted from MacDonald and Langridge-Monopolis (1984).

piled to improve DAMBRK model formulation (MacDonald and Langridge - Monopolis, 1984). These characteristics were not however incorporated in Fread's (1984)
version of DAMBRK and so the comparison remains objective. Given the potential energy in the reservoir and the calculated breach time, the volume of material
eroded from the breach was larger, but not exceptionally larger, than that associated
with most other earth-dams. This fact would normally indicate the presence of
low-cohesion material and structural instability in the dam (MacDonald and Langridge - Monopolis, 1984). The former case is evident at Keppel Cove and, whilst
the presence of piping is speculative it is not uncommon in moraine-dam failures
(Costa, 1985). In any case a relatively low peak rate of discharge would be expected, as observed (Fig. 6), because the relatively narrow, triangular breach 70 to 90 m
in length, would restrict the cross-sectional area and constrain the outflow rate.

DISCUSSION

The primary aim of the project was to ascertain the utility of incorporating
sparse geomorphological data into a dam-break simulation and to assess the model's
potential application to high gradient mountain streams. The model in this respect,
performed well given that, for prescribed calibration limits, initial water levels generated by the routine were not grossly different to levels defined using the geomorphological evidence. Varying parameters within known and realistic limits did not
alter significantly the estimated peak discharge. Problems encountered were similar
to those observed in applying the simulation routine to the Larimer County dam-
bursts (Jarrett and Costa, 1985) in that minor baseflow changes were required and,
in addition, the roughness coefficients needed optimizing. In the present case, Jarrett's regime equation gave good initial estimates of Manning's n, the difference between the average of these and that of the optimized values being 9%. Subcritical
flows were generated through most reaches although modest supercritical flows
were accommodated in some of the gorge sections.
    The boulder berms provided important baseline data concerning probable wa-
ter-levels. Further research is being conducted on the hydraulics of berm formation
but it is appropriate here to check that the simulated velocities were adequate to
transport and deposit the boulders composing major berms.
    Berm deposition occurred mainly at sections 8, 10, 16 and 18. Each deposi-
tional site was associated with major expansions in valley width with contracting
reaches immediately downstream (cf. O'Connor et al., 1986). At each site, valley
slope and water depth might be increasing or decreasing while bed roughness did
not vary significantly, consequently valley width appeared to be the main control
constraining berm location. Average cross-sectional velocities decelerated between
the immediate upstream reaches and depositional reaches with velocities varying between 6.2 and 3.5 ms^{-1} upstream and 3.2 and 1.3 ms^{-1} at berm sites.
    During the flood the largest boulders forming the berms must have been
transported as bedload through the reaches immediately upstream of the depositional

reaches, whilst at the latter sites velocities of necessity must have been close to or equal to the threshold for deposition. Williams (1983) summarized extant boulder threshold data and proposed empirical limiting relationships concerning movement and non-movements with respect to mean velocities. These relationships and the current predicted velocities are shown in Figure 7 wherein the intermediate axes of the ten largest boulders at each section have been averaged to produce representative particle sizes. The predicted velocities would appear to be consistent with Williams' curves.

Finally, in terms of a comparison with naturally generated flood flows the discharge in Glenridding Beck may be compared with recorded and estimated flood discharges in the Lake District and the western flanks of the Pennines. Guaged flood flows in the Lake District are few and are supplemented by estimated discharges (H.R.S., 1968) for the 1967 Forest of Bowland floods (Fig. 8). Also shown in Figure 8 is a proposed envelope curve for peak floods in the U.K. including large floods generated in small basins (Werrity and Acreman, 1985). The discharge in Glenridding village is not exceptional by any means and might be exceeded by a natural flood, being less than half the maximum probable flood discharge estimated using methods in the Flood Studies Report (N.E.R.C., 1975). If only the upper catchment above Red Tarn Beck is considered, once again the discharge,

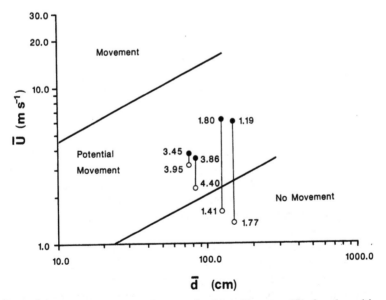

Figure 7. Critical mean velocity versus boulder diameter (O) for deposition sections 10, 8, 18 and 16 (left to right respectively), together with velocities at transport sections immediately upstream (•). Envelope curves enclosing the zone of potential movement are from Williams (1983). The upper curve typicaly represents water depth to particle size ratios of 100 and the ratio for the lower curve is of the order of unity. Numerical values are the ratios associated with the boulders at each section.

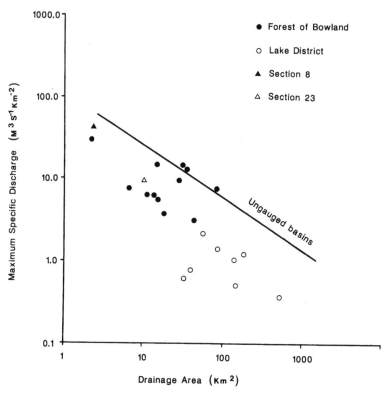

Figure 8. Relationship of the dam-break peak discharges to natural floods. Envelope curve for UK floods in ungauged basins from Werrity and Acreman (1985).

although similar to the FSR maximum probable flood, might be equalled or exceeded by natural floods as it falls within the envelope curve. In making comparisons of this kind it should be remembered that important differences may occur in the nature of dam-break flow propagation and natural flood generation although natural floods in mountainous channels frequently evolve as a roll-wave. The other potential difference is in the nature of the sediment supply to the flood wave. The dam-break provided a concentrated ready source of material whilst in natural storm-flow, sediment sources might be diffuse although debris-flows might concentrate sediment and supply a similar volume of material to a stream-course. With these limitations in mind, dam-break events in some catchments may be useful surrogates for the effects of natural large floods and warrant examination in this context. Further work at this site and other natural flood sites is planned.

ACKNOWLEDGEMENTS

Dr J. Costa (U. S. Geological Survey) and D. Fread (U. S. National Weather Service) are thanked for providing the DAMBRK model computer tape and associated documentation. Dr T. Irish of the F. B. A. freely gave advice and help in implimenting the model. Ms. J. Backhouse of Glenridding village provided contemporary photographs of the flood including Figure. 3a for which we are very grateful, and Mr. J. Thorpe assisted in field survey.

REFERENCES

Anon., 1927, A disaster at the Greenside lead mine: Mining Magazine, 37, p. 360-361.

Carling, P. A., In press, Hydrodynamic interpretation of a boulder-berm and associated debris-torrent deposits: Geomorphology, v.1, (for 1987).

Challis, P. J., 1979, Greenside mine: Ore dressing c. 1900 - The dam disaster 1927: British Mining, v.11, p. 75-81.

Cooper, W. H., 1960, The tarns of Lakeland: Frederick Warne, London, p. 121-122.

Cooper, W. H., 1966, The Lakes: Frederick Warne, London, p. 193.

Costa, J. E., 1984, The physical geomorphology of debris flows. In Costa, J. E. and Fleisher, P. J., eds., Developments and applications of geomorphology: Springer-Verlag, New York, p. 268-317.

Costa, J. E., 1985, Floods from dam failures: U.S. Geological Survey Open-File Report 85-560, 54 p.

Costa, J. E. and Baker, V. R., 1981, Surficial Geology: Wiley, New York, 498 p.

Crichton, J., 1927, The Storm of October 28-29th, 1927: The Meterorological Magazine, v. 62, no. 743, p. 249-252.

Cunge, T. A., Holly, F. M., Jr., Verwey, A., 1980, Practical aspects of computational river hydraulics: Pitman Publishing Ltd., 420 p.

Eastwood, T., 1921, The lead and zinc ores of the Lake District: Memoirs of the Geological Survey. Mineral Resources, No. 22, 56 p.

Fread, D. L., 1977, The development and testing of a dam-break flood forecasting model: Proceedings of the dam-breaks flood routing workshop, U. S. Water Resources Council, p. 164-197.

Fread, D. L., 1984, DAMBRK: The NWS dam-break flood forecasting model: Hydrologic Research Laboratory, National Weather Service, Silver Spring, MD., 60 p.

Gough, D., 1965, Structural analysis of ore shoots at Greenside lead mine, Cumberland, England: Economic Geology, v. 60, p. 1459-1477.

Harvey, A. M., Alexander, R. W. and James, P. A., 1984, Lichens, soil development and the age of Holocene valley floor landforms: Howgill Fells, Cumbria: Geografiska Annaler, v. 66A, no. 4, p. 353-366.

Hay, T., 1928, Glenridding flood of 1927: The Geographical Journal, v. 73, p. 90-91.

Hay, T., 1934, The glaciology of the Ullswater area: The Geographical Journal, v. 84, no. 2, p. 136-148.

Hay, T., 1937, Physiographical notes on the Ullswater area: The Geographical Journal, v. 90, no. 5, p. 426-445.

H. R. S., 1968, Forest of Bowland and Pendle floods of August 1967: Hydraulics Research Station, Wallingford, Berkshire, England. Report EX 382, 40 p.

Innes, J. L., 1985, Lichenometry: Progress in Physical Geography, v. 9, no. 2, p. 187-254.

Jarrett, R . D., 1984, Hydraulics of high gradient streams: Journal of Hydraulic Engineering, v. 110, no. 11, p. 1517-1518.

Jarrett, R. D. and Costa, J. E., 1985, Hydrology, geomorphology and dam-break modeling of the July 15th, 1982 Lawn Lake Dam and Cascade Lake Dam failures, Larimer County, Colorado: U.S. Geological Survey Open-File Report 84-612, 109 p.

Lamb, N. H., 1977, Climate: Present, past and future: v. 1, Methuen and Co., London, 613 p.

Land, L. F., 1981, Computer program documentation user's manual-general purpose dam-break flood simulation model (K-634): U.S. Geological Survey Water-resources Investigations 80-1160, 101 p.

MacDonald, T. C. and Langridge-Monopolis, J. L., 1984, Breaching characteristics of dam failures: Journal of Hydraulic Engineering, v. 110, no. 5, p. 567-586.

Mill, H. R., 1895, Bathymetrical survey of the English lakes: Geographical Journal, v. 6, p. 46-73 and 135-166.

N.E.R.C., 1975, Flood studies report IV. Hydrological Data: Natural Environment Research Council, London. 541 p.

O'Connor, J. E., Webb, R.H. and Baker, V.R., 1986, Paleohydrology of pool-and-riffle pattern development: Boulder Creek, Utah: Geological Society of America Bulletin, v. 97, p. 410-420.

Shaw, W. T., 1983, Mining in the Lake counties: The Dalesman Publishing Company Ltd., Clapham. 128 p.

Sutcliffe, J. V., 1978, Methods of flood estimation; A guide to the flood studies report: Institute of Hydrology, Report 49, 50 p.

Werrity, A. and Acreman, M. C., 1985, The flood hazard in Scotland. in Harrison, S. J., ed., Climatic hazards in Scotland: Geo books, Norwich, p. 25-40.

Williams, G. P., 1983, Paleohydrological methods and some examples from Swedish fluvial environments: Geografiska Annaler, v. 65A, 3-4, p. 227-243.

10

Changes accompaning an extraordinary flood on a sand-bed stream

W. R. Osterkamp and John E. Costa

ABSTRACT

On June 16, 1965, areas south of Denver, Colorado received up to 360 mm of precipitation in four hours, resulting in catastrophic flooding along Plum Creek, a sand-bed stream tributary to the South Platte River. Based on mixed-population flood-frequency analysis, the flood had a recurrence interval between 900 and 1,600 years; the peak discharge of 4,360 m^3s^{-1} was 15 times that of the 50-year flood, the highest ratio ever recorded at a U.S. gaging station. Bottomland damage was extreme, and most geomorphic features and vegetation were extensively modified or destroyed. Along a 4.08-km study reach, average channel width of Plum Creek increased from 26 to 68 m; destruction of vegetation by the flood, followed by heavy spring runoff in 1973, caused increased braiding and a channel averaging 115 m in width. Lack of erosive discharges since 1973 has led to decreasing channel width.

Hydrologic and botanical field investigations suggest that channel narrowing and flood-plain reconstruction along Plum Creek has occurred in two principal manners: (1) On channel bars new vegetation reduced velocities during inundation and caused deposition of sand and relatively rapid aggradation of the recently revegetated surface. (2) Channel islands, which form downstream from channel obstructions, have developed stabilizing woody vegetation that also favors sand deposition and promotes enlargement. Channel islands grow allometrically in the downstream direction by annual increments; they induce channel narrowing by coalescing with other islands and flood-plain edges.

INTRODUCTION

Plum Creek, a sand-bed stream, flows northward between Colorado Springs and Denver, Colorado (Fig. 1). A series of late-spring convectional cells in the area culminated with an intense storm and flood on June 16, 1965. The flood had profound effects on the bottomland features of Plum Creek (Matthai, 1969); processes of geomorphic and vegetative recovery from those changes have continued to the present.

The Plum Creek basin is in a generally arid area of the Great Plains. Through Colorado this area extends about 80 km east of the Rocky Mountains and is subject to intense rainstorms that locally generate large volumes of runoff very rapidly (Hansen and others, 1978). The usual sources of these cloudbursts are warm, moist air masses moving northward from the Gulf of Mexico, coupled with cool, moist air in the northern Great Plains. When moist air rises quickly in an unstable atmosphere, rapid condensation results in tremendous rainfall volumes--300 to 600 mm of rain in 4 to 6 hours. These rainfall intensities are greater than most people have seen, or will see, and the magnitudes are difficult to comprehend.

Above its confluence with the South Platte River, Plum Creek drains nearly 850 km^2 of dissected Tertiary sandstone and siltstone in the heart of the cloudburst region. The southwest quarter of the basin is drained by tributaries to Plum Creek, mostly to West Plum Creek, that head in granitic rocks of the Rocky Mountains (Rampart Range, Fig. 1). Little rainfall and runoff from the mountainous portion of the Plum Creek basin contributed to the 1965 event; instead the flood was largely the result of 4 hours of intense precipitation in the East Plum Creek basin (Matthai, 1969).

The Plum Creek flood, which peaked at 4,360 m^{3}s^{-1}, was a rare event in terms of magnitude, frequency, and alteration of bottomland features. Although unit discharge was less than 6 m^{3}s^{-1}km^{-2}, the flood was extraordinary for this basin. The peak discharge was 15 times greater than the magnitude of the 50-year flood at a streamflow gage near Louviers (Fig. 1); this is the greatest 50-year flood ratio ever measured by the U. S. Geological Survey (Hardison, 1973). The flood occurred in a readily accessible basin where precipitation and streamflow records, aerial photographs, and topographic maps permit comparisons of pre-flood and post-flood conditions. Based largely on hydrologic data starting in 1947, aerial photography starting in 1964, and field investigations begun immediately after the flood, this paper documents some of the flood damage and geomorphic and botanical changes that have occurred in more than two decades following the flood.

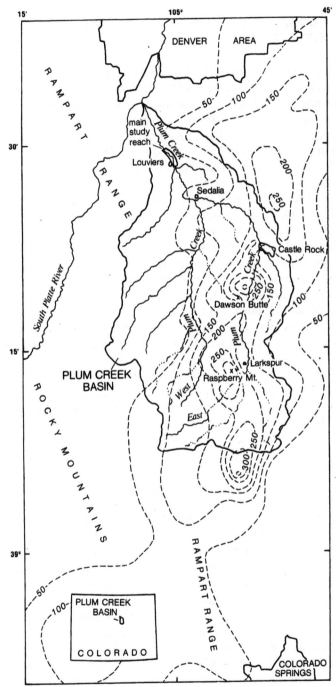

Figure 1. Map of Plum Creek basin and vicinity, showing isohyetals (dashed lines), in millimeters, of June 16 storm.

STORM, FLOOD, AND SUBSEQUENT FLOWS: REVIEW OF EVENTS

Precipitation in early spring, 1965, in the Plum Creek basin and adjacent areas was lower than normal. In the period May 21 through June 13, however, rainfall from several convectional storms totaled roughly twice the normal amount, causing relatively wet antecedent moisture conditions (U. S. Department of Commerce, 1966; Matthai, 1969, p. 12). Thus, relatively moist soils of the basin absorbed little of the intense rain that fell several days later. Rainfall on June 16, 1965, fell unevenly through the length of the Plum Creek basin, and only to the northeast and south was the storm noteworthy in adjacent basins. Heaviest rainfalls were related to topography of the East Plum Creek basin. They were centered (1) near the southern divide of the basin, (2) at Raspberry Mountain, immediately southwest of Larkspur, (3) at Dawson Butte, about 6 km southwest of Castle Rock, and (4) at the basin divide about 6 km southeast of Sedalia (Fig. 1). Maximum precipitation at the four sites was, respectively, about 360, 330, 360, and 255 mm. Most of the East Plum Creek basin received at least 130 mm of rain; precipitation in the West Plum Creek and mainstem Plum Creek valleys (east of the mountains) mostly varied between 30 and 200 mm (Matthai, 1969, pl. 1).

Runoff from the June 16 storm occurred rapidly. Rain began south of Castle Rock in the Raspberry Mountain and Dawson Butte areas about 2 pm. The resulting flood peak reached Louviers before 6 pm, and by 8 pm had traveled another 25 km along lower Plum Creek and the South Platte River into the Denver area (Matthai, 1969). As the flood swept through Denver, it caused damage of roughly $300 million (Matthai, 1969). In the lightly populated Plum Creek basin, however, property losses were much less extensive. Destruction of homes was concentrated at Castle Rock and Sedalia, but elsewhere highways and roads, bridges, and sections of railroad were extensively damaged. Most bridges crossing East Plum Creek and Plum Creek downstream from Larkspur (Fig. 1) failed.

The gaging station near Louviers was destroyed, but reconstruction of the 1965 flood hydrograph is possible using eyewitness accounts, records from undamaged gaging stations downstream, and indirect-discharge estimates. The reconstruction (Fig. 2) suggests that from 5:00 pm, June 16, flow increased from less than 5 $m^3 s^{-1}$ to 4,360 $m^3 s^{-1}$ in about 40 minutes. Recession to base-flow conditions took about 10 hours; thus, the hydrograph is peaked, with a very rapid rise and fall. The valley floor was inundated about 2.5 hours after the flow reached 1,130 $m^3 s^{-1}$, as determined from the stage-discharge relation.

At peak flow, water depths at three valley cross-sections were as great as 5.8 m and averaged from 2.4 to 2.9 m. Velocities computed using flood surveys, the Manning equation, and roughness coefficients of 0.035 to 0.040, ranged from 1.3 ms^{-1} over terraces at the valley sides, to 5.4 ms^{-1} in deeper flows in central parts of the valley. Average velocity for three cross-sections was 3.9 m^{-1}. Average water depths and velocities calculated for the three sections indicate that flows were subcritical, with Froude numbers of 0.72, 0.75, and 0.83. The energy gradient

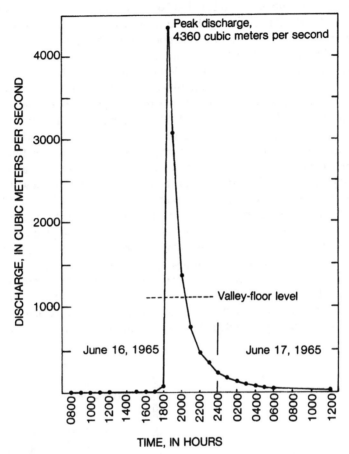

Figure 2. Hydrograph of the June 16, 1965 flood, Plum Creek near Louviers, Colorado.

along the 523-m surveyed reach was 0.00653.

Estimates of recurrence for any very large flood are subject to uncertainty, but are possible if certain assumptions about flood history and frequency distributions are made. The Plum Creek flood inundated terraces of Piney Creek alluvium (Hunt, 1954) to depths of about 0.5 to 1.0 m, but did not deposit identifiable sediment. Piney Creek alluvium is estimated to be 2,500 to 3,000 years old (Scott, 1963). Also inundated were lower, post-Piney Creek terraces formed on alluvium ranging in age from about 1,500 years to modern (Scott, 1963). Because no identifiable sediment was deposited on Piney Creek surfaces in the studied reaches, even though the flood inundated the terraces, earlier floods of comparable magnitude may have occurred without leaving any evidence.

Post-Piney Creek surfaces along Plum Creek received 0.3 to 2.0 m of sediment by the 1965 flood. No other flood deposits were seen on any exposures of post-Piney Creek alluvium along the studied reaches of Plum Creek, thereby suggesting that floods of similar size have not occurred along this reach of stream in roughly 1,500 years. Conventional flood- frequency analysis of very large floods is plagued with difficulties. Using procedures outlined by the U.S. Water Resources Council (Interagency Advisory Committee on Water Data, 1982), however, four possible flood-frequency curves for Plum Creek are plotted in Figure 3. The magnitude of the 1965 flood, and the probable maximum flood for the Louviers site (Bullard, 1986) are also indicated. Curve 1 uses only station data, which may give an unreliable estimate of the true flood frequency because of erratic fluctuations in the station record. The skew coefficient, 1.529, which governs the curvature of the frequency curve on probability paper, is particularly sensitive to flow irregularities like the 1965 flood. Curve 2 uses a generalized skew coefficient (-0.125) for the frequency analysis based on flood data from neighboring gaging stations. Because some nearby stations have different flood populations (such as predominantly from rainfall further to the east, or from snowmelt at higher elevations in the mountains), use of a generalized skew is probably not well advised.

The Interagency Advisory Committee on Water Data (1982) recommends combining the station skew coefficient and generalized skew coefficient by weighting them in inverse proportion to their estimated mean square errors, and averaging the results. The resulting weighted skew coefficient, 0.559, is used to plot curve 3

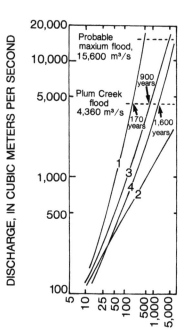

Figure 3. Flood-frequency curves, Plum Creek near Louviers, Colorado. Curves were generated using: Station data and a skew coefficient of 1.529 (1); A generalized skew coefficient for the area of -0.125 (2); A skew coefficient of 0.559 weighted from the first two (3); A foothills-weighted skew coefficient of 0.796 (4).

Table 1. Geomorphic and botanical indicators of bottomland change, Plum Creek Valley, 1964 to 1983.[1]

Date of aerial photography[2]	Oct.4, 1964	June 27, 1967	Aug.7, 1971	June 24, 1975	July 8, 1978	July 9, 1980	Oct. 5 1983
Channel width (ave. in m)	26	68	72	116	70	54	47
Channel gradient	0.0053	0.0063	0.0062	0.0064	0.0064	0.0063	0.0063
Sinuosity	1.22	1.03	1.04	1.01	1.01	1.03	1.03
Channel islands	13	3	15	49	189	171	112
Trees on channel islands[3]	86	2	98	117	580	375	89
Trees, riparian inundated bottomland[3]	1844	913	---	938	---	1030	1420
Trees, (total), inundated bottomland	1930	915	---	1055	---	1405	1509

[1] Data collected from a 4.08-km length of valley centered adjacent to Louvers, CO.

[2] Photographs are enlargements taken at a scale of approximately 1:20,000.

[3] Counts represent trees identifiable on the aerial photography; where island, floodplain, or terrace surfaces are heavily eroded, estimates were made that may include substantial error.

(Fig. 3). Plum Creek and numerous other streams along the eastern Rocky Mountains (foothills) in Colorado, however, are mixed-population flood-frequency stations (Jarrett and Costa, 1983). For these streams, a generalized skew was developed using 26 sets of gage records, and the result was combined with the station skew to produce a foothills-weighted skew for Plum Creek of 0.796. The foothills-weighted skew is the basis of curve 4 (Fig. 3).

Of the four frequency curves in Figure 3, numbers 3 and 4 probably represent the best estimates of recurrence interval for the flood, providing one accepts the extension of a log-Pearson III frequency curve to 2,000 years from 36 years of record. Results suggest the 1965 flood had a recurrence interval of between 900 and 1,600 years. The sediment record in the valley appears consistent with this estimate.

Rampant erosion and undercutting of banks led to failures of locally high cutbanks, and channel widening. These processes were enhanced by debris that caught

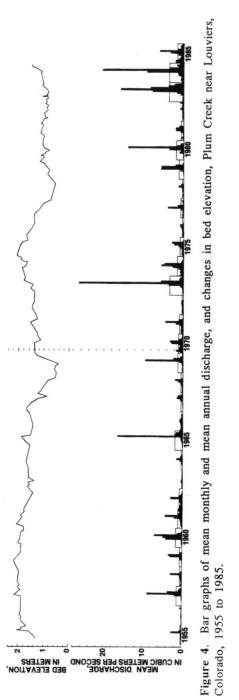

Figure 4. Bar graphs of mean monthly and mean annual discharge, and changes in bed elevation, Plum Creek near Louviers, Colorado, 1955 to 1985.

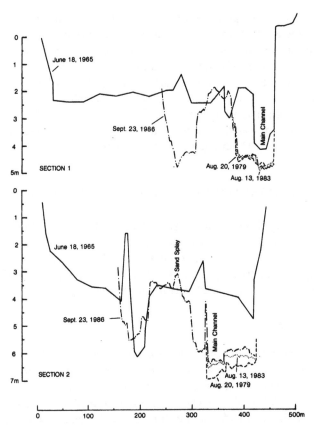

Figure 5. Cross sections showing bottomland changes at two sites near Louviers, Colorado.

on trees and other obstructions, causing them to topple and create sites vulnerable to rapid scour. Near Louviers, deposition of long sand splays occurred on terraces behind large cottonwood trees where debris collected but did not uproot the trees. Where large trees were knocked over by flood debris, however, the flood waters enlarged the freshly exposed pits, in places yielding scalloped channel banks and flood-plain and terrace depressions.

In most places geomorphic evidence of trees uprooted and transported downstream was obliterated. Inspection of aerial photographs, taken in 1964 and 1967 of a 4.08-km reach near Louviers, however, provides an estimate of the flood effects on woody vegetation. Approximately 1,930 trees were identified on lowland surfaces of the reach in the 1964 photograph; in the 1967 photograph about 915 trees (Table 1) were visible in the corresponding area--that covered by the 1965 flood. Thus, an estimated 53 percent of the woody vegetation was destroyed. Some heavily wooded areas that were inundated showed no evidence of toppled trees or else the percentage might have been much higher.

Since June, 1965, the highest discharge recorded at the Louviers gage site was $125 \text{ m}^3\text{s}^{-1}$ on May 8, 1969--less than 3 percent as great as the 1965 flood peak. Bar graphs representing annual and monthly flows (Fig. 4) since 1955, however, show that mean discharges of Plum Creek during the 1973, 1983, and 1984 water years (Oct. 1 through Sept. 30) were larger than in 1965. Sustained flows from snowmelt and spring rains in May of 1980, 1983, 1984, and especially 1973, produced monthly runoffs approaching or exceeding that of June, 1965. Figure 4 also shows that streamflow was low, at times negligible, in Plum Creek during the 1966 through 1969 and 1974 through 1978 water years.

BOTTOMLAND CONDITIONS FOLLOWING THE FLOOD

Few data showing channel conditions of Plum Creek for the period prior to the 1965 flood are available. The distribution and thickness of flood deposits, gage records, photographs of erosion and vegetation damage, and valley sections established to compute peak discharge of the flood, however, permit assessment of conditions near the Louviers site immediately following the flood.

Channel Morphology

An immediate effect of the June 16 flood was movement of large supplies of sand from small tributary basins and bank failures into the valley bottom of Plum Creek. A 31-year record of bed-datum changes (adjusted for changes in gage-site location) of Plum Creek (Fig. 4) suggests that a 6-month trend of increasing bed levels climaxed with the June 16 flood. Better evidence that an influx of coarse sediment caused rapid depositon along Plum Creek is provided by valley sections measured less than 2 days after the flood (solid lines, Fig. 5). Section 1 is 27 m downstream from the present Plum Creek gage site near Louviers; section 2 is 259 m downstream from section 1. At both sections channel deposition was followed by headcutting (right sides of sections, Fig. 5) active during and after the June 18, 1965 measurements (M. S. Petersen, U.S. Geological Survey, unpublished flood records, 1965).

Average channel width, measured on aerial photographs at 22 sites along the 4.08-km study reach, was 68 m in June, 1967, an increase of about 160 percent from an average of 26 m in 1964 (Table 1). In general, the floodwaters carved new lengths of nearly straight channel where Plum Creek had been relatively sinuous, and caused widening of previously less sinuous reaches (Fig. 6) by eroding alluvial channel banks and by the destruction of riparian vegetation. Sinuosity fell from 1.22 to 1.03, accompanied by a 19-percent increase in channel gradient. Occasional channel islands present in 1964 were mostly destroyed, as was almost all woody vegetation on those islands (Table 1).

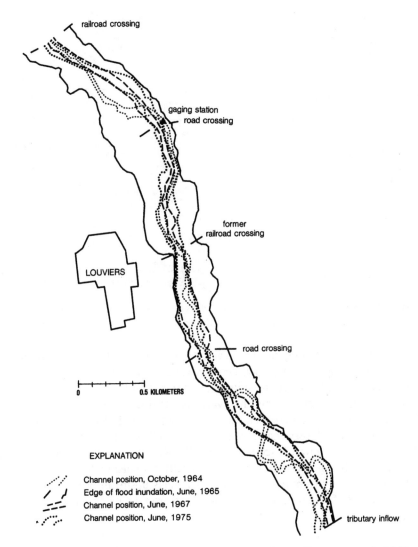

railroad crossing

gaging station
road crossing

former
railroad crossing

LOUVIERS

road crossing

0   0.5 KILOMETERS

EXPLANATION

Channel position, October, 1964
Edge of flood inundation, June, 1965
Channel position, June, 1967
Channel position, June, 1975

tributary inflow

Figure 6. Map showing changes in position of the Plum Creek channel, 1964 to 1975, and area inundated, June, 1965.

Flood Deposits

Sediment deposited by the flood on terraces and the flood plain is varied and diversely distributed. At valley edges, deposits of fine- to strongly-fine skewed (Folk, 1974, p. 47), fine to medium sand were deposited (Table 2). The massive to strongly horizontally laminated sediment accumulated to thicknesses of 1.3 m. Facies changes across the valley bottom are pronounced. At the valley sides, up to

Figure 7. Sequences of coarsening-upwards sand and granules that were deposited rapidly on a low terrace adjacent to Plum Creek downvalley from Louviers, Colorado.

1.5 m of fine- to medium-grained sand was deposited on the flood plain and low terraces. In the valley middle, flood deposits varied in thickness from 0.2 m at the downstream end of a surviving bottomland forest to 4.6 m filling the pre-flood channel of Plum Creek. In this area waves of (1) fine-skewed to symmetrically-distributed, medium to coarse sand, and (2) nearly symmetrically-distributed, poorly to moderately sorted, very coarse sand to granules (Table 2) were deposited in successive coarsening-upwards sequences (Fig. 7). Repetition of the depositional pattern suggests that the flood flow was periodically enriched with large volumes of sediment, probably from collapse at high cutbanks upstream of the study reach. Various scars of failed terrace sediment upstream from Louviers support this possibility.

Vegetation

About half of the bottomland woody vegetation near Louviers was removed by the 1965 flood (Table 1). Tree ring studies demonstrate that at flood plain and terrace levels some species, particularly willows, were sheared along the main axis of the stem, and subsequently resprouted from the intact roots as multi-trunked trees. Many other medium-sized trees were uprooted as debris piled against them. Consequently, most surviving trees either (1) grew where the depth of inundation, based on datable trunk and limb scars, was less than about 2 m, (2) were large enough to withstand the floodwater and debris accumulation, or (3) were on the protected lee

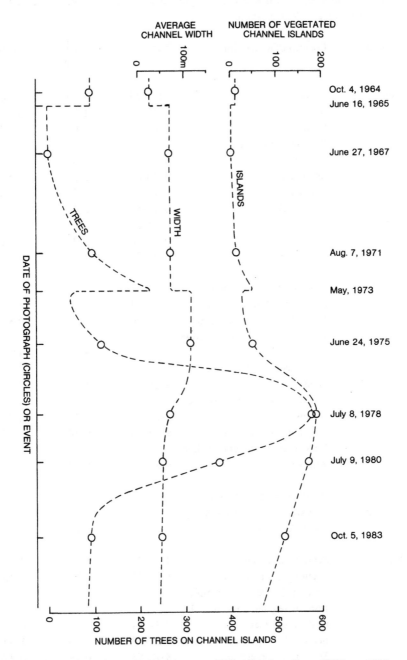

Figure 8. Channel and vegetation changes, Plum Creek valley, 1964 to 1983.

Table 2. Summary of particle-size analyses of sediment samples collected from Plum Creek area. 1975 to 1986.

Surface	Location/Description	Date	d_{50}[1] (mm)	Sorting Index [2]
Bed	Section 1	11-1-75	1.75	3.34
	Section 1	1-25-77	1.5	3.99
	Section 2	8-25-77	2.0	4.17
	Section 2, ave. of 2 samples	4-5-82	1.8	4.22
	Section 2	4-30-83	2.1	3.54
	Section 1	8-13-83	2.5	2.88
	Section 2	8-14-83	2.4	2.86
	Section 2	7-24-84	1.5	3.58
	Section 1	9-24-86	4.0	4.72
	Section 2	9-22-86	2.1	3.85
Island	Ave. of ten unvegetated to poorly vegetated surfaces	8-14-83	1.7	5.17
	Vegetated downstream end	9-25-86	1.5	2.82
	Unvegetated; interior of downstream half	9-25-86	3.0	4.27
	Poorly vegetated lee end; grasses	9-25-86	1.0	3.05
	Vegetated stoss, grasses and willows	8-13-83	0.22	6.07
	Vegetated stoss, grasses	9-25-86	0.40	3.62
	Vegetated center; grasses and willows	9-25-86	0.97	4.37
Island attachment	Poorly vegetated channel deposits recently connecting island to bank	9-25-86	2.1	4.61
Bar	Section 2; unvegetated	8-25-77	1.0	4.13
	Section 2; unvegetated	8-13-83	2.9	4.33
	Section 2; ave. of 2 samples from unvegetated recently deposited sediment	7-24-84	2.3	4.59
	Section 2; vegetated, willows	8-25-77	0.16	8.88
	Section 1; vegetated	8-13-83	0.14	3.87

Table 2. continued

Surface	Location/Description	Date	d_{50}[1] (mm)	Sorting Index [2]
Banks-	Section 2, left side	4-5-82	0.10	2.50
root zone	Section 1, right side	8-13-83	0.11	1.80
	Section 2, left side	8-13-83	0.11	1.92
	Section 2, right side	9-22-86	0.16	3.91
Flood plain	Section 1, left side	4-5-82	0.46	3.97
	Section 2, right side	4-5-82	0.33	7.24
	Section 1, left side	4-30-83	0.37	5.31
	Section 2, right side	9-22-86	0.20	6.60
Low terrace	Section 2, left side; unvegetated	1-25-77	1.4	3.50
High terrace	Section 2, left side; poorly vegetated	1-25-77	1.1	3.35
	Section 2, left side; poorly vegetated	9-22-86	0.93	3.14
	Section 2, left side vegetated	8-13-83	0.75	3.54
Terrace[3]	Valley edge;	7-24-84	0.34	1.58
	vertical	7-24-84	2.3	1.98
	exposure	7-24-84	0.70	1.72
		7-24-84	1.4	1.81
Channel	Valley middle,	8-14-85	0.35	1.54
fill &	vertical	8-14-85	2.3	1.88
flood	exposure	8-14-85	0.70	1.80
plain [4]		8-14-85	1.5	1.77

[1]Median particle diameter of sediment sample

[2] $\frac{1}{2}\left[\frac{d_{84}}{d_{50}}+\frac{d_{50}}{d_{16}}\right]$, where d_{84} and d_{16} are particle diameters at the 84 and 16 percentiles.

[3] Descending order, at equal intervals from a 1.3-m thickness of flood deposits.

[4] Descending order, at equal intervals from a 0.7-m thickness of two coarsening-upwards sequences of flood deposits.

of large surviving trees that collected much of the floating debris. Principal areas of surviving trees were along old meander loops cutting into valley walls that offered protection from the main flow, and where fan deposits of tributaries to Plum Creek provided sufficient relief to minimize depth of inundation. Trees were especially protected where these conditions combined.

Measurements of seven surviving trees with debris piles on a terrace downstream from and within 300 m of the gage (Fig. 6), and within 150 m of the present left bank of Plum Creek, show that all had trunk diameters of at least 68 cm reconstructed to the time of the flood. The average trunk diameter of the trees, six cottonwoods and a willow, in 1983 was 98 cm. The average height above the present terrace surface of the piles was 2.6 m, the highest being 4.6 m, and the average volume of the piles was about 75 m^3. Although other data necessary to compute the force exerted on the trees by stacked debris are not available, these figures are suggestive of the extreme shear to which many trees along Plum Creek were exposed.

Flood-documentation photographs and aerial photography of June, 1967, show that, even 2 years after the flood, most inundated areas were largely devoid of shrubs and herbs. Except near valley margins, the depth of flood deposits was generally great enough that little pre-flood, non-woody vegetation survived. The lack of fine sizes and nutrients in the flood deposits was probably a principal cause for slow re-establishment of shrubs and herbs on the higher depositional surfaces.

CHANGES THROUGH TIME

Normal annual precipitation in the Plum Creek basin averages about 400 mm, and soils and exposed rocks are largely sandy. As reported by Schumm and Lichty (1963) and Burkham (1972) for roughly comparable basin conditions, these characteristics limit redeposition of fine sediment on floodplain surfaces and thus slow the development of stable vegetation. In the Plum Creek valley channel morphology, characteristics of flood plain sediment, and flood-plain vegetation continue to change in response to the 1965 flood.

Channel Morphology

Two valley sections measured for slope-area computations of peak discharge two days after the flood were remeasured in 1979 and 1983; complete resurveys (to areas of post-flood construction on the left half of the valley floor) were made in 1986 (Fig. 5). Two principal channels, through which all non-flood discharge now flows, are apparent in the 1986 sections. Prior to 1965, the main channel at section 1 was in about its present position on the right side of the valley. Downstream the channel turned left and occupied the position now shown in the middle of section 2 (figs. 5 and 6). Both channel areas were covered with as much as 4.6 m of sand and gravel, but headcutting during and following recession reestablished the channel incisions. In the main channel, degradation continued slowly during the following four years, a period of deficient rainfall (Fig. 4).

Increasing bed levels correspond to several relatively wet years starting in 1969 (Fig. 4), but aerial photography suggests that renewed channel widening did not occur until the high spring runoff of May, 1973 (Table 1). This runoff destroyed most post-flood channel vegetation and probably most channel islands. From August, 1971 through June, 1975, average channel width increased about 60 percent, and channel gradient increased about 3 percent with an accompanying decrease in sinuosity (Table 1). Interpretations of bottomland changes through 1983 are given in Figure 8. Reductions in average channel width, gradient, and bed level accompanied normal discharge rates in Plum Creek following the erosive flows of 1973. By 1978 the number of channel islands stabilized by vegetation had increased several fold (Table 1; Fig. 8), but only locally along Plum Creek is there evidence of flood plain reconstruction by narrowing from the excessive channel widths of 1973. One example is the right side of section 2 (Fig. 5), where aggradation in the channel and on a large bar began about 1978. Botanical evidence, including the burial of organic litter and root collars of trees dating to 1978, confirms the changes depicted by Figure 5.

Figure 6 illustrates channel positions near Louviers in 1964, 1967, and 1975, and shows that extensive changes in stream course caused by the flood were followed by only minor changes of lateral migration and widening during the next decade. Spring runoff in 1973 caused widening but little else. Although not shown, channel-position changes since 1975 also have been minor, and are mostly post-1973 trends of channel narrowing (Fig. 8).

Of particular pertinence to the recovery of the Plum Creek channel following flood damage is variation in number of channel islands. Since the 1965 flood, the establishment, growth, and eventual adjoining of channel islands to riparian surfaces has been the predominant geomorphic process by which narrowing, gradient reduction, and revegetation of Plum Creek has progressed. Not included in this grouping are larger, generally sheltered bars, such as that described at the right side of section 2. These bars initially may be channel islands by virtue of stormrunoff anabranches that separate them from higher geomorphic surfaces, but they generally show little tendency for growth in size and evolve into attached flood- plain surfaces as the anabranch becomes vegetated and filled with fluvial sediment.

Numbers of channel islands and trees on the islands in the 4.08-km study reach were estimated for each date of available photography (Table 1); yielded are interpretations of changes from 1964 through 1983 (Fig. 8). Table 2, summarizing sediment-size data, shows large differences in median sizes of sediment from vegetated and unvegetated surfaces. Comparing Tables 1 and 2, the modest increases in channel width and number of vegetated islands (Fig. 8) during the 8-year period up to spring, 1973, suggest steady but limited accumulation of fine sizes on lowland surfaces. The discharges of 1973 winnowed little fine sediment from the alluvium of Plum Creek and were not followed by erosive flows in succeeding years, thus allowing the number of vegetated islands to increase rapidly through 1977 or 1978

(Fig. 8). A 40-percent decrease in average channel width during the period, supported by field studies and aerial photography, strongly suggests that as the number of channel islands peaked, the attachment rate to the banks to cause channel narrowing also peaked. Since 1981, the number of islands in the reach near Louviers has been too small to favor rapid channel narrowing by attachment processes. As channel islands have attached to flood-plain areas, of course, the remaining trees on islands have decreased and the number of trees on riparian surfaces has increased (Table 1; Fig. 8).

Gradient and Bed Level

Thalweg gradient and sinuosity of Plum Creek have changed little since the channel was straightened by the 1965 flood. Minor changes occurred in 1973 when fine-grained alluvium and riparian vegetation were inadequate to prevent additional bank erosion. By 1983 alluvial banks were sufficiently fine grained and vegetated (Tables 1 and 2) to prevent similar spring discharges (Fig. 4) from reversing the trend of channel narrowing and gradient reduction. Minor gradient reduction owing to recent channel aggradation at section 2 (Fig. 5) may be related to sedimentation caused by a flood-control reservoir 7 km downstream. Comparison of bed levels and monthly discharges (Fig. 4) generally shows aggradation during high-runoff periods and scour during low flow. This pattern may be typical of sand channels with perennial discharge, and implies input of sand by ephemeral tributaries during periods of flow, but unreplenished removal of sand from the trunk channel during base-flow periods.

Sediment Characteristics

Selected sampling of alluvial deposits in the Louviers area began in 1975. Initially expected trends of decreasing median particle size following the flood and high flows of 1973 are not apparent in integrated bed samples collected at sections 1 and 2 (Fig. 5). In addition, samples collected from other unvegetated surfaces are inadequate to identify trends (Table 2). Samples collected from unvegetated channel islands, recent channel fill connecting an island to flood plain, and unvegetated bars show similar sizes and sorting as the channel material from which they were derived.

Sharp differences are apparent between samples collected from vegetated and unvegetated surfaces. The association of vegetation with deposits of fine sediment on sandy bars and channel islands strongly suggests that flow-resistant vegetation traps and stores coarser sizes of suspended fluvial sediment during periods of nonerosive inundation, and thus channel narrowing and flood-plain construction following highly erosive flooding are largely dependent on the establishment of permanent vegetation. Highest values for sorting index (poor sorting) are from vegetated surfaces (Table 2) and reflect a bimodal distribution of coarse channel sands veneered by silt and fine sand trapped by vegetation.

CHANNEL, FLOOD-PLAIN, AND VEGETATION DEVELOPMENT

Narrowing of alluvial channels typically progresses by a variety of sorting processes including formation of bars, encroachment of riparian vegetation, and accretion of suspended sediment on banks. Along the flood-widened sand channel of Plum Creek, narrowing has occurred through bar development, but a more important process has been narrowing by the growth and attachment of islands to the channel sides.

Flood-plain Development

Documentation of post-flood channel narrowing along Plum Creek by bar formation, revegetation, and conversion to flood plain is provided by changes at section 2 (Fig. 5) and botanical data. Similar processes at section 1 have been modified by a road embankment. Aerial photography of June, 1975, verified by field observations 4 months later, shows 80 m of unvegetated channel- bar at the right side of section 2. By August, 1979, the bar surface, no longer part of the channel, averaged 0.5 m higher than the thalweg (Fig. 5). Photography from the following summer showed most of the bar surface to be vegetated with herbaceous plants and tree seedlings. Spring runoff in 1983 and 1984 (Fig. 4) caused additional deposition on the bar surface. Root collars of three willows on the bar, up to 5 years old by summer, 1984, were buried to an average depth of 0.6 m; at various sites along the section-2 bar, depth to the first buried organic zone, indicating 1984 deposition, ranged between 19 and 32 cm. These observations, combined with section data of 1979 and later (Fig. 5), leave little doubt that channel narrowing by deposition and subsequent development of flood plain from bar surfaces, depends on the presence of vegetation that persists during all but the largest floods. Hereford (1984) reached closely comparable conclusions in a study of the Little Colorado River in Arizona.

Channel Islands

Vegetation similarly influences deposition on channel islands of Plum Creek, and also appears to be a principal cause of island growth through time. The numbers of channel islands and trees visible on aerial photographs (Table 1) help summarize the relation between the islands and bottomland morphology. Further analysis of the photographs, however, shows that channel islands resistant to spring runoff steadily grow downstream with allometric increases in maximum width. Typically, island size increases until an anabranch separating the island from a nearby bank or island is narrowed to the degree that filling occurs and the island is attached to the bank or other island. Where two or more islands are closely spaced along flow lines, lee-side deposition of bed sediment also leads to their coalescing. When the rate of island attachments begins to exceed that of island initiation, as occurred along the Plum Creek study reach in about 1978, the rate of channel nar-

Figure 9. Channel island that adjoined to left bank of Plum Creek in about 1984. Debris (logs) of unknown age form the upstream end of the island.

Figure 10. Channel island, Plum Creek near Sedalia, Colorado, that is expanding downstream from 1965 flood debris.

FLOOD PLAIN

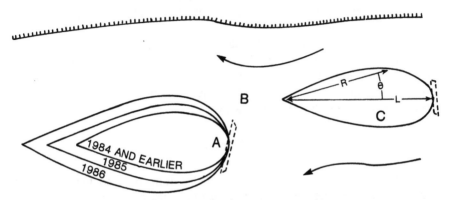

Figure 11. Schematic diagram of a channel island, Plum Creek valley. Islands form behind partially-buried logs (dashed figures). Based on field observations near Sedalia, Colorado, the ages of annual growth increments are indicated by year; arrows show directions of streamflow, A is the site of oldest woody vegetation on the island, and B suggests the site where, assuming continuing incremental growth, the island will be adjoined with island C. The compound island later is expected to attach to the flood plain.

rowing probably is at or near its maximum (Fig. 8).

In September, 1986, 58 channel islands along a 1-km reach starting 1 km downstream from Sedalia were surveyed. Most were unattached, but 18 were in various stages of adjoinment, including six showing evidence of 1984 to 1986 attachment to a bank or another island (Fig. 9). At most islands the upstream end was protected by a relatively immobile obstruction, resulting in sediment deposition and the subsequent establishment of vegetation. All obstructions were either logs or used tires. The tires were carried by the 1965 flood from a large repository upvalley on the Plum Creek flood plain. Many obstructions, therefore, appeared to predate an island and to be the origin of its formation (Fig. 10).

Island lengths ranged from 0.5 to 120 m, and widths were 0.4 to 27 m. Except for very small, relatively young islands, the highest part of the island was immediately downstream from the upper end and tailed smoothly into the channel, reflecting lee-end deposition. The smaller islands, those less than 5 m long, had variable shapes largely dependent on size and position of the obstructions behind which they formed. Therefore, they showed an inconsistent relation between length and width. Twelve single unattached islands with lengths exceeding 5 m, however, defined a length (L)-width (W) relation of:

$$L = 3.1 \ W^{1.05} \qquad\qquad (1)$$

with a coefficient of determination (r^2) of 0.97 and a standard error of estimate of 0.082. A similar analysis for 38 larger islands of the Columbia, Mississippi, and Missouri Rivers yielded coefficient, exponent, and r^2 values, respectively, of 4.0, 0.96, and 0.96 (Komar, 1984).

Equation 1, relating length and width, suggests that Plum Creek islands evolve in a predictable manner. Komar (1983; 1984) showed that channel islands are one of several geomorphic features, both erosional and depositional, that assume a streamlined, minimum-drag form. In plan, these features are described by the equation, in polar coordinates, for a lemniscate loop:

$$R = L \cos(k\theta) \qquad (2)$$

where R is distance from the origin, at the tailpoint of an island, to a point on the perimeter; θ is the angle with the horizontal bisecting the island, and k is a function of length and area (Komar, 1984, p. 134) (island C, Fig. 11).

Most islands surveyed along Plum Creek, particularly those longer than 5 m, showed a well-defined lemniscate form (equation 2). Moreover, the well-shaped islands exhibited growth patterns conforming to periods of recent runoff. Figure 11 idealizes a Plum Creek island 55 m long that showed incremental growth in the downstream direction. Growth bands up to 1.5 m wide and extending from the sides to the tailpoint showed stepped increases of 1 to 4 cm in elevation above water level, and botanical observations suggested that each band represented annual or less-frequent growth. Coalescing with a nearby island or the flood plain had not yet occurred, and both islands grew downstream from a partially-buried log (figs. 9, 10, and 11).

The outermost growth band had an estimated 50-percent cover of grasses and other herbs; the next step or band had an estimated 60-percent cover that included willow and cottonwood saplings in the first year of growth. Inside of the second band, growth increments could not be distinguished, but some willows and cottonwoods were in the second year of growth and vegetative cover was nearly total (Fig. 11). Most root collars of saplings were buried several centimeters. Thus, it is inferred that incremental enlargement occurred during spring-runoff aggradational periods of 1986, 1985, 1984, and probably 1983 (Fig. 4). Earlier, poorly defined incremental growth on this island and others was indicated by decreasing tree ages, up to 17 years in two instances, from the upper to lower ends of an island. Low flows and limited deposition in water years 1981 and 1982 (Fig. 4) may have resulted in little or no incremental growth.

CONCLUSIONS

Extraordinary floods can affect far-reaching, long-term, and complex responses in stream channels. Even in sand-bed channels, where low flows can transport most of the available bed-material sizes, recovery of channel and valley morphology can extend over decades. The record of an extraordinary flood in Plum Creek is preserved in stream morphology, vegetation damage and new growth, and deposits on low terraces not normally reached by floods. Processes of channel recovery are dependent on the ability of normal flows to transport sand that is trapped by juvenile vegetation on island bars. These bars grow in size in a downstream direction as streamlined channel islands until they coalesce with other islands and ultimately with the channel bank; the result is flood-plain expansion and channel narrowing. Recovery is also controlled by the frequency of high flows in years following the disrupting flood. Such flows can slow the regrowth of sediment-trapping vegetation, and thus retard the recovery process. The historic record following the 1965 Plum Creek flood documents the interrelated processes and products that lead to the recovery or relaxation of this particular landscape.

REFERENCES

Bullard, K. L., 1986, Comparison of estimated maximum flood peaks with historic floods: Hydrology Branch, Engineering and Research Center, U.S. Bureau of Reclamation, Denver, Colo.,165 p.

Burkham, D. E., 1972, Channel changes of the Gila River in Safford Valley. Arizona, 1846-1970: U.S. Geological Survey Professional Paper 655-G, 24 p.

Folk, R. L., 1974, Petrology of sedimentary rocks: Hemphill Publishing Co., Austin, Texas, 182 p.

Hansen, W. R., Chronic, John, and Matelock, John, 1978, Climatography of the Front Range Urban Corridor and vicinity, Colorado: U.S. Geological Survey Professional Paper 1019, 59 p.

Hardison, C. H., 1973, Probability distribution of extreme floods: Highway Research Record, no. 479, Highways and catastrophic floods of 1972, Highway Research Board, National Research Council, Washington, D.C., p. 42-45.

Hereford, Richard, 1984, Climate and ephemeral-stream processes: twentieth-century geomorphology and alluvial stratigraphy of the Little Colorado River, Arizona: Geological Society of America Bulletin, v. 95, no. 6, p. 654-668.

Hunt, C. B., 1954, Pleistocene and Recent deposits in the Denver area, Colorado: U. S. Geological Survey Bulletin 996-C, p. 91-140.

Interagency Advisory Committee on Water Data, 1982, Guidelines for determining flood flow frequency: Hydrology subcommittee, Bulletin 17B, U. S. Geological Survey, Reston, Va.

Jarrett, R. D., and Costa, J. E., 1983, Multidisciplinary approach to the flood hydrology of foothill streams in Colorado: International Symposium on Hydrometeorology, American Water Resources Association, Bethesda, Md., p. 565-569.

Komar, P. D., 1983, The shapes of streamlined islands on earth and Mars; experiments and analysis of the least-drag form: Geology, v. 11, p. 651-654.

Komar, P. D., 1984, The lemniscate loop--comparisons with the shapes of streamlined landforms: Journal of Geology, v. 92, p. 133-145.

Matthai, H. F., 1969, Floods of June 1965 in South Platte River basin, Colorado: U.S. Geological Survey Water-Supply Paper1850-B, 64 p.

Schumm, S. A., and Lichty, R. W., 1963, Channel widening and flood-plain construction along Cimarron River in southwestern Kansas: U.S. Geological Survey Professional Paper 352-D, p. 71-88.

Scott, G. R., 1963, Quaternary geology and geomorphic history of the Kassler Quadrangle, Colorado: U. S. Geological Survey Professional Paper 421-A, 70 p.

U.S. Department of Commerce, 1966, Climatological Data, Colorado: Annual Summary 1965, v. 70, no. 13, Environmental Science Services Administration, Weather Bureau, Asheville, p. 237-247.

11

A medieval catastrophic flood in central west Iran

Ian A. Brookes

ABSTRACT

The stratigraphic record of Holocene alluviation in the intermontane Qara Su Basin, central west Iran (34°N, 47°E) is dominated by four overbank mud units and four soils, labelled II-V and A-D, in order of increasing age. Unit I is modern alluvium. In numerous lowland cutbank sections, unit V, with thick soil D on it, is truncated by a paleochannel up to 2 m deep, occasionally flanked by a level, eroded, channel-border zone. The paleochannel contains a structureless, poorly sorted, matrix-supported, muddy sand gravel, unit IVa, interpreted as the product of a single, brief, basin-wide, catastrophic flood. There is no evidence of comparable earlier or later Holocene events in the basin.

Chronological evidence, drawn from associated archeological occurrences and comparison of the alluviation record with palynological records of Holocene vegetation-climate change in the Zagros Mountains, indicates that deposition of the exposed alluvium began in the early Holocene, and converges on a date for the flood between 200 and 1850 A.D., with justifiable preference given to a date between ca. 950 and ca. 1250 A.D.

The flood is most simply attributed to the stalling of a Mediterranean winter 'low' over the basin, the hydrological effect of which may have been enhanced by basin conditions changed as a consequence of a sudden political disruption, perhaps the Mongol conquest.

INTRODUCTION

Background to the Study

The subject of this paper arose from a study of the geomorphology and Quaternary geology of the Qara Su (Black River) watershed in central west Iran, undertaken in 1975 and 1978. The basin is approximately centred on 34°20'N, 47°00'E. This study was one facet of a multi-disciplinary, regional archeological project directed from the West Asian Department, Royal Ontario Museum. The role of project geomorphologist was initially seen as providing information on land resources available to past human populations, and erecting a scheme of environmental changes which could potentially be applied to the archeological evidence of changes in population densities, settlement patterns, and adaptive strategies in the study area.

Surface archaeological survey located 1200 sites in roughly 50 percent of the 4900 km^2 watershed area, 80 percent of them within the combined landform categories of alluvial fan and alluvial lowland. Geological studies thus focused on those zones. They early revealed that fluvial activities occurring very late in the geomorphic history of the basin (i.e., within the last 1000 years) have so altered its surface that no direct information on previous land resources can be provided. More significantly, fluvial activities have affected the distribution and the size of archeological sites to a greater degree than cultural ones, particularly the effects of a basin-wide catastrophic flood, the only such event recorded in the Holocene alluvial sequence, and subsequently renewed overbank sedimentation of muds. A detailed account of the geological research is given in Brookes (in press), and archeological implications of the alluvial history are found in that work, and more briefly in Brookes et al. (1982).

The Study Area

The Qara Su (Fig. 1) is a headwater tributary of the Saimarreh-Karkheh River system which drains the central sector of the Zagros Mountains into the Hawr-al Hammar marshes of southern Iraq. The river drains the largest intermontane basin in the Zagros.

Geologically, the basin straddles three tectonic zones of the Zagros Orogen (Fig. 2): (1) thrust slices of Middle Cretaceous crystalline limestone to the NE; (2) a crush zone on Cretaceous melange (ophiolites and cherts), cherts, and limestone breccias, transitional to ; (3) a folded zone of Jurassic to Miocene shelf carbonates and flysch clastics to the SW.

Topographically, these belts are reflected as (1) a NE highland rim, with several discrete, massive ranges, with serrated summit profiles at 2350-3350 m (Fig. 3a) with one major valley (Ab-i-Razawar) and two smaller ones transecting them, guided by faults; (2) smooth-crested central uplands at 1500-1800 m (Fig. 3b), flanked southwestward by the Kuh-e Sefid scarp at 2000-2500 m; (3) a serrated boundary

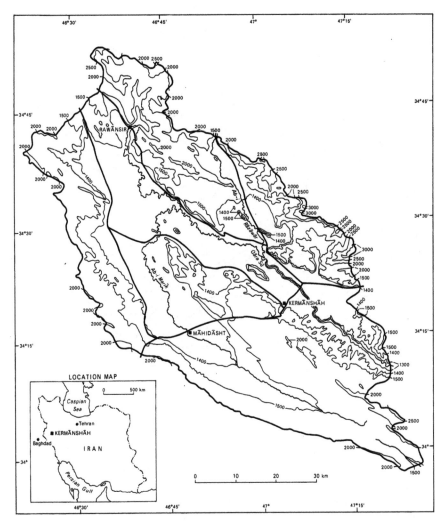

Figure 1. Qara Su basin, showing contours at 100 m intervals below 1500 m, 500 m above 1500 m, drainage net, paved roads, places named in text.

ridge at 1800-2300 m to the SW, developed on SW-dipping Oligocene limestone, and fronted by pediments planed across folds on older rocks, sloping to 1500 m at the lowland edge.

Denudation, initially activated by Late Cretaceous orogeny and rejuvenated by Pliocene movements, has maintained high relief on strong rocks, with deep valleys such as Qara Su, Ab-i-Razawar, and Ab-i-Marik in structurally weak fault and fold-axial zones. Thus, the highlands are flanked by alluvial lowlands drained by peren-

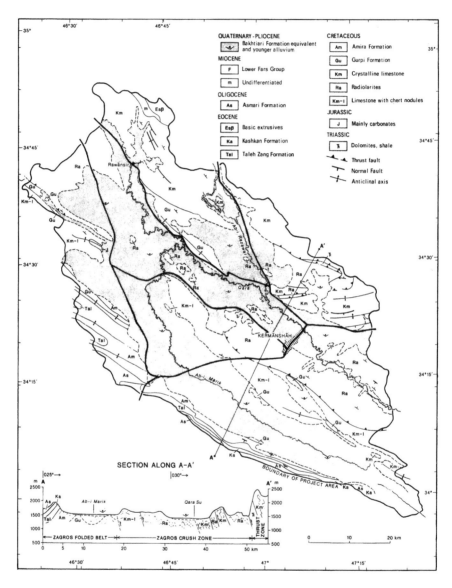

Figure 2. Generalized geology of Qara Su basin and SW-NE geological section.

Table 1. Stratigraphy of fan sediments and soils, pit wall 3.5 km west of Mahidasht.

_____

Unit IVb-II:	0.25 - 1.0 m dark brown, stony loam with immature Ah (or ? Ap) horizon. Gradational contact with
Unit IVa:	0.5 m pebbly and cobbly, subrounded and subangular limestone gravel, with modal clast size 5-10 cm, maximum 30 cm. Sharp, undulating contact with
Unit V:	0.5-2.0 m reddish brown, massive silty clay, with pedogenic prismatic structure and powdery $CaCO_3$ nodules up to 3 cm. Gradational contact with
Unit VI:	1.5 cm crudely bedded, subrounded and subangular, clast-supported, limestone gravel, with pockets of muddy sand matrix. Upper 25-35 cm impregnated by powdery $CaCO_3$ ('caliche'). Base below pit floor.

nial trunk streams following a NW-SE parallel pattern, with connections via fault- and plunge-controlled valleys. Lower-order intermittent tributaries adopt slope-controlled dendritic patterns, flowing from highlands across pediments and (or) fans to alluvial lowlands.

The climate of the region is of montane, Mediterranean type (Fig. 4) with winter precipitation from east-moving depressions. Kermanshah, near the lowest part of the Qara Su basin at 1322 m, receives 480 mm, with an average of only 9 days of snowfall and 80 frost-days. These presumably increase with altitude. No meteorological data are available to confirm this, but Maire (1978) reported perennial firn in shaded karstic hollows at 2500 m. Mean annual temperature at Kermanshah is 13.8°C, with a January mean of 1.8°C, and a July mean of 26.8°C (Iran, Ministry of Water and Power).

The probability distribution of daily precipitation at Kermanshah is shown in Fig. 5. The median value is 4 mm, whereas 10 percent of precipitation days (4.4 days annually) receive more than 24 mm. No recording rain gauge data are available, but a crude measure of annual precipitation intensity shows no correspondence between high-intensity years and years of high annual run-off or flood peaks on the Qara Su.

The basin is thus well-watered, especially at altitude, yet the mountain and upland zones have no perennial drainage, due to highly porous and fractured carbonate

Table 2.   Stratigraphy of alluvium and soils, Ab-i-Marik, 1 km NW of Mahidasht

----

Unit I: discontinous, variably textured modern channel bars in mid- and side-channel positions.

Soil A: 20 cm Ah (?Ap) horizon over unaltered C horizon, presently ploughed and grazed.

Unit II: 1 m silty clay alluvium, structureless, friable. Forms top of section.

Soil B: strongly developed through all of Unit III with thick, strong Ah and Bca horizons.

Unit III: absent from downstream section. Upstream, 1 m of massive silty clay alluvium.

Soil C: impregnates upper 75 cm of Unit IVb. Moderately developed Ah over weakly developed Bca horizon. Present only in upstream section.

Unit IVb: 1.5 m of fining-upward silty sand to clay silt alluvium; structureless, rare potsherds sporadically distributed. A coarser base may represent decline of flood which emplaced IVa in palaeochannel, but most of it represents later, repetitive alluviation over an abandoned channel and the flanking floodplain. Upstream section shows Unit V passing up into Unit IVb through a 15 cm transition of lightening colour with structure changing from blocky to massive.

Unit IVa: up to 75 cm of poorly sorted, matrix-supported, subangular and sub-rounded, limestone and chert gravel with clasts from 1 cm to 5 cm; matrix of red-brown muddy sand (<2 mm fraction contains up to 43% clay). Clasts unoriented. Structureless, but 2-3 faint partings visible when surface washed. Contains variably abraded potsherds up to 7 cm long, including abundant plain "Yellow Ware" believed to have appeared no earlier than ca. 200 A.D. Also contains bone fragments, bovid teeth, articulated and broken freshwater bivalve shells. Deposit floors a wide, shallow paleochannel cut across unit V. IVa absent from above unit V upstream.

Soil D: impregnates entire 3 m of Unit V and presumably extends below section. Structure blocky with abundant slickensides and Mn stains on block faces. Represents B-horizon of strongly developed soil, with A-horizon and probably top of B-horizon eroded. Elsewhere, often powdery $CaCO_3$ nodules present.

Unit V: up to 3 m fine-grained (modal texture, silty clay, ranging from clay to silt loam), reddish brown alluvium with no sedimentary structures, possibly due to complete impregnation by soil D. Base not exposed, top erosional in downstream sub-section, gradational to IVb upstream.

Figure 3. **a.** Rugged front of thrust slice of Middle Cretaceous crystalline limestone, 100 m high above 1400 m alluvial lowland. Fans and springs (marked by
village) at mountain foot; 30 km NW of Kermanshah (IAB 17 May 1978).
b. Smooth-crested hills at 1900 m on cherty limestone in central uplands, 10 km
W. of Kermanshah (IAB, 10 Aug. 1975).

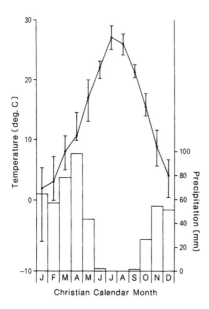

Figure 4. Climograph for Kermanshah, based on 1956-1972 data (Iran, Ministry of Water and Power).

bedrock. The drainage net is fed from scores of springs, a majority perennial, aligned at geological contacts coinciding with the lowland border.

The mean annual discharge of the Qara Su near its outlet is 24 m^3s^{-1}, ranging from 6 m^3s^{-1} to 60 m^3s^{-1}. Hydrographs for the water years with the lowest and highest flood peaks are shown in Fig. 6, with daily precipitation. Winter precipitation feeds a steady rise in baseflow by percolating into bedrock (heavily karstified in the NE mountain zone--Maire 1978, Waltham and Ede 1973), with discharge in mountain foot springs. On the baseflow curve discharge spikes respond to individual storms, with a 3-4 day delay at the basin outlet--a lag accordant with an estimate of flowpath length and mean flow velocity. No snowmelt discharge peak occurs, probably because snow yields from separate storms are largely melted between storms, so that snowpack does not accumulate to great depths over the winter.

Runoff is rapid to low-order intermittent streams across mountain and upland slopes, which have been cleared of native oak parkland and steppe scrub (Zohary 1973) during rougly 5000 years of occupation by sedentary lowland agriculturalists. Alluvial lowlands with level topography, deep absorptive sediments, and water diversion/impoundment for irrigation, merely sustain rather than generate runoff; no streams arise within those zones.

Vegetation history is known from spectra of pollen types in lacustrine sediments only from two sites, 160 km NW and SE of this area (van Zeist and Bottema, 1977). Parallel changes in these spectra, however, permit reasonable interpolation of a similar history for the Qara Su basin. Pollen zones at these sites and inferred vegetation and climates are shown in Fig. 7, as well as correlative phases in the alluvial chronology of that basin.

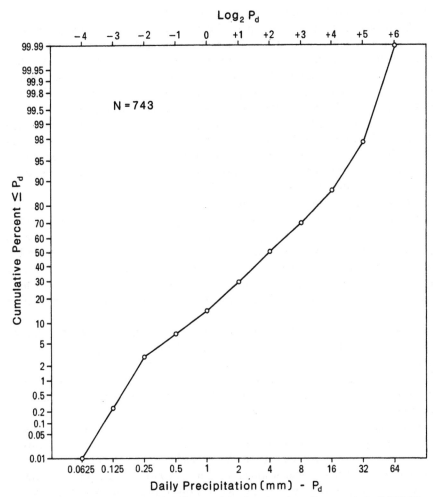

Figure 5. Probability plot of daily precipitation (mm) at Kermanshah (1956-72, Iran Ministry of Water and Power).

ALLUVIAL SEQUENCE

Stratigraphy

Alluvial fans occupy 30 percent, and alluvial lowlands 24 percent of the basin area. Sediments in the fans are rarely exposed, except in shallow banks of intermittent streams crossing them en route to lowland trunks. Fortunately, one man-made cut in a gravel pit exposed a sedimentary sequence which could be correlated with nearby exposures in lowland river cutbanks. The section is shown diagrammatically in Fig. 8 and described in Table I. Units are designated to accord with correlative alluvial units in lowland cutbanks (see Fig. 9 and Table II).

Lowland streams are incised up to 12 m along hundreds of kilometres of sinuous channel. While accessibility and burial by slump and slope wash, (not to mention the attentions of packs of semi-domestic dogs!) limited study of much of this potential exposure, roughly 40 sections were logged and sampled. They show a uniformity of lithology and stratigraphy which struck me as tedious initially, but, as the chronology became clearer, this uniformity became remarkable.

The section exposing most alluvial units, soils, and a disconformity is located 1 km of Mahidasht (Fig.1). It is given the status of a "standard section" with which those showing fewer environmental changes could be compared and correlated. The local setting of the section is shown in Fig. 9a, the section itself is schematized in Fig. 9b and described in Table 2.

Alluvial units in the 40 or so other cutbank sections logged can be correlated with the numbered units in the standard section by, (1) the distinctive blocky structure of unit V with soil D impregnating it, always appearing low in the sections, never exposing its base; (2) the disconformity truncating it (Fig. 10a), often but not always overlain by gravels of unit IVa, which contains sherds of recent date (Fig. 10b); (3) the common occurrence of a prominent calcic horizon on or within an alluvial unit above this surface, correlated with soil B at the standard section (Fig. 10c).

The texture of the sampled units is compared by using probability plots of grain size (Fig. 11). These show the broad similarity of the muddy units II-IVb, and V, and also distinguish gravelly unit IVa. Differences amongst the muddy units reflect source materials. Unit V, also present in the fan section, is interpreted as the fluvially redeposited finer fraction of *terra rossa* soils developed over undulating upland interfluves, and still intact there, in contrast to the skeletal loam rubbles on valley-side slopes subject to erosion. Since the *terra rossa* in this region is dominated by far-travelled aeolian dusts, unit V, as a second-cycle deposit, closely reflects the texture of that dust. Units IVb, III and II, on the other hand, have variable admixtures of sand, some of which was winnowed from coarse pediment and fan gravels on the piedmont, with some sand reworked from unit IVa. Modern suspended load deposits (Unit I) are the sandiest of the muddy units since, unlike the others, they are being deposited within the modern channels rather than by overbank sedimentation.

Overbank sedimentation is the preferred mode of emplacement of pre-modern, muddy alluvium in the lowlands. No traces of extensive lateral stream migration are associated with units IVb-II underlying the lowland surface. Point-bar structures were recognized at one locality, but they underlie a fragmentary terrace formed during very recent incision of the drainage system. Nor are levees recognizable, but in such fine-grained sediments they would not be expected. Aggradation probably resulted from flooding in axial lowland streams, as well as in sub-parallel tributaries joining at small angles, and in transverse tributaries. Water table lowering, which has resulted both from stream incision and from population increase in recent time, has desiccated those tributaries, but their traces remain, often bordered by

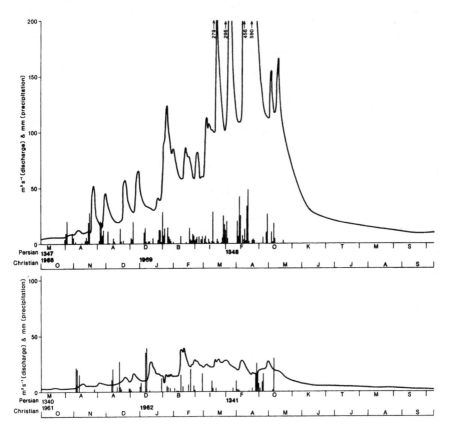

Figure 6. Hydrograph of Qara Su at Gharbaghestan for (a) low flood-peak water year (1961/62), and (b) high flood-peak year (1968/69), with daily precipitation bar graphs (Iran Ministry of Water and Power).

mounds marking abandoned villages.

Clearly, then, the gravels of Unit IVa, and the often channelled disconformity on which they lie, are anomalous in this alluvial sequence. Particle sizes up to 12 cm (one potsherd 20 cm high, weighing about 1 kg was found in them at one locality), extremely poor sorting, absence of cross-bedding or cut-and-fill structures, and the erosion surface below, which truncated soil D, together point to a brief flood within a drainage system not very different in position (but perhaps in density) from the modern one. Floodwaters eroded gravels from the only available sources on fans bordering the lowland, and from pediments above them, as well as perhaps from more distant, steeper upland and mountain slopes. These gravels were routed through channels into the lowland axes, a minimum distance of 5 km in the widest lowland, and thence along axial streams for unknown distances.

POLLEN ZONE	POLLEN TYPES	VEGETATION & CLIMATE	ALLUVIAL HISTORY QARA SU BASIN	AGE ^{14}Cy. BP
7	arboreal pollen (AP) 25–65% grass important	increasing human influence on vegetation through deforestation & agriculture	Units IVb–I —flood (Unit IVa)— Soil D	~1000
6	AP up to 70%	max. tree cover; warm & moist		5500 — 6200
5	AP incr. to 30% low % Artemisia	forest-steppe; incr. trees, esp. Pistacia; grasses important; increasing moisture	Unit V alluvium	
4	AP incr. to ≤ 5% Artemisia decr.	Increasing trees; warming, drying, but ppt. poss. increasing	caliche on Unit VI	10500 — 14000
3	AP ≤ 3%; Artemisia 10–30% Chenopod.~30% Umbellif. ≤ 30%	treeless steppe; cool, dry. too cool for trees at high elevations, too dry at low elevations	Unit VI fan gravel ?	33500
2	high non-arboreal pollen (NAP); AP fluctuating ≤ 7%	steppe; fluctuating warmer-cooler or moister-drier		39300
1	AP ≤ 5% NAP ≥ 95%	treeless steppe; cool, increasingly dry		

Figure 7. Pollen zones, inferred vegetation and climate, Lake Zeribar and Lake Mirabad sediment cores (after van Zeist and Bottema 1977), with inferred correlative phases of Qara Su basin alluvial history (this study).

Evidence of this brief and intense disruption of relatively placid geomorphic regime (soil D was forming on unit V) was recovered widely enough within the Qara Su basin that a regional flood with a regional cause must be invoked. It certainly was a catastrophe ("a sudden, widespread, and single disaster"), but discharges cannot be determined due to inadequate opportunity for geometric measurements. Even if paleochannel geometry was estimated from fragmentary evidence, the necessary manipulation of bed shear formulae, using maximum particle size, would be inappropriate, since sediment structure suggests transport as a flow of high but unknown density and viscosity.

Chronological Evidence

This flood was not just catastrophic: the stratigraphic record contains evidence of only one such event in the entire Holocene epoch. Consideration must therefore be given to chronological evidence before speculation on cause(s). Such evidence is not as secure as one would like, but that from six sources (five providing a maximum, one a minimum age) converges rather than conflicts, and indicates that the flood occurred between the mid-7th century A.D. and the mid-13th century A.D., with a possibly narrower lower limit of mid-10th century A.D. These sources are discussed below in approximate order of confidence placed in them.

First, articulated valves of a riverine, Unioid bivalve contained in the flood gravels at the standard section were radiocarbon-dated at 3520±95 years BP (I-9821). More reliable evidence from the age of potsherds also contained in the gravels indicates that this date is at least 2000 years too old. Contamination of such shells by 'old' carbon has been reported by Keith and Anderson (1963).

Second, 6 km SE of Ravansir (Fig. 1) unit IVa is exposed low in the Qara Su

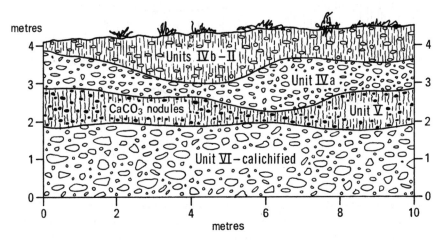

Figure 8. Stratigraphy of sediments and soils in fan section 3 km west of Mahidasht.

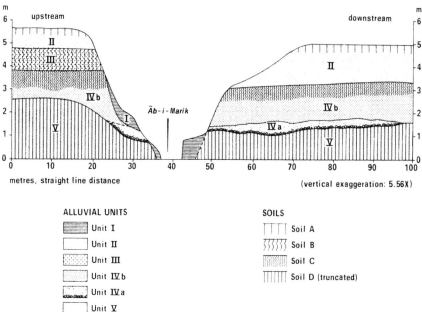

Figure 9 a. Ab-i-Marik channel incised into alluvial lowland 1 km NW of Mahidasht. Downstream cutbank of standard section at left (mostly obscured). Upstream cutbank out of view to right. Note prominent calcic soil B horizon (to right), modern channel deposits, and highly erodible channel border areas. (IAB, 21 May 1978). **b.** Stratigraphy of alluvial units and soils at standard section, 1 km NW of Mahidasht.

Figure 10. **a**. Paleochannel floored with Unit IVa gravel (pick rests on this), cut into Unit V below, filled with undifferentiated Unit IVb-II. (IAB, 5 Aug. 1975) **b**. Potsherd in Unit IVa gravel, standard section, 1 km NW of Mahidasht. (IAB, 5 Aug. 1975)

Figure 10 c. Upstream cutbank, standard section, showing unit V in lower half, [Vb-II and soils C, B (calcic), and A in upper half. (IAB, 5 Aug 1975)

cutbanks, overlain by undifferentiated units IVb to II. The latter units extend inland beneath a terrace which appears to bury the distal edge of a small fan issuing from a valley cut into the piedmont behind. The fan is thus tentatively correlated with unit IVa. Its surface, only where plowed, is littered with small, abraded potsherds, most likely turned up by plowing. None was diagnostic, except for a sliver of blue glass, a material unknown in this region before Islamic times (beginning 640 A.D.). If this glass was indeed once contained in the fan gravels, a maximum age of the fan, and by extension unit IVa, would be mid-7th century A.D.

Third, 33 km NNW of Kermanshah, the Razawar River has but a 12 m cliff in the side of a large village mound. The lowest cultural debris in the mound rests on sterile, blocky-structured, silty clay alluvium which is identified as unit V. The upper 30 cm of the alluvium is darker than material below, is penetrated by mud-filled tubules (root or worm-burrow casts), and contains 0.28 percent organic carbon, compared to 0.10 percent in lower material. This horizon is therefore believed to be the A-horizon of soil D on unit V, eroded from all other sections logged, but preserved here by the cultural overburden. Since the lowest cultural horizon contains pottery no older than Late Parthian in age (1-220 A.D.), the flood which eroded soil D at other sites occurred later than this period.

Fourth, unit IVa contains potsherds in nearly all exposures. Although usually worn and undiagnostic, they include many of a ware found on abandoned village mounds first occupied in the Sasanian period (220-640 A.D.) and which is still manufactured today. Unit IVa must therefore post-date Sasanian occupation.

Fifth, one sherd of this ware from unit IVa at the standard section was dated by thermoluminescence at 975±120 A.D. (Archeometry Laboratory, Oxford University, 1976). While dates on several sherds from several sections would be ideal in determining the youngest of this ware, some confidence can be placed in this date. While the sherd was not carefully collected for purposes of TL dating, the internal material would not be optically bleached by exposure to sunlight. If any bleaching had occurred it would decrease the TL age, but the sherd cannot be younger than mid-17th century (see next paragraph). In fact, it must be older than that to permit accumulation of alluvial units IVb-II, formation of soils B, C, and D, and subsequent incision of streams through the sequence--all following emplacement of unit IVa.

Sixth, at Mahidasht, 1 km upstream from the standard section (Fig. 1), a bridge over Ab-i-Marik contains stonework characteristic of mid-17th century A.D. architecture. This was buried by the muddy alluvium beneath the present surface, referable to units IVb-II, and is now re-exposed by recent river incision. Unit IVa must considerably pre-date bridge construction to allow for deposition of units IVb-II and later incision.

The deposition of unit IVa flood gravels can therefore be placed within the broadest limits imposed by Late Parthian occupation (1-220 A.D.) at the Razawar valley site (3, above) and mid-17th century bridge construction at Mahidasht. The blue glass find would raise the lower limit to post-640 A.D., the sherd TL date to post-975±120 A.D. The upper limit is imprecisely set at mid-17th century A.D., but must be lowered several centuries to allow for deposition of IVb-II, formation of soils B, C, and D, and recent incision. The age limits of unit IVa could therefore reasonably be set at mid-10th century to mid-13th century A.D.

Speculation on Cause(s) of the Flood

The simplest explanation of the Medieval Qara Su flood would call upon a rare, large precipitation event, such as would be generated by the stalling of a frontal system over the basin. The Colorado Big Thompson River flood of 1976 was associated with a deep low stalled for 24 hours over mountainous terrain, and has been assigned a minimum recurrence interval of 1000 years (Costa 1978). Winter synoptic patterns over the central Zagros would certainly permit this situation to develop; the area is closer to Mediterranean moisture sources than eastern Colorado

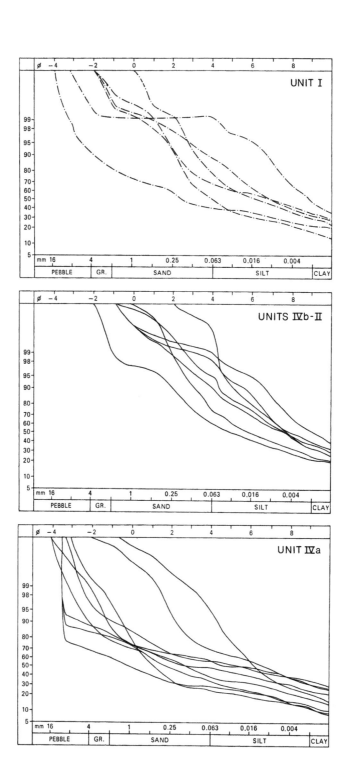

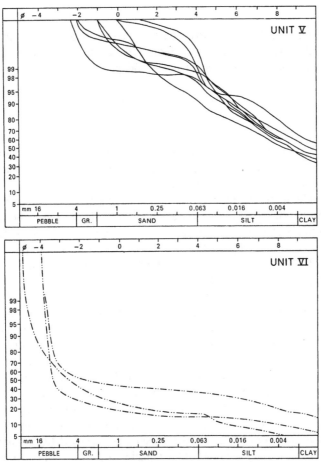

Figure 11. Probability plots of grain size for alluvial units samples, Qara Su basin.

is to Pacific ones, while the abrupt break between the Mesopotamian plain and the western Zagros would potentially induce more powerful orographic lift.

The very predisposition of this region to such events, however, leads to the expectation that they would be commoner than recorded in the Holocene alluvial sequence of the Qara Su basin. That sequence is likely not freakish, because the extensive alluvial lowland would act as a trap for sediments moved into them by meteorologically driven hydrologic events. In two other nearby Zagrosian intermontane basins (Kangavar and Holailan), eroded sediment is stored in broad fans, the opposing toes of which are close to the axial stream, so that the alluvial record is fragmentary at best, often absent altogether.

Statistical considerations also predict higher probabilities of such an event. The probability (r) of an event one or more events with a return period, T, in an n-year period is given by (Bruce and Clark, 1966, p. 149) as

$$r = 1 - (1-1/T)^n$$

If T for the case under discussion is given a value of 1000 years and n is also 1000 years, r = 0.63. Or, there is a 63 percent chance that a flood larger than one which occurred 1000 years ago has occurred in the last 1000 years.

One might, therefore, propose that, while large storms have been common over the Qara Su basin, runoff response was once, and only once, greatly magnified by basin conditions changed briefly as a result of one or more cultural factors. The most obvious, widely reported, and appropriately timed cultural discontinuity which could potentially have had the appropriate effects on the surface condition of this landscape is the Mongol conquest.

Western Iran lay at the hub of conflicts amongst Arabs, Turks, Iranians, and Mongols between the mid-11th century, when the Turkish Saljuks took Baghdad, and the mid-13th century, when Mongols under Genghiz Khan sacked that city. In 1197-98, Kermanshah province "was laid waste by the amir Miyadjuk" (Lambton 1980, p. 169), during his rout of Saljuks. Then, in 1220/21, his dominions were overrun by Mongols, under Genghiz Khan (Lambton 1980, p. 170). An Arab historian is quoted by Lambton (1980, p. 169) as referring to Kermanshah town being reduced to a mere village, as well as to declining rural production in wider Kurdistan asa direct consequence this invastion. Examples of drastic population and economic decline all over Iran at this time are summarized by Petroshevsky (1968).

While contemporary accounts of the rapacity of the Mongol conquest are probably exaggerated, its effects on the landscape would nevertheless have been severe. A highly developed, finely tuned system of intensive irrigation agriculture in lowlands which, archaeological records demonstrate, were densely populated, was almost overnight abandoned in favour of extensive pastoralism. Mongols had neither experience of, nor interest in the maintenance of canals, dams, and embankments which could have previously mitigated the effects of high floods. If this sounds too dramatic, it is relevant to add that, prior to Mongol invasion, Saljuk administration in the neighbouring Diyala basin to the west, in what is now Iraq, had deteriorated to a point at which similar irrigation systems had fallen into disrepair, so that "The Mongol invasion seems to have dealt no more than a *coup de grâce* to an urban civilization whose roots in rural agriculture had already withered" (Adams, 1965).

CONCLUSION

The stratigraphic record of lowland sediments in the Qara Su intermontane basin of central west Iran is dominated by several units of silty clay alluvium deposited by overbank floods, separated by soils formed when alluviation slowed or ceased. The only exception to this is a unit of rapidly deposited channel gravels which record a brief and intense flood. Chronological evidence is imprecise, but can reasonably be interpreted to show that the entire exposed alluvial record spans the Holocene epoch, and that the flood most probably occurred between the mid-10th century and mid-13th century A.D.

The orographic and meteorological environments of the region lead to an expectation that such intense floods would be more frequent. Absence of their effects from the alluvial sequence prompts speculation that a cultural discontinuity lowered the threshold of hydrologic response. If such was necessary, the most likely discontinuity would have been the Mongol invasion of this area in the 1220's A.D.

Whether entirely natural, or culturally magnified, this Medieval flood extensively erased a land surface which had previously been aggraded by placid lowland streams (unit V), then stabilized roughly 5000 years ago to allow formation of a deep soil (soil D). After the flood, sedimentation of muddy alluvium by overbank floods was resumed, continuing through the mid-17th century A.D., perhaps to the mid-19th century, when channel incision began to expose the record discussed here.

ACKNOWLEDGMENTS

I am deeply indebted to Louis D. Levine, West Asian Department, Royal Ontario Museum, who, as director of the Mahidasht Project, invited my participation and provided support of all kinds to my work. The project was funded by the Canada Council (now Social Sciences and Humanities Research Council of Canada), and the Museum. The Iran Centre for Archaeological Research assisted in the support and conduct of the project. R. A. Finnie very competently assisted in the field and laboratory work. York University has provided essential support for publication, especially the assiduous cartographers in my home department, and typing of Secretarial Services.

REFERENCES

Adams, R. Mc., 1965, Land behind Baghdad: Chicago, University of Chicago Press.

Brookes, I. A., in press, The physical geography and geomorphology of the Mahidasht Project area, Qara Su basin, central west Iran: Royal Ontario Museum, Publications in Archaeology. Toronto, Royal Ontario Museum.

Brookes, I. A., Levine, L. D. and Dennell, R. W., 1982, Alluvial sequence in central west Iran and implications for archeological survey: Journal of Field Archeology, v. 9, p. 285-299.

Bruce, J. P. and Clark, R. H., 1966, Introduction to Hydrometeorology: New York, Pergamon Press.

Costa, J. E., 1978, Colorado Big Thompson flood: geologicla evidence of a rare hydrologic event: Geology, v. 6, p. 617-620.

Iran, Ministry of Water and Power, 1956-1972: Meteorological Yearbooks, Teheran.

Iran, Ministry of Water and Power, 1954-1970, Hydrographic Yearbooks, Teheran.

Keith, M. L., and Anderson, G. M., 1963. Radiocarbon dating: fictitious results with mollusk shells: Science, v. 141, p. 634-637.

Lambton, A. K. S., 1980. "Kirmanshah", p. 169-170 *in*, The Encyclopedia of Islam: Leiden, E. J. Brill.

Maire, R., 1978. Première reconnaissance sur un grand karst d'altitude des Zagros: le plateau de Ravansar, Province de Kermanshah, Loristan, Iran: Bulletin de l'Association des Géographes Français, No. 449, p. 51-58.

Petroshevsky, I. P., 1968. The socio-economic conditions of Iran under the Il-Khans, p. 483-537 *in* J.A. Boyle, editor, The Cambridge History of Iran, v. 5: The Saljuq and Mongol Periods: Cambridge, Cambridge University Press.

van Zeist, W. and Bottema, S., 1977, Palynological investigations in Western Iran: Palaeohistoria, v. 19, p. 19-85.

Waltham, A. C. and Ede, D. P., 1973. The karst of the Kuh-e Parau: Transactions of the Cave Research Group, Great Britain, v. 15, p. 27-40.

Zohary, M., 1973, Geobotanical Foundations of the Middle East.: Geobotanica Selecta, Band III. Stuttgart, G. Fischer Verlag.

12

Occurrence and geomorphic effects of streamflow and debris flow floods in northern Arizona and southern Utah

Robert H. Webb

ABSTRACT

Extreme meteorological events can cause at least two types of catastrophic floods in northern Arizona and southern Utah. Streamflow floods occur in large drainage basins composed of Mesozoic sedimentary rocks or Tertiary volcanic rocks in southern Utah. A storm in the summer of 1909 triggered a streamflow flood in the Escalante River basin that initiated the erosion of a deep, rectangular-shaped arroyo into a previously unincised flood plain. Debris flows occur in small, high-relief basins developed in Paleozoic strata in the Grand Canyon National Park, Arizona. A winter frontal storm in December 1966 caused debris flows in the Lava-Chuar and Crystal Creek drainage basins that reached the Colorado River. The debris flows, triggered by slope failures during intense rainfall, caused bed scour, lateral migration of channels into older terraces, and deposition of levees of poorly sorted sediments.

Rectangular channels form in response to flooding regardless of the type of flood although channels affected by streamflow floods require more time to completely form. Channels change after the floods, by local bank collapse and channel widening, sedimentation on flood plains that form between the channel banks or walls, and reinvasion of riparian vegetation. The net effect of these changes is the creation of a low-water channel and flood-plain system within the arroyo walls or channel banks formed by the large flood.

INTRODUCTION

Catastrophic floods occur infrequently but cause profound changes in channel geometry. Use of the word *catastrophic* in describing a flood implies that the flood caused either a loss of human life, destruction of buildings or property, or extreme channel change. In southern Utah, for example, all major rivers had large floods between 1862 and 1909 that were catastrophic because they initiated the formation of incised channels locally known as arroyos (Webb and Baker, 1987). The floods on many of these rivers destroyed entire towns or settlements, such as the destruction of five towns during the Virgin River floods of 1862 (Larson, 1957). Much of the damage caused by these floods occurred because the channels widened and (or) deepened to transport the unusually large discharges.

Floodflow in natural channels can be separated into three categories depending on the sediment concentration. Streamflow, or Newtonian flow, occurs when the sediment concentration is less than 40 percent by weight (Beverage and Culbertson, 1964). Hyperconcentrated flow occurs when the sediment concentration is between 40 and 80 percent by weight (Beverage and Culbertson, 1964). Debris flows, or cohesive flows, occur when the sediment concentration exceeds 80 percent by weight and interactions between particles become a major source of friction during flow. All three types of flow can occur under natural conditions. The geomorphic effects of streamflow floods are generally well known in the Southwest (Baker, 1977; Roeske and others, 1978). The effects of debris or hyperconcentrated flow on channel morphology are not well documented in the Southwest, although Cooley and others (1977) documented debris flows in the Grand Canyon that occurred in 1966.

The types of storms that initiate streamflow floods include incursions of tropical storms over the continent (Costa, 1974), torrential thunderstorms (Baker, 1977), and winter convergent storms (Nolan and Marron, 1985). Channel widening is the most common geomorphic effect of large floods (Burkham, 1972; Costa, 1974; Stewart and LaMarche, 1967). Channel beds may erode or aggrade depending on the channel gradient and location within the drainage basin (Nolan and Marron, 1985; Stewart and LaMarche, 1967). Large streamflow floods are considered to have a more lasting impact on fluvial systems than more frequent discharges in the western United States (Baker, 1977; Nolan and Marron, 1985). The effects of debris flows on channel morphology have been well documented for areas devastated by volcanic eruptions (Janda and others, 1981) and for otherwise undisturbed areas subjected to intense or prolonged precipitation (Blackwelder, 1928; Costa and Jarrett, 1981; Gallino and Pierson, 1985; Pierson, 1980). Streamflow floods can follow debris flows in the same channel (Blackwelder, 1928); therefore, the separation of the geomorphic effects of the two types of floods is difficult. Janda and Meyer (1986) found that streamflow floods on the Toutle River in Washington widen channel,s whereas debris or hyperconcentrated flows create narrower channels. The differences in geomorphic effects were attributed to the different sediment-transport processes of the two types of flows (Janda and Meyer, 1986).

The purpose of this paper is to contrast the occurrence and effects of large streamflow and debris-flow floods in northern Arizona and southern Utah. A storm in August and September 1909 caused a large streamflow flood that deeply eroded the flood plain of the Escalante River in southern Utah (Fig. 1). A storm in December 1966 initiated debris flows in small drainage basins in the Grand Canyon of northern Arizona (Fig. 1). The two types of floods had similar effects on channel geometry during the event, and the channels had a similar recovery following the event. However, the channel changes were significantly different for the channels affected by the two types of flows.

STORM TYPES INITIATING FLOODS

The climate of northern Arizona and southern Utah is characterized by an ambiguous seasonality and strong orographic components of precipitation. For example, precipitation at higher altitudes in the Grand Canyon (Fig. 2) has a definite bimodal seasonality, whereas lower altitudes have a peak in precipitation only during the summer months. Cold air masses from the North Pacific are the dominant circulation features controlling winter precipitation (Mitchell, 1976). Several interacting features control summer rainfall; the "summer monsoon" airmass boundary (Mitchell, 1976) bisects the region leading to a combination of continental-interior and tropical-Pacific sources of moisture for summer rainfall (Fig. 1).

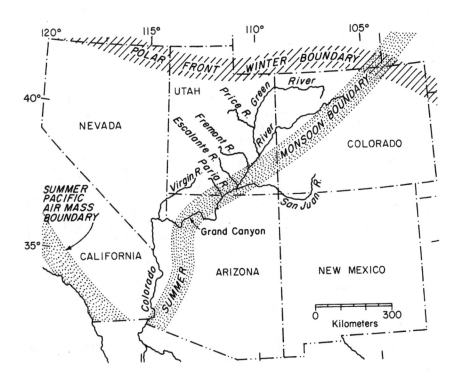

Figure 1. Airmass boundaries of the western United States (after Mitchell, 1976) and the location of study sites discussed in this paper.

Storms that cause large floods in the Southwest may include convergent storms from the North Pacific in winter (Hansen and Shwarz, 1981), incursions of hurricanes or tropical storms during the summer (Smith, 1986), or cutoff lows during the spring or fall (Douglas, 1974). Descriptions of the floods of December 1862 and December 1889 on the Virgin River (Larson, 1957) and the floods of March 1884 on the Paria and San Juan Rivers (Chidester and Bruhn, 1949; Perkins and others, 1957) suggest that successive convergent storms from the North Pacific were the cause of flooding. In December 1966, a flood on the Virgin River and debris flows in the Grand Canyon were caused by convergent storms that originated in the North Pacific and moved west to east across southern Utah (Butler and Mundorff, 1970; Cooley and others, 1977). The three largest floods on the Paria River,

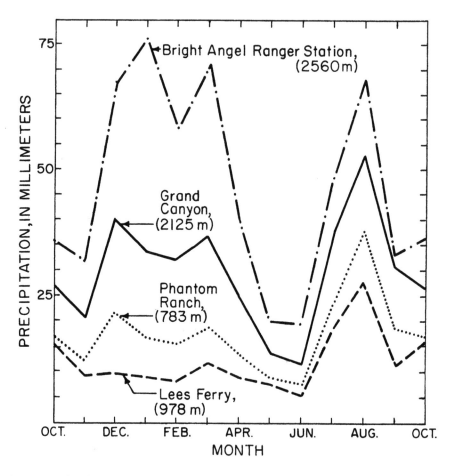

Figure 2. Precipitation as affected by elevation in the Grand Canyon of northern Arizona (after Warren and others, 1982).

Table 1. Seasonality of historic floods in southern Utah (Webb, 1985).

River	Total drainage area (km^2)	Percentage of Reported Floods, 1850 - 1980		
		Nov-June	July-Aug	Sept-Oct
Virgin River	9,920	20	50	30
Kanab Creek	510	6	51	33
Paria River	4,070	7	71	21
Escalante River	810	0	92	8
Fremont River	3,130	11	53	37
San Rafael River	2,410	6	72	22
Price River	3,990	0	76	24
Pack and Mill Creeks	340	14	57	29
San Juan River and tributaries	59,570	11	15	74

recorded in 1925, 1926, and 1939, were related to the penetration of moisture from tropical cyclones off the west coast of Mexico (Smith, 1986).

Floods can occur in northern Arizona and southern Utah in any season but may be related to the location of the drainages with respect to the summer-monsoon boundary (Fig. 1). The Virgin River in southwestern Utah is susceptible to flooding in December and January (Table 1). The largest floods on the Paria and Escalante Rivers have occurred primarily in late August and early September (Webb, 1985). The San Juan River in southeastern Utah and southwestern Colorado has large floods in late September and early October (Table 1).

STREAMFLOW FLOODS ON THE ESCALANTE RIVER

The Flood of August-September 1909

Historic channel changes on the Escalante River in south-central Utah (Figs. 1 and 3) provide an excellent example of geomorphic change caused by large streamflow floods. The bedrock in the headwaters of the Escalante River consists of Mesozoic sandstones and shales and Tertiary volcanic rocks (Hackman and Wyant, 1973). Upper Valley Creek, the headwater channel of the Escalante River, flows through an alluvial valley upstream from Escalante, Utah (Fig. 3). The river flows into a bedrock canyon below Escalante. A discontinuous ephemeral stream was present with two isolated arroyo segments (Fig. 3) before 1894, and the average width of the channel at section-line crossings was 8.4 m in 1893-94 (Table 2; Webb, 1985).

In late August and early September 1909, a mid-latitude low-pressure system moved west to east across southern Utah (Brandenberg, 1909). The storm stalled over the Escalante River basin between August 31 and September 1 and dropped a total of 91 mm of rainfall at Escalante. The slow movement of the late-summer storm suggests that the storm was a cutoff low. The storm has been the only storm that has caused widespread flooding in the history of southern Utah (Webb, 1985).

The peak discharge for the flood of 1909 was estimated to be 570 m^3s^{-1} at a site 3.2 km downstream from the gaging station (Webb and others, 1987a). The drainage area is 820 km^2. Uncertainty introduced by the reported channel scour during the flood (U.S. Geological Survey, 1910) was evaluated using photogrammetric methods, and a discharge of 500 m^3s^{-1} was estimated for channel conditions before the flood (Webb and others, 1987a). These discharges are comparable on flood-envelope curves with the largest recorded discharges in southern Utah and northern Arizona (Webb, 1985).

The recurrence interval for the flood is questionable because the annual flood series for the Escalante River is not necessarily stationary (Webb and Baker, 1987). Historic floods like the flood of 1909 may represent a different population of flood events. A recurrence interval of 50-100 yrs has been estimated for a discharge of 570 m^3s^{-1} by incorporating four historic floods, including the flood of 1909, in a flood-frequency analysis of the gaging record for the Escalante River (Webb and Baker, 1987). On the basis of the gaging record only, the recurrence interval for 570 m^3s^{-1} is in excess of 1000 yrs (Webb and Baker, 1987).

Geomorphic Effects of the Flood of 1909
and Subsequent Floods

The flood of 1909 initiated the incision of a new channel into the nearly level flood plain that was composed of layered sediments that ranged in size from fine sand to clay. The gaging station on the Escalante River below the confluence of Pine Creek and the Escalante River (Fig. 3) was destroyed when the channel entrenched 1 m and widened on August 31, 1909 (U.S. Geological Survey, 1910). Wallace Roundy, a pioneer of Escalante, stated that the flood of 1909 "dug out a big hole" near the town of Escalante, and that subsequent floods "began coming often and cut the hole back farther and farther" (Edson Alvey, resident of Escalante, written commun., 1985).

The erosion coalesced discontinuous channel segments into a continuously incised channel within 30 yrs. Floods between 1910 and 1930 and two large floods in the summer of 1932 resulted in significant erosion of agricultural land and destruction of bridges and buildings (Webb, 1985). Comparison of aerial photographs taken in 1940 and 1974 revealed little change in the location of the arroyo walls along Upper Valley Creek (Fig. 4). Floods of the same magnitude as the flood of 1909 have not occurred after 1940 (Webb and Baker, 1987).

Table 2. Widths of Upper Valley Creek and the Escalante River in 1893-94 (Lewman, 1893-94) and 1984-85.

Township range	Between sections	1893-94 width (m)	1984-85 width (m)	Notes
	E of 9	5.2	36	
	8-9	6.7	199	Swamp in 1894
T.35S.,	7-8	7.1	>75	
R.3E.	W of 7	6.9	139	
	11-12	7.5	61	
	10-11	10.1	165	
	10-15	8.0	132	
T.35S.,	15-16	8.6	300	Arroyo in 1872
R.2E.	16-17	7.7	46	
	17-20	6.0	22	
	20-29	6.0	149	
	29-30	15.5	40	
	E of 25	4.3	48	
T.35S.,	25-26	10.8	63	
R.1E.	26-35	17.9	--	
	34-35	36.1	142	
	S of 34	6.5	52	
	3-4	5.4	60	Arroyo in 1894
	4-9	4.7	98	Arroyo in 1894
	9-10	4.5	79	
T.36S.,	8-9	5.2	38	
R.1E.	8-17	6.4	33	Swamp in 1894
	17-18	4.5	29	Swamp in 1894
	18-19	4.5	23	
	19-30	4.5	16	Meadow in 1872
	W of 30	---	11	
		---	--	
Average		8.4	82	

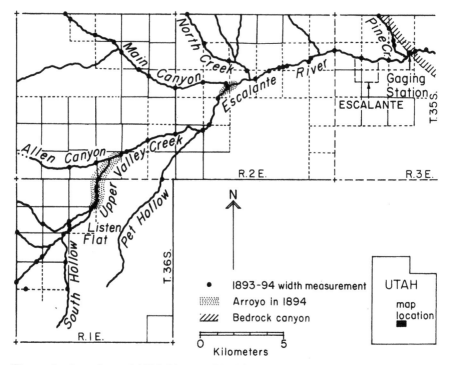

Figure 3. Locations of 1893-94 and 1984-85 width measurements on Upper Valley Creek and the Escalante River, south-central Utah.

Land-survey notes recorded in 1893-94 (Lewman, 1893-94) and aerial photographs taken in 1940 were used to identify relict segments of the channel as it appeared before 1909 (Webb, 1985). A 220-m reach on Listen Flat was surveyed to compare dimensions between the relict and modern channels (Fig. 5). The relict channel ranged in width from 7 to 44 m in this reach and had a range of 6.6 to 34.7 in width-depth ratio. In contrast, the modern arroyo in the same reach ranged from 29 to 79 m in width and ranged from 3.5 to 5.3 in the width-depth ratio. The hydraulic radius of the relict channel ranged from 0.2 to 1.9 m in five cross sections, whereas the hydraulic radii of three modern arroyo cross sections ranged from 4.4 to 9.1 m (Webb, 1985). Other channel reaches have similar differences (Table 2).

The dimensional contrasts between the relict and modern channels serves to underscore the effect of channel changes on bankfull discharge. By using the slope-area method and assuming no scour, the discharge required to overtop for the relict channel banks was estimated to be 40 m^3s^{-1} compared with greater than 11,000 m^3s^{-1} for the modern arroyo walls. The channel flood plain prior to 1909 must have sustained periodic overbank floods as the bankfull discharge was exceeded, whereas only an extremely long recurrence interval flood could possibly overtop the

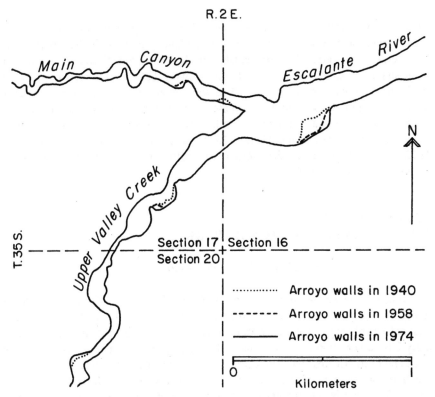

Figure 4. The confluence of Main Canyon and Upper Valley Creek showing bank erosion between 1940 and 1974.

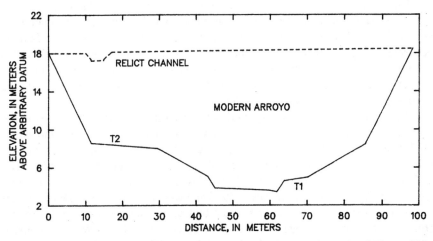

Figure 5. Comparison of adjacent relict and modern-arroyo channels of Upper Valley Creek at Listen Flat (T.36S., R.1 between sections 4 and 9).

modern arroyo walls. The decrease in large floods in Upper Valley Creek resulted in the formation of a small channel and flood-plain system within the arroyo walls.

Aerial and ground photographs taken in the 1930's and 1940's show a rectangular-shaped channel with no vegetation growing between the arroyo walls. In 1985, the channel was as narrow as 3 m wide locally because riparian vegetation has colonized the flood plain between the arroyo walls. Low terraces--as many as three at some sites--have formed within the arroyo walls (Fig. 5), either as a result of the encroaching vegetation (Hadley, 1961), decreased flood frequency (Burkham, 1972), or adjustments in the storage of sediment within the drainage basin (Schumm and others, 1984).

DEBRIS FLOWS IN
GRAND CANYON NATIONAL PARK

Magnitudes of the Debris Flows of December 1966

Debris flows on two tributaries of the Colorado River in Grand Canyon National Park (Fig. 6) illustrate the effects of a second type of catastrophic flood on channel morphology. Debris flows occurred during a winter convergent storm in early December 1966 that also caused flooding on many major rivers in southern Utah and northern Arizona. The storm lasted from December 3-6 and resulted in as much as 300-360 mm of total precipitation in southwestern Utah (Butler and Mundorff, 1970) and in the vicinity of the Grand Canyon (Fig. 6; Cooley and others, 1977). Peak discharges in the Virgin River basin (Fig. 1) are the largest on record for most of the gaging stations, although the flood of 1862 may have been larger (Butler and Mundorff, 1970).

The intense precipitation on the Kaibab Plateau caused slope failures in the Paleozoic Supai Group and overlying Hermit Shale, which are common formations in Grand Canyon National Park (Huntoon and others, 1986). Ninety slope failures occurred during the storm of December 1966, ten of which traveled short distances as debris flows (Cooley and others, 1977). Debris flows in the Lava-Chuar and Crystal Creek drainage basins (Fig. 6) were of sufficient magnitude to reach the Colorado River (Cooley and others, 1977).

Field evidence and laboratory work were used to verify that the floods in the Lava-Chuar and Crystal Creek drainages were debris flows. Cooley and others (1977) describe mudlines and poorly sorted deposits in both drainages that they attributed to debris-flow origins; similar features were still apparent in 1986 (Webb and others, 1987b). Samples of the debris-flow matrix were taken from the Lava-Chuar and Crystal Creek drainages in 1986 and were reconstituted to determine the moisture content using methods presented by Gallino and Pierson (1985). Reconstituted water contents ranged from 22 to 32 percent by volume (Webb and others, 1987b), and sediment concentration by weight was 85 to 90%. The debris flow in the Lava-Chuar Creek drainage had a higher sediment concentration than the debris

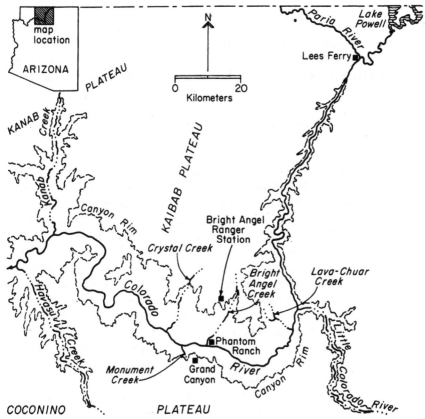

Figure 6. Selected streams and major geographical landmarks in the Grand Canyon of northern Arizona.

flow in the Crystal Creek drainage basin (Cooley and others, 1977).

Magnitudes of the debris flows in the Lava-Chuar and Crystal Creek drainages have been calculated with uncertain accuracy. Cooley and others (1977) used the slope-area method to estimate a peak discharge of 820 m^3s^{-1} for the debris flow at one site on Dragon Creek, which is a tributary of Crystal Creek. The slope-area method, however, was developed for streamflow floods and resulted in substantial overestimation of discharges for debris flows in Colorado (Costa and Jarrett, 1981). Webb and others (1987b) used superelevation and velocity-head equations to estimate discharges that ranged from 250 to 400 m^3s^{-1} for the debris flow of 1966 at three sites in the Crystal Creek drainage. Webb and others (1987b) also estimated a peak discharge of 110 m^3s^{-1} at one site for the debris flow in the Lava-Chuar Creek drainage.

Recurrence intervals for debris flows are not known for the ungaged and isolated drainages. Webb and others (1987b) used historical evidence and radiocarbon dating of deposits in these canyons to determine a chronology of debris flows. Debris flows that reach the Colorado River may recur every 20-30 yrs in the Lava-Chuar Creek drainage and every 50 yrs in the Crystal Creek drainage. The streamflow flood of December 1966 on Bright Angel Creek, which is the only comparable, gaged drainage in the Grand Canyon, had an estimated recurrence interval of 50 yrs (Cooley and others, 1977).

Geomorphic Effects of the Debris Flows of 1966
and Subsequent Flows

The debris flows of 1966 significantly changed the geometry of channels in the Lava-Chuar and Crystal Creek drainages. Because of their isolation, little is known of the exact conditions of the channels before 1966. Geomorphic and historical evidence along with aerial and ground photography were used to determine quantitatively the extent of changes (Cooley and others, 1977; Webb and others, 1987b).

The debris flow in the Lava-Chuar Creek drainage, which traveled 10.5 km to the Colorado River, caused extensive erosion of the channel and deposition of levees of poorly sorted sediment. Before December 1966, the channel and flood plain were filled with riparian vegetation, and the debris flow of 1966 stripped vegetation from a 50-m-wide area of the flood plain. The post-flow morphology of the channel at a given site was a function of the erosiveness of the debris flow and the width of the channel. At some sites, a rectangular-shaped channel was eroded into the flood plain (Fig. 7) with a width-depth ratio ranging from 4 to 8. At other sites, clay- to boulder-sized sediment as large as 1.5 m in diameter was deposited in lobes and levees that completely filled the channel and covered the flood plain.

Subsequent fluvial events, notably a smaller debris flow that occurred between 1973 and 1978, changed the channel of the Lava-Chuar Creek drainage (Webb and others, 1987b). A cross section taken in 1967 (Cooley and others, 1977) was compared with a cross section taken in 1986 at about the same site and datum (Fig. 7). The walls of the rectangular channel have been eroded laterally 3-5 m and the angle of the banks has decreased (Fig. 7). The debris flow that occurred between 1973 and 1978 locally deposited a small levee along most of the length of the drainage. Low discharge caused a narrow and shallow channel to form within the channel created by the debris flow of 1966 (Fig. 7). The shallow channel is continuous through the drainage despite the variability in channel form that resulted from the debris flow of 1966. Riparian vegetation reinvaded the flood plain by 1986, although not to the density present before 1966, and has locally encroached upon the channel.

In the Crystal Creek drainage, the debris flow of December 1966 traveled 20.9 km before it reached the Colorado River and caused extensive channel changes and erosion. Much of the erosion may have occurred during recessional streamflow instead of during the peak of the debris flow (Cooley and others, 1977). Recessional flow, however, may have consisted of pulses of hyperconcentrated or debris flow (Webb and others, 1987b). Cottonwood trees 0.6 to 0.9 m in diameter were broken 0.3 to 0.9 m above the ground in narrow sections where the stage of the debris flow exceeded 12 m. As much as 3.7 m of downcutting in the channel occurred during the debris flow on the basis of compared elevations of a tributary channel and the main channel at one site (Cooley and others, 1977). Local aggradation also occurred, particularly in expanding reaches downstream from severe contractions. Cooley and others (1977) reported lateral cutting generally of less than 3 m during the flood. The resulting alluvial channel had a width-depth ratio between 5 and 10. Boulders as much as 2 m in diameter were transported in the debris flow (Webb and others, 1987b). The channel in the Crystal Creek drainage experienced a slight widening, the invasion of riparian vegetation onto the flood plain, and the formation of a narrow channel within the enlarged channel. A resurvey of three cross sections used for a slope-area determination of discharge in 1967 (site 46 of Cooley and others, 1977) revealed a 2-5 m widening of the channel and a decrease in the bank angle (Fig. 8). The bed of the channel at this site, reported as bedrock in 1967, was irregularly covered with a veneer of alluvium in 1986. Near the Colorado River, riparian vegetation choked the channel in 1986, and a 1-3 m wide channel formed to convey the perennial base-flow discharge.

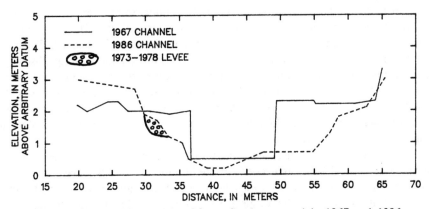

Figure 7. Comparison of channels of Lava Creek surveyed in 1967 and 1986.

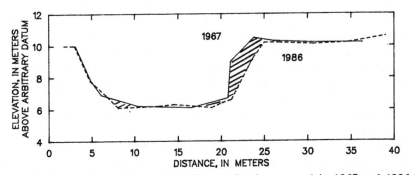

Figure 8. Comparison of channels of Dragon Creek, surveyed in 1967 and 1986.

COMPARISON OF STREAMFLOW
AND DEBRIS FLOW FLOODS

Characteristics of the three drainage basins (Table 3) may explain the difference in the type of flood that occurred in Upper Valley Creek in comparison to the floods that occurred in the Grand Canyon. The drainage area of Upper Valley Creek is more than three times larger than the two drainages in the Grand Canyon but Upper Valley Creek has a lower topographic relief and ratio of relief to length of channel. The bedrock in the headwaters of Upper Valley Creek is easily weathered to sand or smaller-sized particles, whereas the bedrock in the headwaters of the two Grand Canyon drainages is resistant to weathering. Although debris-flow deposits that extend downslope from landslides occur at the bases of cliffs in the Escalante River basin (Williams, 1984), there is no evidence of past debris flows along Upper Valley Creek. The lower potential for debris flows in Upper Valley Creek compared with the potential for debris flows in drainages in the Grand Canyon is attributed to the lower relief and greater weathering rates of bedrock in the headwaters of Upper Valley Creek.

Streamflow floods and debris flows have had similar short-term effects on channel morphology in southern Utah and northern Arizona (Table 4). Channels became rectangular in shape with width-depth ratios between 3.5 and 10. Riparian vegetation that lined the channels and induced deposition during low-stage discharges was stripped away. Channels affected by streamflow floods generally tend to be smooth and composed of well-sorted sands and gravels with infrequent bars of cobble-size particles. Channels affected by debris flows tend to have levees composed of poorly sorted clay- to boulder-sized particles and bars composed of boulders as much as 2 m in diameter.

The meteorological causes are similar for both types of floods; however, an additional geological factor is needed for the initiation of debris flows. A winter-convergent storm in December 1966 caused both debris flows in the Grand Canyon and streamflow floods in the Virgin River drainage. Debris flows also can result from other types of storms. An isolated thunderstorm in July 1984 initiated a debris flow in Monument Creek, a small drainage northwest of Grand Canyon (Figure 6; Webb and others, 1987b). Tropical Storm Norma in 1970 caused a debris flow in a tributary of Tonto Creek in central Arizona and also caused streamflow floods throughout the Southwest (Roeske and others, 1978).

Debris flows also require slope failures to provide sediment to sustain the flow of debris. Historic debris flows in Grand Canyon National Park were initiated in only a few of the geologic formations (Webb and others, 1987b). Debris flows require an unusual storm event, high relief in the drainage basin, and bedrock conditions conducive to slope failure, whereas streamflow floods only require an extreme storm event.

Upper Valley Creek recovered slowly after the floods because of the semiarid environment and the large differences between the discharges of ordinary and large floods (Webb and Baker, 1987). The arroyo that was formed by streamflow floods

Table 3. Comparison of drainage basin characteristics for catastrophic floods in Upper Valley Creek, Utah, and Lava-Chuar and Crystal Creek, Grand Canyon National Park, Arizona.

Drainage name	Drainage area (km^2)	Relief (m)	Relief divided by channel length	Bedrock formation of headwaters	Particle size of channel banks	Type of flood
Upper Valley Creek, UT	391	1250	0.06	Tertiary volcanic and Cretaceous sedimentary rock	sand to clay	streamflow
Lava-Chuar Creek, AZ	55	1615	0.13	Paleozoic sedimentary rocks	boulders to clay	debris flow
Crystal Creek, AZ	112	1980	0.07	Paleozoic sedimentary rocks	boulders to clay	debris flow and streamflow

Table 4. Comparison of required conditions and responses for catastrophic floods in Upper Valley Creek, Utah, and Lava-Chuar and Crystal Creeks, Grand Canyon National Park, Arizona

Drainage name	Required conditions	Response to flooding	Channel changes after flooding
Upper Valley, UT	Intense or prolonged rainfall	Initiated arroyo formation that lasted 31 years.	Minimal changes in arroyo walls. Low-water channel narrows within arroyo wall; vegetation reinvades
Lava-Chuar Creek, AZ	Intense rainfall and slope failure	Rectangular channel formed.	Large local changes in channel banks. Low-water channel narrows within arroyo wall; vegetation reinvades.
Crystal Creek, AZ	Intense rainfall and slope failure	Rectangular channel formed.	Minimal changes in channel banks. Low-water channel narrows within channel banks; vegetation reinvades.

has undergone little erosion of the arroyo walls after cessation of the large floods (Fig. 4). Arroyos initiated by streamflow floods may not fully form until decades after the flood. Channels created by catastrophic debris flows, however, can either widen appreciably (Fig. 7) or remain unchanged with time (Fig. 8). In both cases, formation of a low-water channel and low terraces within the channel or arroyo walls (Figs. 5 and 7) and invasion of riparian vegetation occur after the flood. Large floods can either quickly change the morphology of a channel or initiate an erosional pattern that results in a completely different channel morphology with time.

REFERENCES

Baker, V. R., 1977, Stream-channel response to floods, with examples from central Texas: Geological Society of America Bulletin, v. 88, p. 1057- 1071.

Beverage, J. P., and Culbertson, J. K., 1964, Hyperconcentrations of suspended sediment: American Society of Civil Engineers, Journal of the Hydraulics Division, v. 90, p. 117-126.

Blackwelder, E., 1928, Mudflow as a geologic agent in semiarid mountains: Geological Society of America Bulletin, v. 39, p. 465-494.

Brandenberg, F. H., 1909, Climatological data for September,1909, District No. 9, Colorado Valley: Monthly Weather Review, v. 37, p. 648-649.

Burkham, D. E., 1972, Channel changes of the Gila River in Safford Valley, Arizona, 1846-1970: U. S. Geological Survey Professional Paper 655-G, 24 p.

Butler, Elmer, and Mundorff, J. C., 1970, Floods of December 1966 in southwestern Utah: U. S. Geological Survey Water-Supply Paper 1870-A, 40 p.

Chidester, I., and Bruhn, E., 1949, Golden nuggets of pioneer days, a history of Garfield County: Panguitch, Utah, The Garfield County News, 374 p.

Cooley, M. E., Aldridge, B. N., and Euler, R. C., 1977, Effects of the catastrophic flood of December, 1966, North Rim area, eastern Grand Canyon, Arizona: U.S. Geological Survey Professional Paper 980, 43 p.

Costa, J. E., 1974, Response and recovery of a Piedmont watershed from Tropical Storm Agnes, June 1972: Water Resources Research, v. 10, p. 106-112.

Costa, J. E., and Jarrett, R. D., 1981, Debris flows in small mountain stream channels of Colorado and their hydrologic implications: Bulletin of the Association of Engineering Geologists, v. 18, p. 309-322.

Douglas, A. V., 1974, Cutoff lows in the southwestern United States and their effects on the precipitation of this region: Tucson, Arizona, University of Arizona, Laboratory of Tree-Ring Research, Final Report on Department of Commerce Contract 1-35241-No. 3, 40 p.

Gallino, G. L., and Pierson, T. C., 1985, Polallie Creek debris flow and subsequent dam-break flood of 1980, East Fork Hood River basin, Oregon: U. S. Geological Survey Water-Supply Paper 2273, 22 p.

Hackman, R. J., and Wyant, D. G., 1973, Geology, structure, and uranium deposits of the Escalante quadrangle, Utah and Arizona: U. S. Geological Survey Miscellaneous Investigations Series map I-744, scale 1:250,000, 2 sheets.

Hadley, R. F., 1961, Influence of riparian vegetation on channel shape, northeastern

Arizona, *in* Geological Survey Research, 1961: U. S. Geological Survey Professional Paper 424-C, p. 30-31.

Hansen, E. M., and Shwarz, F. K., 1981, Meteorology of important rainstorms in the Colorado River and Great Basin drainages: U. S. Department of Commerce, National Oceanic and Atmospheric Administration, Hydrometeorological Report No. 50, 167 p.

Huntoon, P. W., Billingsley, G. H., Jr., Breed, W. J., Sears, J. W., Ford, T. D., Clark, M. D., Babcock, R. S., and Brown, E. H., 1986, Geologic map of the eastern part of the Grand Canyon National Park, Arizona: Flagstaff, Museum of Northern Arizona Special Publication Map, scale 1:62,500, 1 sheet.

Janda, R.J., Scott, K.M., Nolan, K.M., and Martinson, H.A., 1981, Lahar movement, effects, and deposits: U. S. Geological Survey Professional Paper 1250, p. 461-478.

Janda, R. J., and Meyer, D. F., 1986, Sediment concentration and channel morphology, *in* Proceedings of the Fourth Federal Interagency Sedimentation Conference: Interagency Advisory Committee on Water Data, Subcommittee on Sedimentation Conference, March 24-27, 1986, Las Vegas, Nevada, v. 1, p. 3-83 to 3-92.

Larson, A. K., 1957, The red hills of November: Salt Lake City, Utah, The Deseret News Press, 330 p.

Lewman, William, 1893-94, Unpublished field notes: Salt Lake City, Bureau of Land Management, General Land Office Survey, Book A-238, p. 46-461.

Mitchell, V. L., 1976, The regionalization of climate in the western United States: Journal of Applied Meteorology, v. 15, p. 290-297.

Nolan, M. K., and Marron, D. C., 1985, Contrast in stream-channel response to major storms in two mountainous areas of California: Geology, v. 13, p. 135-138.

Perkins, C. A., Nielson, M. G., and Jones, L. B., 1957, Saga of San Juan: Blanding, Utah, San Juan County Daughters of Utah Pioneers, 367 p.

Pierson, T. C., 1980, Erosion and deposition by debris flows at Mt. Thomas, North Canterbury, New Zealand: Earth Surface Processes, v. 5, p. 227-247.

Roeske, R. H., Cooley, M. E., and Aldridge, B. N., 1978, Floods of September 1970 in Arizona, Utah, Colorado, and New Mexico: U.S. Geological Survey Water-Supply Paper 2052, 135 p.

Saarinen, T. F., Baker, V. R., Durrenberger, Robert, and Maddock, Thomas, Jr., 1984, The Tucson, Arizona flood of October 1983: Washington, D.C., National Academy Press, 112 p.

Schumm, S. A., Harvey, M. D., and Watson, C. C., 1984, Incised channels, morphology, dynamics, and control: Littleton, Colorado, Water Resources Publications, 200 p.

Smith, Walter, 1986, The effects of eastern north Pacific tropical cyclones on the southwestern United States: Salt Lake City, Utah, National Oceanic and Atmospheric Administration Technical Memorandum NWS WR-197, 229 p.

Stewart, J. H., and LaMarche, V. C., Jr., 1967, Erosion and deposition produced by the flood of December 1964 on Coffee Creek, Trinity County, California: U.S. Geological Survey Professional Paper 422-K, 22 p.

U.S. Geological Survey, 1910, Surface water supply, part IX, Colorado River basin: U.S. Geological Survey Water-Supply Paper 269, p. 183-184.

Warren, P. L., Reichhardt, K. L., Mouat, D. A., Brown, B. T., and Johnson, R. R., 1982, Vegetation of Grand Canyon National Park: Tucson, Arizona, National Park Service and University of Arizona Cooperative National Park Resources Studies Unit, Contribution No. 017/06, Technical Report No. 9, 140 p.

Webb, R. H., 1985, Late Holocene flooding on the Escalante River, south-central Utah [Ph.D. thesis]: Tucson, Arizona, University of Arizona, 204 p.

Webb, R. H., and Baker, V. R., 1987, Changes in hydrologic conditions related to large floods on the Escalante River, south-central Utah, *in* Singh, V., editor, Proceedings of the International Symposium on Flood Frequency and Risk Analysis, May 14-17, 1986: Dordrecht, Holland, D. Reidel Publishing, in press.

Webb, R. H., O'Connor, J. E., and Baker, V. R., 1987a, Paleohydrologic reconstruction of flood frequency on the Escalante River, south-central Utah, *in* Baker, V.R., Patton, P. C., and Kochel, R. C., editors, Flood geomorphology: New York, John Wiley and Sons, in press.

Webb, R. H., Pringle, P. T., and Rink, G. R., 1987b, Debris flows from tributaries of the Colorado River in Grand Canyon National Park, Arizona: U.S. Geological Survey Open-File Report 87-118.

Williams, V. S., 1984, Pedimentation versus debris-flow origin of plateau-side desert terraces in Southern Utah: Journal of Geology, v. 92, p. 457-468.

13

Techniques used by the U. S.Geological Survey in estimating the magnitude and frequency of floods

Wilbert O. Thomas, Jr.

ABSTRACT

The U.S. Geological Survey is actively involved in the development and test-ing of procedures for estimating flood-peak discharges at selected probabilities of exceedance. This paper describes the techniques utilized by the U.S. Geological Survey in estimating flood-peak discharges (1) at gaging stations, (2) at both rural and urban ungaged sites, and (3) at ungaged sites on streams with one or more gag-ing stations. The U.S. Geological Survey uses Bulletin 17B, Guidelines for Deter-mining Flood Flow Frequency, for estimating flood-peak discharges at gaging sta-tions. At ungaged sites on streams with one or more gaging stations, weighted es-timates are made based on data at the nearby gaging station and regional estimates at the ungaged site. The regional estimates for ungaged sites are generally based on multiple-regression techniques that relate flood discharges of various exceedance probabilities to watershed, climatic, and channel-geometry characteristics. The ac-curacy of U.S. Geological Survey procedures for flood estimation for ungaged sites is also compared to six frequently used rainfall-runoff estimation techniques.

INTRODUCTION

The mission of the Water Resources Division (WRD) of the U.S. Geological Survey (USGS) is to provide the hydrologic information and understanding needed for the best use and management of the Nation's water resources for the benefit of the people of the United States. In support of this mission, WRD offices across the country collect, analyze, and interpret basic water-resources data, conduct prob-lem-oriented research in hydrology to better understand hydrologic systems, dissem-inate water data and results of investigations and research, and maintain a water-resources data base. The collection and analysis of flood data are a major part of the surface-water activities of the WRD, USGS. The purpose of this paper is to discuss the techniques used by the USGS in estimating flood-peak discharges at (1) gaging stations, (2) rural and urban ungaged sites, and (3) ungaged sites on streams with one or more gaging stations. In addition, the procedures for estimating flood-peak discharges for ungaged sites are compared to six frequently used rainfall-runoff estimating techniques. These comparisons are taken from a U.S. Water Resources Council report (1981) that compared several techniques for estimating flood-peak discharges for ungaged rural watersheds.

ANALYSIS OF FLOOD-PEAK DATA AT GAGING STATIONS

The USGS maintains a Peak Flow File as part of the National Water Data Storage and Retrieval System known as WATSTORE (Lepkin and DeLapp, 1979). This file contains data on annual instantaneous peak discharges and partial-duration peaks (peaks above a base value) for both active and discontinued stations. As of December 1986, this file contained about one-half million flood peaks at over 22,000 stations with an average record length of 22 years. A station must have a minimum of 5 years of annual peak data before it is added to the file. At over 8,300 stations partial-duration peaks are also available. There are approximately three partial-duration peaks per year of record for these stations. Those flood peaks affected by regulation, backwater, dam breaks, etc., are coded so that the analyst will be aware they do not represent natural flow conditions. The flood-peak data can be retrieved from the Peak Flow File using procedures documented by Lepkin and DeLapp (1979).

The annual flood peaks and partial-duration peaks are published each year in the annual data reports for each State (e.g. Water Resources Data, West Virginia, Water Year 1984). The partial-duration peaks are computed and published at all unregulated continuous-record stations. Annual peak discharges are published for all stations including crest-stage stations. Crest-stage stations are those stations where only peak stages greater than some predetermined level are recorded. Only the annual maximum peak discharge is published for these stations.

Federal agencies in the United States, including the USGS, use Bulletin 17B, Guidelines for Determining Flood-Flow Frequency (Hydrology Subcommittee, Interagency Advisory Committee on Water Data, 1982), for analyzing flood-peak data at gaging stations. Bulletin 17B guidelines recommend fitting the logarithms of the annual maximum peak discharges to a Pearson Type III distribution using the method of moments approach for parameter estimation as the standard method. Deviations from the standard method are allowed with proper documentation of the analysis. The objective of the analysis is to determine those flood discharges that have, for example, a 50-, 10-, 2-, or 1-percent chance of being exceeded in any given year. Bulletin 17B guidelines include outlier tests for identifying high and low outliers, procedures for including historic and paleoflood information in the analysis and procedures for weighting station skew (based on observed annual peaks) with a regional value of skew. In the future, the USGS plans to utilize nonparametric statistical tests for evaluating if the annual peak data meet the requirements of frequency analysis. These tests would evaluate, for example, if the annual peaks are independent, homogeneous, and free of trends. Thomas (1985) provides a description of the historical development of Bulletin 17B and discusses the current methodology used in it.

Since this paper is being published as part of a geomorphology symposium, it is worthwhile to comment further on the procedures in Bulletin 17B for including historic and paleoflood information in the analysis. Historic information implies knowledge of the magnitude and the year that a major flood occurred outside of the systematic gaging record. This information can often be obtained from local residents or newspaper accounts. Paleoflood information is similarly defined except that this information is obtained by (1) tree ring analysis of floodplain trees destroyed or scarred by flood waters, (2) radiocarbon dating of organic materials in flood deposits, or (3) dating of landforms eroded or buried by floods based on changes in quantitative soil properties versus time (Costa, 1978). Both approaches attempt to determine the time period since the major or catastrophic flood occurred. Historic information generally extends the systematic record by 50 to 200 years while the paleoflood information may add a few thousand years to the systematic record.

The technique presently used in Bulletin 17B computes improved estimates of the logarithmic mean and standard deviation and the skew coefficient of the annual peak data using the historic period and the magnitude of the historic flood. In this approach the systematic annual peaks are replicated to create a data set equivalent to the number of years in the historic period. For example, suppose there are 25 years of systematic record and it is known that the largest flood occurred 51 years prior to the systematic record. This implies the historic flood is the highest in 76 years. In order to produce a 76-year record, each systematic annual peak must be duplicated twice to "fill-in" the missing 50-year period. The logarithmic mean and standard deviation and the skew coefficient are then computed giving a weight of three to each systematic peak and a weight of one to the historic peak. Cohn and Stedinger (1983) have shown that the weighted moments procedure in Bulletin 17B does not efficiently use the historic information. They proposed a method using censoring theory and maximum likelihood estimation and demonstrated how this method provided improved estimates of design floods. The Cohn-Stedinger method is particularly useful when analyzing paleofloods because the historic period may be a few thousand years. Under these conditions, the weighted moments procedure in Bulletin 17B gives very little weight to the paleoflood. As the historic period increases, more weight is given to the systematic record in an effort to fill in the missing historic period and as a result less weight is given to the paleoflood. Since the use of paleoflood information is very important in computing improved estimates of design floods, the present technique in Bulletin 17B should be revised to more efficiently use this paleoflood information.

Often the annual flood-peak discharges are caused by two different hydrologic events such as rainfall or snowmelt runoff. In these instances it may not be possible to adequately fit the annual peak discharges with a three-parameter distribution such as the Pearson Type III. For these mixed population problems, the USGS utilizes the combination of probability approach as described by Crippen (1978). These procedures involve computing separate curves for the two different hydrologic events, such as rainfall and snowmelt flood peaks, and then combining them into a single curve using the following equation

$$P3 = P1 + P2 - (P1)(P2) \qquad (1)$$

where P3 = probability of exceedance of a given flood discharge from either hydrologic event,

P1 = probability of exceedance of a given rainfall peak,

P2 = probability of exceedance of a given snowmelt peak, and
(P1)(P2) = probability of exceedance from both events.

The USGS has developed a computer program for analyzing at-site flood-peak data according to Bulletin 17B guidelines (Kirby, 1979). Many copies of this program (J407) have been provided to consulting engineers, universities, other Federal and State agencies, etc. In addition, flood-frequency estimates determined using Bulletin 17B guidelines are available from the various WRD District Offices of the USGS. These analyses generally are based on weighted skew using the skew map in Bulletin 17B and the available information in the USGS Peak Flow File. If alternate analyses are required using different data/assumptions, then these analyses must be made at the user's expense. As an example of the demand for this type of information, the USGS office in Jackson, Mississippi, answers about 100 requests per year for flood-frequency estimates from engineers and planners. The requests come from both public and private agencies for purposes ranging from highway and culvert design to flood-plain management planning. The most up-to-date estimates of flood frequency at gaging stations can be obtained by contacting the local USGS office. A Water Resources Division Information Guide (U.S. Geological Survey, 1984) gives a listing of addresses, telephone numbers, and office hours for the USGS offices. However, the desired flood-frequency estimates can also be obtained from (1) published reports of the USGS describing flood-peak estimation techniques for ungaged sites (discussed later in this paper) and (2) the USGS Streamflow and Basin Characteristics File described by Dempster (1983).

ESTIMATING FLOOD-PEAK
DISCHARGES AT UNGAGED SITES

A major effort of the USGS related to flood frequency is the development of techniques for estimating flood discharges (such as the 1-percent chance flood) for both rural and urban ungaged sites. These estimating techniques are generally multiple-regression equations relating a flood characteristic, such as the 1-percent chance flood, to watershed and climatic characteristics such as watershed size, slope of the main channel, an index of precipitation, percent of the watershed forested, mean elevation of the watershed, etc. Multiple-regression techniques are described in several textbooks, such as Draper and Smith (1981), so no description is provided herein. Benson (1962, 1964) and Thomas and Benson (1970) describe some of the earlier applications of multiple-regression techniques for estimating floodflow characteristics. Thomas and Benson (1970) and others have shown that the following model form is appropriate in hydrologic regression analysis

$$Q = B_0 \, X_1^{B1} \, X_2^{B2} \; \ldots \; X_k^{Bk} \tag{2}$$

where B_i are regression coefficients determined by the principle of least squares, X_i are watershed and climatic characteristics, and Q is a floodflow characteristic such as the 1-percent chance flood. By taking the logarithms of both sides of equation 2, the model form becomes linear allowing the analyst to use linear regression techniques.

Since 1973 every USGS District Office has published a report for their State for estimating flood-peak discharges for ungaged sites. Appendices 1-3 are listings by State of the most recent reports (December 1986). These reports are constantly being updated, so the lists in these tables will naturally change with time. For instance, regional flood-frequency studies are underway in Illinois, Alaska, Minnesota, Kentucky, Kansas, New Jersey, and undoubtedly other States. Therefore, the reports for those States will be updated in the near future. General guidelines are that these reports should be updated every 5 to 10 years. Copies of reports in Appendices 1-3 can be obtained by writing to U.S. Geological Survey, Books and Open-File Reports, Federal Center, Building 41, Box 25425, Denver, Colorado 80225.

For those reports in Appendix 1, the flood-peak discharges are estimated on the basis of watershed and climatic characteristics. The most frequent watershed and/or climatic characteristics utilized are watershed size, an index of watershed topography such as slope of the main channel, mean basin slope, or mean watershed elevation, and an index of precipitation such as the mean annual precipitation or the 2-year, 24-hour precipitation intensity.

Most analysts divided a given State into hydrologic regions to improve the accuracy of the estimating technique. In addition, many USGS District Offices have developed multiple-regression equations utilizing channel-geometry characteristics, such as active channel width, to estimate flood characteristics. In many arid or semiarid areas, flood characteristics cannot be estimated accurately using traditional watershed characteristics such as watershed size and channel slope therefore the channel-geometry approach is a suitable alternative. Riggs (1978), Wahl (1977), and Hedman and Osterkamp (1982) discuss the application and accuracy of the channel-geometry technique and define the channel characteristics needed for application at ungaged sites. The method does usually require a field visit to the site. The reports utilizing channel-geometry techniques are listed in Appendix 2.

Appendix 3 is a listing of reports for estimating flood-peak discharges for urban watersheds. These flood-peak discharges are estimated on the basis of watershed and climatic characteristics with at least one characteristic indicative of the urbanization effect. The percentage of impervious cover in the watershed is the most frequently used urbanization factor, but some investigators have used bank-full channel conveyance (Liscum and Massey, 1980) and basin development factor (Sauer and others, 1983). Bank-full channel conveyance as defined by Liscum and Massey (1980) is the conveyance of the channel in a reach downstream from the gaging station when the water-surface elevation is equal to that of the lower bank. It is computed using Manning's equation and is indicative of the channel capacity or relative efficiency of the channel. The basin development factor as described by Sauer and others (1983) is an index of the drainage features of the watershed such as the extent of channel improvements, curb and gutter streets, and storm sewers.

For many of the reports shown in Appendices 1-3, the annual peak data base was enhanced by using a rainfall-runoff model (Dawdy et al., 1972). Concurrent rainfall and runoff data were used to calibrate a rainfall-runoff model. Using long-term climatic data at a nearby station, a long series of annual-peak discharges were simulated at the site of interest. The logarithms of the simulated-peak discharges were then fit to a Pearson Type III frequency distribution to define at-site flood characteristics. These flood characteristics were then related to watershed and climatic characteristics as part of the regional analysis. These record extension techniques were used for small rural and urban watersheds (generally less than 50 square kilometers) where the annual-peak record length was too short for frequency analysis (less than 10 years).

Traditionally ordinary least squares regression has been used by the USGS in developing the multiple-regression equations. In this approach all stations are given equal weight regardless of record length and the possible correlation between flood estimates at nearby sites. However, recent research by Stedinger and Tasker (1985, 1986) indicates that generalized least squares regression analysis may be better for regionalization of streamflow characteristics. The generalized least squares approach takes into consideration the variance of flood estimates at a given site (function of record length and at-site variability) and the correlation structure of flood estimates at nearby sites when estimating the parameters of the multiple-regression equation. In addition, the variability about the regression equation is partitioned into model error and sampling error. The model error results from assuming an incomplete model form and this error cannot be reduced by collecting more data. The sampling error includes both time- and space-sampling errors and can be reduced by obtaining longer records at the present stations, possibly install-

ing additional stations or some combination of both. This attribute of the generalized least squares technique makes it useful in evaluating the worth of additional
data collection.

As noted earlier, most analysts divided their State into hydrologic regions to
improve the accuracy of the estimating technique. These regions were generally determined by examining the areal distribution and magnitude of the residuals
(deviations between estimated and gaging station data) of the regression analysis.
Although this procedure does improve the accuracy of the estimating technique, it
is somewhat subjective and sometimes the defined flood boundaries cross watershed
divides making application of the technique difficult. Recently the USGS has begun advocating the use of more objective techniques for regionalizing such as clus-
ter and discriminant analysis as described by Tasker (1982) and Wiltshire and Beran
(1986). In this approach, the hydrologic regions may not be geographical regions
but groupings of stations with similar watershed and climatic characteristics.

ESTIMATING FLOOD-PEAK DISCHARGES AT UNGAGED SITES ON GAGED STREAMS

Oftentimes, flood-peak estimates are needed at ungaged sites on streams having
one or more gaging stations. In these cases, it is desirable to use information from
both the gaging station and that provided by the regional regression equation to obtain a more accurate estimate. There are several ways of combining the station data
with the regional data, and only a few of the more popular techniques are described.
The USGS does not have a recommended way of transferring data from a gaging
station to an ungaged site, so a study comparing frequently used techniques is
worthwhile.

Sauer (1973) suggested the following technique for estimating flood-peak discharges for selected exceedance probabilities for ungaged sites on a gaged stream in
Oklahoma. The drainage area of the ungaged site must be within 50 percent of the
drainage area of the gaging station for the technique to be applicable. Sauer suggests the flood-peak discharge be estimated as

$$Qu = R' (Qru) \qquad (3)$$

where Qu = the final discharge (for a selected exceedance probability) at the
ungaged site,

Qru = the estimated discharge at the ungaged site using the regional regression
equations, and

R' = an adjustment factor defined as

$$R' = R - (D/0.5 \ Ag)(R - 1.00) \qquad (4)$$

where R = the ratio of the weighted estimate at the gaging station to the regional regression estimate at the gaging station,

Ag = drainage area of gaging station, and

D = absolute value of the difference in drainage area between the gaging station and ungaged site.

The weighted estimate at the gaging station is a weighted average of the regional regression estimate and the estimate based on station data. This concept of weighting the gaging station estimate with the regional regression estimate is recommended by the USGS to obtain more accurate estimates at gaging stations. The two estimates are weighted inversely proportional to their variances. Sauer's technique essentially gives a weight of one to the weighted estimate at the gaging station, and this weight decreases to zero when the drainage area of the ungaged site is 50 percent more or less than the gaging station. At this point, the regional regression estimate is given a weight of 1.00. Carpenter (1980), Lee (1985), and others have used Sauer's technique.

Hannum (1976) suggested a different technique for estimating flood discharges at ungaged sites on gaged streams in Kentucky. Hannum's estimate for the ungaged site (for a selected exceedance probability) is

$$Qu = Ks \, (Qru) \tag{5}$$

where Qu and Qru are the same as described earlier and

Ks = an adjustment factor defined as

$$Ks = (Kg - 1)((2 \, Ag - Au)/Ag) + 1 \tag{6a}$$

if ungaged site is downstream from gaging station or

$$Ks = (Kg - 1)((2 \, Au - Ag)/Ag) + 1 \tag{6b}$$

if ungaged site is upstream from gaging station

where Kg = estimate based on gaging station data divided by regional regression estimate at gaging station,

Au = drainage area of ungaged site, and

Ag = drainage area of gaging station.

Hannum's technique is applicable if the drainage area of the ungaged site is within one-half to two times the drainage area of the gaging station. Otherwise, the regional regression equations are used. Webber and Bartlet (1977) used this technique in their study in Ohio. Zembrzuski and Dunn (1979) and Bridges (1982) used the same technique in studies in New York and Florida with a slight modification. They computed Kg using a weighted estimate at the gaging station rather than the gaging station data alone.

Colson and Hudson (1976) suggested yet another technique for use in Mississippi. Their estimate for the ungaged site (for a selected exceedance probability) on the gaged stream is

$$Qu = Qru \ (2 \ D/Ag) + Qut \ (1 - 2 \ D/Ag) \qquad (7)$$

where all terms are previously defined except

$$Qut = Qg \ (Au/Ag)^n \qquad (8)$$

where n is the regression coefficient (exponent) for drainage area in the regional regression equation. Qut is an estimate at the ungaged site made by transferring the gaging station estimate (Qg) on the basis of drainage area. This estimate is weighted with an estimate based on the regional regression equation (Qru) to obtain the final estimate (Qu) at the ungaged site. Simmons and Carpenter (1978), Conger (1980), and Olin (1984) also used this technique in their studies in Delaware, Wisconsin, and Alabama. This technique is also applicable if the drainage area of the ungaged site is within one-half to two times the drainage area of the gaging station.

Finally, Jordan (1984) suggested another technique for use in Kansas. Jordan's estimate for the ungaged site (for a selected exceedance probability) on the gaged stream is

$$Qu = We \ (Qru) + Wg \ (Qg) \qquad (9)$$

where Qru and Qg are previously defined and

$$We = 0.5 - 0.5 \cos \ (4.53 \ln \ (Au/Ag)) \qquad (10)$$

and Wg is then computed as 1 - We. This technique gives a weight of zero to the regional regression estimate (Qru) when Au = Ag and a weight of one to Qru when the Au/Ag exceeds 2.0 or is less than 0.5. The cosine function defined in equation 10 allows a nonlinear interpolation between the gaging station and the limits of the technique. The weighting technique is applicable if the drainage area of the ungaged site is one-half to two times the drainage area of the gaging station.

All of these techniques are similar but obviously not exactly the same. Using hypothetical input, it can be shown that equations 3, 5, 7, and 9 give different estimates. An interesting study would be to apply the four techniques to actual gaging station data and determine which technique is most accurate and if there are significant differences among the four techniques.

COMPARISON OF SELECTED TECHNIQUES FOR ESTIMATING FLOOD-PEAK DISCHARGES

In 1978 a work group of the Hydrology Committee of the U.S. Water Resources Council (WRC) undertook the comparison of nine different techniques for estimating flood-peak discharges for rural ungaged watersheds. This study culminated with a report in 1981 (U.S. Water Resources Council, 1981). Subsequent to this study, Newton and Herrin (1982) further analyzed the data collected by the work group. Selected results will be abstracted from the WRC report to illustrate that the USGS regional regression equations are very accurate and reproducible when compared to six frequently used rainfall-runoff procedures.

In the WRC study, five different people applied the nine techniques at 42 watersheds in the Midwestern United States and 28 watersheds in the Northwestern United States. These techniques were applied as if the watershed were ungaged. In reality, each of the 70 watersheds had at least 20 years of unregulated peak-flow record to compute a flood-frequency curve using Bulletin 17B guidelines. The criteria for comparison were bias, reproducibility, and time to apply the techniques in estimating the flood peak discharges with exceedance probabilities of 0.50, 0.10, and 0.01.

Bias is the systematic variation of the estimated discharges from the estimates based on gaging station data and is defined as

$$Bik = (1/m \sum_{j=1}^{m} (Yijk - Yoi))/Yoi \qquad (11)$$

where Bik is the bias for watershed i and technique k computed across m (m = 5) testers, Yijk is an estimate of selected frequency by tester j on watershed i using procedure k, and Yoi is the estimate of the same frequency based on gaging station data for watershed i. Reproducibility is a measure of how well different testers can reproduce the same results at a site (variance of five estimates about mean of estimate) with the same procedure and is defined as

$$REik = \sum_{j=1}^{m} ((Yijk - \overline{Yik})/Yoi)^2/m - 1 \qquad (12)$$

where REik is the reproducibility for watershed i and technique k computed across the (m=5) testers, Yijk and Yoi are defined above, and Yik is the mean of the m estimates for a selected frequency for watershed i using technique k. Both bias and reproducibility are standardized by the gage estimate (Yoi) in an attempt to remove the effect of watershed size. Finally, the time to apply (TA) the technique was simply the time in hours it took the tester to make the required flood estimates using a given technique. Only a small portion of the WRC test results will be given here. The interested reader should refer to the WRC report (U.S. Water Resources Council, 1981) or Newton and Herrin (1982). Results for the Midwest Region (Illinois, Indiana, Ohio, and Missouri) and the Northwest Region (Washington, Oregon, Idaho, and Montana) will be shown separately. Only the results for the USGS regional regression equations and the following six rainfall-runoff techniques are shown--the Reich technique (Reich, 1968), the rational formula (Schaake and others, 1967), TR 55 chart technique (Soil Conservation Service, 1975), TR 55 graphical technique (Soil Conservation Service, 1975), TR 20 (Soil Conservation Service, 1969), and HEC-1 (U.S. Army Corps of Engineers, 1973). The interested reader should consult the WRC report (U.S. Water Resources Council, 1981) for further details and a description of these techniques. The bias (equation 11), reproducibility (Equation 12), and time to apply values for the two regions for the seven techniques are shown in Table 1.

The reader should be aware that not all techniques were applied for all watersheds (28 in Northwest, 42 in Midwest). Certain techniques (such as the rational formula) are only applicable for certain drainage area sizes and therefore were applied to watersheds within these limits. If the bias (B) values in Table 1 are multiplied by 100, they become the percent differences between the estimated flood discharges and the estimates based on gaging station data (Bulletin 17B guidelines). The results in Table 1 are averaged across the flood estimates for the three exceedance probabilities (0.50, 0.10, 0.01).

Table 1. Summary of bias (B), reproducibility (RE), and time to apply (TA) for the various techniques

Technique	Midwest Region			Northwest Region		
	B	RE	TA (hours)	B	RE	TA (hours)
USGS equations	0.092	0.012	3.09	0.206	0.058	3.80
Reich	1.809	0.367	3.68	*	*	*
Rational formula	1.393	0.254	2.19	3.231	2.091	3.51
TR 55 (charts)	0.201	0.156	2.46	0.672	0.552	3.96
TR 55 (graphical)	1.086	0.504	4.08	1.557	1.591	4.25
TR 20	0.631	0.218	25.16	0.479	0.352	22.26
HEC-1	0.301	0.239	40.76	1.223	0.648	34.32

*Technique not applicable.

The smaller the values in Table 1, the more unbiased, reproducible, and easier to apply the technique is for estimating flood-peak discharges for rural ungaged watersheds. As shown in Table 1, the USGS regression equations have the smallest bias and reproducibility. In general, these bias and reproducibility values are statistically different from the values for the six rainfall-runoff procedures. The time to apply values are all essentially the same with the exception of TR 20 and HEC-1. These two techniques are rather complex computer programs that take a long time to apply for a given watershed. The values for B, RE, and TA in Table 1 illustrate that putting more effort (time) into making the flood-peak estimate does not necessarily improve the accuracy of the estimate. It should be noted, however, that TR 20 and HEC-1 provide estimates of the entire flood hydrograph and not just the flood peak. In some studies, the entire flood hydrograph may be needed.

SUMMARY AND CONCLUSIONS

Techniques utilized by the USGS for estimating the magnitude and frequency of flood-peak discharges for gaging stations and rural and urban ungaged watersheds were described. Generally, these techniques are developed and published on a statewide basis. Listings of these reports by State and information on how to obtain these reports were provided. Also, the accuracy of the USGS regional regression equations was compared to six frequently used rainfall-runoff procedures. This comparison, abstracted from a WRC report (1981), indicated that the USGS regional regression equations are more unbiased and reproducible than the six rainfall-runoff procedures. The USGS regression equations are also easy to apply and therefore are useful in estimating flood peak discharges for ungaged sites.

REFERENCES

Benson, M. A., 1962, Factors influencing the occurrence of floods in a humid region of diverse terrain: U.S. Geological Survey Water-Supply Paper 1580-B, 62 p.

Benson, M. A., 1964, Factors affecting the occurrence of floods in the Southwest: U.S. Geological Survey Water-Supply Paper 1580-D, 72 p.

Bridges, W. C., 1982, Technique for estimating magnitude and frequency of floods on natural-flow streams in Florida: U.S. Geological Survey Water-Resources Investigations 82-4012.

Carpenter, D. H., 1980, Technique for estimating magnitude and frequency of floods in Maryland: U.S. Geological Survey Water-Resources Investigations Open-File Report 80-1016.

Cohn, T. A., and Stedinger, J. R., 1983, The use of historical floods in flood frequency analysis: Invited paper presented at the December 1983 Fall American Geophysical Union Meeting in San Francisco, California.

Colson, B. E., and Hudson, J. W., 1976, Flood frequency of Mississippi streams: Mississippi State Highway Department.

Conger, D. H., 1980, Techniques for estimating magnitude and frequency of floods for Wisconsin streams: U.S. Geological Survey Water-Resources Investigations Open-File Report 80-1214.

Costa, J. E., 1978, Holocene stratigraphy in flood frequency analysis: Water Resources Research, v. 14, no. 4, p. 626-632.

Crippen, J. R., 1978, Composite log-Pearson type III frequency-magnitude curve of annual floods: U.S. Geological Survey Open-File Report 78-352, 4 p.

Dawdy, D. R., Lichty, R. W., and Bergmann, J. M., 1972, A rainfall-runoff simulation model for estimation of flood peaks for small drainage basins: U.S. Geological Survey Professional Paper 506-B, 28 p.

Dempster, G. R., Jr., 1983, Streamflow/basin characteristics retrieval (program E796): U.S. Geological Survey WATSTORE User's Guide, v. 4, chapter II, section B, 31 p.

Draper, N. R., and Smith, H., 1981, Applied regression analysis: New York, New York, John Wiley and Sons, 2d ed., 709 p.

Hannum, C. H., 1976, Technique for estimating magnitude and frequency of floods in Kentucky: U.S. Geological Survey Water-Resources Investigations 76-62 (PB-263 762/AS).

Hedman, E. R., and Osterkamp, W. R., 1982, Streamflow characteristics related to channel geometry of streams in Western United States: U.S. Geological Survey Water-Supply Paper 2193, 17 p.

Hydrology Subcommittee of the Interagency Advisory Committee on Water Data, 1982, Bulletin 17B, Guidelines for determining flood-flow frequency: U.S. Geological Survey, Office of Water Data Coordination, Reston, Virginia.

Jordan, P. R., 1984, Magnitude and frequency of high flows of unregulated streams in Kansas: U.S. Geological Survey Open-File Report 84-453.

Kirby, W. H., 1979, Annual flood frequency analysis using U.S. Water Resources Council guidelines (program J407): U.S. Geological Survey Open-File Report 79-1336-I, WATSTORE User's Guide, v. 4, chapter I, section C, 56 p.

Lee, F. N., 1985, Floods in Louisiana, Magnitude and frequency, Fourth edition: Department of Transportation and Development, Water Resources Technical Report No. 36.

Lepkin, W. D., and DeLapp, M. M., 1979, Peak flow file retrieval (program J980): U.S. Geological Survey Open-File Report 79-1336-I, WATSTORE User's Guide, v. 4, chapter I, section B, 64 p.

Liscum, F., and Massey, B. C., 1980, Technique for estimating the magnitude and frequency of floods in the Houston, Texas, metropolitan area: U.S. Geological Survey Water-Resources Investigations 80-17.

Newton, D. W., and Herrin, J. C., 1982, Assessment of commonly used flood frequency methods: Transportation Research Record 896, Transportation Research Board, Washington, D.C.

Olin, D. A., 1984, Magnitude and frequency of floods in Alabama: U.S. Geological Survey Water-Resources Investigations Report 84-4191.

Reich, B. M., 1968, Rapid flood-peak determination on small watersheds: American Society of Agricultural Engineers Transactions, 11(2), p. 291-295.

Riggs, H. C., 1978, Streamflow characteristics from channel size: American Society of Civil Engineers, Journal of Hydraulics Division, v. 104, no. HY1, p. 87-96.

Sauer, V. B., 1973, Flood characteristics of Oklahoma streams: U.S. Geological Survey Water-Resources Investigations 52-73.

Sauer, V. B., Thomas, W. O., Jr., Stricker, V. A., and Wilson, K. V., 1983, Flood characteristics of urban watersheds in the United States: U.S. Geological Survey Water-Supply Paper 2207.

Schaake, J. C., Jr., Geyer, J. C., and Knapp, A. W., 1967, Experimental examination of the rational method: American Society of Civil Engineers Proceedings, Journal of the Hydraulics Division, v. 93, no. HY6, p. 353-70.

Simmons, R. H., and Carpenter, D. H., 1978, Technique for estimating the magnitude and frequency of floods in Delaware: U.S. Geological Survey Water-Resources Investigations Open-File Report 78-93.

Soil Conservation Service, 1969, Computer program for project formulation-hydrology: Technical Release No. 20, Supplement No. 1, Central Technical Unit, Washington, D.C.

Soil Conservation Service, 1975, Urban hydrology for small watersheds: Technical Release No. 55, Washington, D.C.

Stedinger, J. R., and Tasker, G. D., 1985, Regional hydrologic analysis -- ordinary, weighted and generalized least squares compared: Water Resources Research, v. 21, no. 9, p. 1421-1432.

Stedinger, J. R., and Tasker, G. D., 1986, Regional hydrologic analysis, 2, model-error estimators, estimation of sigma and log-Pearson type 3 distributions: Water Resources Research, v. 22, no. 10, p. 1487-1499.

Tasker, G. D., 1982, Comparing methods of hydrologic regionalization: Water Resources Bulletin, v. 18, no. 6, p. 965-970.

Thomas, D. M., and Benson, M. A., 1970, Generalization of streamflow characteristics from drainage-basin characteristics: U.S. Geological Survey Water-Supply Paper 1975, 55 p.

Thomas, W. O., Jr., 1985, A uniform technique for flood frequency analysis: American Society of Civil Engineers, Journal of Water Resources Planning and Management, v. 111, no. 3, July 1985, p. 321-337.

U.S. Army Corps of Engineers, 1973, HEC-1, flood-hydrograph package user's manual: Hydrologic Engineering Center, Davis, California.

U.S. Geological Survey, 1984, Water Resources Division (WRD) Information Guide: November 1984, 15 p.

U.S. Water Resources Council, 1981, Estimating peak flow frequencies for natural ungaged watersheds--A proposed nationwide test: U.S. Water Resources Council, Washington, D.C.

Wahl, K. L., 1977, Accuracy of channel measurements and the implications in estimating streamflow characteristics: U.S. Geological Survey Journal of Research, v. 5, no. 6, p. 811-814.

Webber, E. E., and Bartlett, W. P., Jr., 1977, Floods in Ohio--Magnitude and frequency: State of Ohio, Department of Natural Resources, Division of Water, Bulletin 45.

Wiltshire, S., and Beran, M., 1986, Multivariate techniques for the identification of homogeneous flood frequency regions: Paper presented at the International Symposium on Flood Frequency and Risk Analysis, Baton Rouge, Louisiana, May 1986.

Zembrzuski, T. J., and Dunn, B., 1979, Techniques for estimating magnitude and frequency of floods on rural unregulated streams in New York, excluding Long Island: U.S. Geological Survey Water-Resources Investigations 79-83 (PB-80 201 148).

Appendix 1. List of reports for estimating rural flood-peak discharges using watershed and climatic characteristics

Alabama:
Olin, D. A., 1984, Magnitude and frequency of floods in Alabama: U.S. Geological Survey Water-Resources Investigations 84-4191.

Alaska:
Lamke, R. D., 1978, Flood characteristics of Alaskan streams: U.S. Geological Survey Water-Resources Investigations 78-129.

Arizona:
Eychaner, J. H., 1984, Estimation of magnitude and frequency of floods in Pima County, Arizona, with comparisons of alternative methods: U.S. Geological Survey Water-Resources Investigations 84-4142.
Roeske, R. H., 1978, Methods for estimating the magnitude and frequency of floods in Arizona: U.S. Geological Survey Open-File Report 78-711.

Arkansas:
Neely, B. L., Jr., 1986, Magnitude and frequency of floods in Arkansas: U.S. Geological Survey Water-Resources Investigations Report 86-4335.

California:
Waananen, A. O., and Crippen, J. R., 1977, Magnitude and frequency of floods in California: U.S. Geological Survey Water-Resources Investigations 77-21 (PB-272 510/AS).

Colorado:
Kircher, J. E., Choquette, A. F., and Richter, B. D., 1985, Estimation of natural streamflow characteristics in Western Colorado: U.S. Geological Survey Water-Resources Investigations Report 85-4086.
Livingston, R. K., 1980, Rainfall-runoff modeling and preliminary regional flood characteristics of small rural watersheds in the Arkansas River Basin in Colorado: U.S. Geological Survey Water-Resources Investigations 80-112 (NTIS).
McCain, J. R., and Jarrett, R. D., 1976, Manual for estimating flood characteristics of natural-flow streams in Colorado: Colorado Water Conservation Board, Technical Manual No. 1.

Connecticut:
Weiss, L. A., 1975, Flood flow formula for urbanized and non-urbanized areas of Connecticut: Watershed Management Symposium of ASCE Irrigation and Drainage Division, p. 658-675, August 11-13, 1975.

Delaware:
Simmons, R. H., and Carpenter, D. H., 1978, Technique for estimating the magnitude and frequency of floods in Delaware: U.S. Geological Survey Water-Resources Investigations Open-File Report 78-93.

Florida:
Bridges, W. C., 1982, Technique for estimating magnitude and frequency of floods on natural-flow streams in Florida: U.S. Geological Survey Water-Resources Investigations 82-4012.

Georgia:

Price, M., 1979, Floods in Georgia, magnitude and frequency: U.S. Geological Survey Water-Resources Investigvations 78-137 (PB-80 146 244).

Hawaii:

Matsuoka, I., 1978, Flow characteristics of streams in Tutuila, American Samoa: U.S. Geological Survey Open-File Report 78-103.

Nakahara, R. H., 1980, An analysis of the magnitude and frequency of floods on Oahu, Hawaii: U.S. Geological Survey Water-Resources Investigations 80-45 (PB-81 109 902).

Idaho:

Kjelstrom, L. C., and Moffatt, R. L., 1981, Method of estimating flood frequency parameters for streams in Idaho: U.S. Geological Survey Open-File Report 81-909.

Thomas, C. A., Harenburg, W. A., and Anderson, J. M., 1973, Magnitude and frequency of floods in small drainage basins in Idaho: U.S. Geological Survey Water-Resources Investigations 7-73 (PB-222 409).

Illinois:

Curtis, G. W., 1977, Technique for estimating magnitude and frequency of floods in Illinois: U.S. Geological Survey Water-Resources Investigations 77-117 (PB-277 255/AS).

Indiana:

Glatfelter, D. R., 1984, Techniques for estimating magnitude and frequency of floods in Indiana: U.S. Geological Survey Water- Resources Investigations 84-4134.

Iowa:

Lara, O., 1973, Floods in Iowa: Techniques manual for estimating their magnitude and frequency: State of Iowa Natural Resources Council Bulletin no. 11.

Kansas:

Jordan, P. R., and Irza, T. J., 1975, Magnitude and frequency of floods in Kansas, unregulated streams: Kansas Water Resources Board Technical Report no. 11.

Kentucky:

Hannum, C. H., 1976, Technique for estimating magnitude and frequency of floods in Kentucky: U.S. Geological Survey Water-Resources Investigations 76-62 (PB-263 762/AS).

Louisiana:

Lee, F. N., 1985, Floods in Louisiana, Magnitude and frequency, Fourth edition: Department of Transportation and Development, Water Resources Technical Report No. 36.

Lowe, A. S., 1979, Magnitude and frequency of floods for small watersheds in Louisiana: Louisiana Department of Transportation and Development, Office of Highways, Research Study No. 65-2H.

Maine:

Morrill, R. A., 1975, A technique for estimating the magnitude and frequency of floods in Maine: U.S. Geological Survey open-file report.

Maryland:

Carpenter, D. H., 1980, Technique for estimating magnitude and frequency of floods in Maryland: U.S. Geological Survey Water-Resources Investi- gations Open-File Report 80-1016.

Massachusetts:
Wandle, S. W., 1983, Estimating peak discharges of small rural streams in Massa-
    chusetts: U.S. Geological Survey Water-Supply Paper 2214.
Michigan:
Holtschlag, D. J., and Croskey, H. M., 1984, Statistical models for estimating flow
    characteristics of Michigan streams: U.S. Geological Survey Water Resources
    Investigations 84-4207.
Minnesota:
Guetzkow, L. C., 1977, Techniques for estimating magnitude and frequency of floods
    in Minnesota: U.S. Geological Survey Water-Resources Investigations 77-31
    (PB-272 509/AS).
Mississippi:
Colson, B. E., and Hudson, J. W., 1976, Flood frequency of Mississippi streams:
    Mississippi State Highway Department.

Missouri:
Hauth, L. D., 1974, A technique for estimating the magnitude and frequency of Mis-
    souri floods: U.S. Geological Survey Open-file report.
Montana:
Omang, R. J., Parrett, C., and Hull, J. A., 1986, Methods of estimating magnitude
    and frequency of floods in Montana based on data through 1983: U.S. Geolog-
    ical Survey Water-Resources Investigations Report 86-4027.
Nebraska:
Beckman, E. W., 1976, Magnitude and frequency of floods in Nebraska: U.S. Geo-
    logical Survey Water-Resources Investigations 76-109 (PB-260 842/AS).
Nevada:
Moore, D. O., 1976, Estimating peak discharges from small drainages in Nevada ac-
    cording to basin areas within elevation zones: Nevada State Highway Depart-
    ment Hydrologic Report no. 3.
New Hampshire:
LeBlanc, D. R., 1978, Progress report on hydrologic investigations of small drain-
    age areas in New Hampshire--Preliminary relations for estimating peak dis-
    charges on rural, unregulated streams: U.S. Geological Survey Water-
    Resources Investigations 78-47 (PB-284 127/AS).
New Jersey:
Stankowski, S. J., 1974, Magnitude and frequency of floods in New Jersey with ef-
    fects of urbanization: New Jersey Department of Environmental Protection
    Special Report 38.
New Mexico:
Hejl, H. R., Jr., 1984, Use of selected basin characteristics to estimate mean annual
    runoff and peak discharges for ungaged streams in drainage basins containing
    strippable coal resources, northwestern New Mexico: U.S. Geological Survey
    Water-Resources Investigations Report 84-4260.
Waltmeyer, S. D., 1986, Techniques for estimating flood-flow frequency for unregu-
    lated streams in New Mexico: U.S. Geological Survey Water- Resources In-
    vestigations Report 86-4104.
New York:
Zembrzuski, T. J., and Dunn, B., 1979, Techniques for estimating magnitude and fre-

quency of floods on rural unregulated streams in New York, excluding Long Island: U.S. Geological Survey Water-Resources Investigations 79-83 (PB-80 201 148).

North Carolina:

Jackson, N. M., Jr., 1976, Magnitude and frequency of floods in North Carolina: U.S. Geological Survey Water-Resources Investigations 76-17 (PB-254 411/ AS).

North Dakota:

Crosby, O. A., 1975, Magnitude and frequency of floods in small drainage basins in North Dakota: U.S. Geological Survey Water-Resources Investigations 19-75 (PB-248 480/AS).

Ohio:

Webber, E. E., and Bartlett, W. P., Jr., 1977, Floods in Ohio magnitude and frequency: State of Ohio, Department of Natural Resources, Division of Water, Bulletin 45.

Oklahoma:

Tortorelli, R. L., and Bergman, D. L., 1984, Techniques for estimating flood peak discharge for unregulated streams and streams regulated by small floodwater retarding structures in Oklahoma: U.S. Geological Survey Water-Resources Investigations 84-4358.

Oregon:

Harris, D. D., and Hubbard, L. E., 1982, Magnitude and frequency of floods in eastern Oregon: U.S. Geological Survey Water-Resources Investigations 82-4078.

Harris, D. D., Hubbard, L. L., and Hubbard, L. E., 1979, Magnitude and frequency of floods in western Oregon: U.S. Geological Survey Open-File Report 79-553.

Pennsylvania:

Flippo, H. N., Jr., 1977, Floods in Pennsylvania: A manual for estimation of their magnitude and frequency: Pennsylvania Department of Environmental Resources Bulletin no. 13.

Puerto Rico:

Lopez, M. A., Colon-Dieppa, E., and Cobb, E. D., 1978, Floods in Puerto Rico: magnitude and frequency: U.S. Geological Survey Water-Resources Investigations 78-141 (PB-300 855/AS).

Rhode Island:

Johnson, C. G., and Laraway, G. A., 1976, Flood magnitude and frequency of small Rhode Island streams--Preliminary estimating relations: U.S. Geological Survey Open-file report.

South Carolina:

Whetstone, B. H., 1982, Floods in South Carolina--Techniques for estimating magnitude and frequency of floods with compilation of flood data: U.S. Geological Survey Water-Resources Investigations 82-1.

South Dakota:

Becker, L.D., 1974, A method for estimating the magnitude and frequency of floods in South Dakota: U.S. Geological Survey Water-Resources Investigations 35-74 (PB-239 831/AS).

Becker, L.D., 1980, Techniques for estimating flood peaks, volumes, and hydrographs on small streams in South Dakota: U.S. Geological Survey Water-Resources Investigations 80-80 (PB-81 136 145).

Tennessee:
Randolph, W. J., and Gamble, C. R., 1976, A technique for estimating magnitude and frequency of floods in Tennessee: Tennessee Department of Transportation.

Texas:
Schroeder, E. E., and Massey, B. C., 1977, Techniques for estimating the magnitude and frequency of floods in Texas: U.S. Geological Survey Water-Resources Investigations Open-File Report 77-110.

Utah:
Thomas, B. E., and Lindskov, K. L., 1983, Methods for estimating peak discharges and flood boundaries of streams in Utah: U.S. Geological Survey Water-Resources Investigations 83-4129.

Vermont:
Johnson, C. G., and Tasker, G. D., 1974, Flood magnitude and frequency of Vermont streams: U.S. Geological Survey Open-File Report 74-130.

Virginia:
Miller, E. M., 1978, Technique for estimating the magnitude and frequency of floods in Virginia: U.S. Geological Survey Water-Resources Investigations Open-File Report 78-5.

Washington:
Cummans, J. E., Collins, M. R., and Nassar, E. G., 1974, Magnitude and frequency of floods in Washington: U.S. Geological Survey Open-File Report 74-336.

Haushild, W. L., 1978, Estimation of floods of various frequencies for the small ephemeral streams in eastern Washington: U.S. Geological Survey Water-Resources Investigations 79-81.

West Virginia:
Runner, G. S., 1980, Technique for estimating magnitude and frequency of floods in West Virginia: U.S. Geological Survey Open-File Report 80-1218.

Wisconsin:
Conger, D. H., 1980, Techniques for estimating magnitude and frequency of floods for Wisconsin streams: U.S. Geological Survey Water-Resources Investigations Open-File Report 80-1214.

Wyoming:
Craig, G. S., Jr., and Rankl, J. G., 1977, Analysis of runoff from small drainage basins in Wyoming: U.S. Geological Survey Water-Supply Paper 2056.

Lowham, H. W., 1976, Techniques for estimating flow characteristics of Wyoming streams: U.S. Geological Survey Water-Resources Investigations 76-112 (PB-264 224/AS).

Regional Reports:
Landers, M. N., 1985, Floodflow frequency of streams in the alluvial plain of the Lower Mississippi River in Mississippi, Arkansas and Louisiana: U.S. Geological Survey Water-Resources Investigations 85-4150.

Wetzel, K. L., and Bettandorff, J. M., 1986, Techniques for estimating streamflow characteristics in the Eastern and Interior Coal Provinces of the United States: U.S. Geological Survey Water-Supply Paper 2226.

Appendix 2. List of reports for estimating rural flood-peak discharge using channel-geometry characteristics

Colorado:

Hedman, E. R., Moore, D. O., and Livingston, R. K., 1972, Selected streamflow characteristics as related to channel geometry of perennial streams in Colorado: U.S. Geological Survey open-file report.

Idaho:

Harenberg, W. A., 1980, Using channel geometry to estimate flood flows at ungaged sites in Idaho: U.S. Geological Survey Water-Resources Investigations 80-32 (PB-81 153 736).

Kansas:

Hedman, E. R., Kastner, W. M., and Hejl, H. R., 1974, Selected streamflow characteristics as related to active-channel geometry of streams in Kansas: Kansas Water Resources Board Technical Report no. 10.

Montana:

Omang, R. J., 1983, Mean annual runoff and peak flow estimates based on channel geometry of streams in southeastern Montana: U.S. Geological Survey Water-Resources Investigations Report 82-4092.

Parrett, C., 1983, Mean annual runoff and peak flow estimates based on channel geometry of streams in northeastern and western Montana: U.S. Geological Survey Water-Resources Investigations Report 83-4046.

Nevada:

Moore, D. O., 1974, Estimating flood discharges in Nevada using channel-geometry measurements: Nevada State Highway Department Hydrologic Report no. 1.

New Mexico:

Scott, A. G., and Kunkler, J. L., 1976, Flood discharges of streams in New Mexico as related to channel geometry: U.S. Geological Survey open-file report.

Ohio:

Roth, D. K., 1985, Estimation of flood peaks from channel characteristics in Ohio: U.S. Geological Survey Water-Resources Investigations Report 85-4175.

Webber, E. E., and Roberts, J. W., 1981, Floodflow characteristics related to channel geometry in Ohio: U.S. Geological Survey Open-File Report 81-1105.

Utah:

Fields, F. K., 1974, Estimating streamflow characteristics for streams in Utah using selected channel-geometry parameters: U.S. Geological Survey Water-Resources Investigations 34-74 (PB-241 541/AS).

Wyoming:

Lowham, H. W., 1976, Techniques for estimating flow characteristics of Wyoming streams: U.S. Geological Survey Water-Resources Investigations 76-112 (PB-264 224/AS).

Regional Reports:

Hedman, E. R., and Kastner, W. M., 1977, Streamflow characteristics related to channel geometry in the Missouri River Basin: U.S. Geological Survey Journal of Research, v. 5, no. 3, p. 285-300.

Hedman, E. R., and Osterkamp, W. R., 1982, Streamflow characteristics related to channel geometry of streams in Western United States: U.S. Geological Survey Water-Supply Paper 2193.

Appendix 3. List of reports for estimating urban flood-peak discharges

Alabama:

Olin, D. A., and Bingham, R. H., 1982, Synthesized flood frequency of urban streams in Alabama: U.S. Geological Survey Water-Resources Investigations 82-683.

California:

Waananen, A. O., and Crippen, J. R., 1977, Magnitude and frequency of floods in California: U.S. Geological Survey Water-Resources Investigations 77-21 (PB-272 510/AS).

Connecticut:

Weiss, L. A., 1975, Flood flow formula for urbanized and non-urbanized areas of Connecticut: Watershed Management Symposium of ASCE Irrigation and Drainage Division, p. 658-675, August 11-13, 1975.

Florida:

Franklin, M. A., 1984, Magnitude and frequency of floods from urban streams in Leon County, Florida: U.S. Geological Survey Water-Resources Investigations 84-4004.

Lopez, M. A., and Woodham, W. M., 1982, Magnitude and frequency of flooding on small urban watersheds in the Tampa Bay area, west-central Florida: U.S. Geological Survey Water-Resources Investigations 82-42.

Georgia:

Inman, E. J., 1983, Flood-frequency relations for urban streams in metropolitan Atlanta, Georgia: U.S. Geological Survey Water-Resources Investigations 83-4203.

Illinois:

Allen, H. E., Jr., and Bejcek, R. M., 1979, Effects of urbanization on the magnitude and frequency of floods in northeastern Illinois: U.S. Geological Survey Water-Resources Investigations 79-36 (PB-299 065/AS).

Iowa:

Lara, O., 1978, Effects of urban development on the flood-flow character- istics of Walnut Creek basin, Des Moines metropolitan area, Iowa: U.S. Geological Survey Water-Resources Investigations 78-11 (PB-284 093/AS).

Kansas:

Peek, C. O., and Jordan, P. R., 1978, Determination of peak discharge from rainfall relations for urbanized basins, Wichita, Kansas: U.S. Geological Survey Open-File Report 78-974.

Missouri:

Becker, L. D., 1986, Techniques for estimating flood-peak discharges for urban streams in Missouri: U.S. Geological Survey Water-Resources Investigations Report 86-4322.

Spencer, D. W., and Alexander, T. W., 1978, Technique for estimating the magnitude and frequency of floods in St. Louis County, Missouri: U.S. Geological Survey Water-Resources Investigations 78-139 (PB-298 245/AS).

New Jersey:

Stankowski, S. J., 1974, Magnitude and frequency of floods in New Jersey with effects of urbanization: New Jersey Department of Environmental Protection Special Report 38.

North Carolina:

Martens, L. A., 1968, Flood inundation and effects of urbanization in metropolitan Charlotte, North Carolina: U.S. Geological Survey Water-Supply Paper 1591-C.

Putnam, A. L., 1972, Effect of urban development on floods in the Piedmont province of North Carolina: U.S. Geological Survey Open-file report.

Ohio:

Sherwood, J. M., 1986, Estimating peak discharges, flood volumes, and hydrograph stages of small urban streams in Ohio: U.S. Geological Survey Water-Resources Investigations Report 86-4197.

Oklahoma:

Sauer, V. B., 1974, An approach to estimating flood frequency for urban areas in Oklahoma: U.S. Geological Survey Water-Resources Investigations 23-74 (PB-235 307/AS).

Oregon:

Laenen, Antonius, 1980, Storm runoff as related to urbanization in the Portland, Oregon-Vancouver, Washington, area: U.S. Geological Survey Water-Resources Investigations Open-File Report 80-689.

Tennessee:

Neely, B. L., Jr., 1984, Flood frequency and storm runoff of urban areas of Memphis and Shelby County, Tennessee: U.S. Geological Survey Water-Resources Investigations 84-4110.

Robbins, C. H., 1984, Synthesized flood frequency for small urban streams in Tennessee: U.S. Geological Survey Water-Resources Investigations 84-4182.

Wibben, H. C., 1976, Effects of urbanization on flood characteristics in Nashville-Davidson County, Tennessee: U.S. Geological Survey Water-Resources Investigations 76-121 (PB-266 654/AS).

Texas:

Land, L. F., Schroeder, E. E., and Hampton, B. B., 1982, Techniques for estimating the magnitude and frequency of floods in the Dallas-Fort Worth Metropolitan Area, Texas: U.S. Geological Survey Water-Resources Investigations 82-18.

Liscum, F., and Massey, B. C., 1980, Technique for estimating the magnitude and frequency of floods in the Houston, Texas, metropolitan area: U.S. Geological Survey Water-Resources Investigations 80-17 (ADA-089 495).

Veenhuis, J. E., and Garrett, D. G., 1986, The effects of urbanization on floods in the Austin metropolitan area, Texas: U.S. Geological Survey Water-Resources Investigations Report 86-4069.

Virginia:

Anderson, D. G., 1970, Effects of urban development on floods in Northern Virginia: U.S. Geological Survey Water-Supply Paper 2001-C.

Wisconsin:

Conger, D. H., 1986, Estimating magnitude and frequency of floods for ungaged urban streams in Wisconsin: U.S. Geological Survey Water- Resources Investigations Report 86-4005.

Regional Reports:

Sauer, V. B., Thomas, W. O., Jr., Stricker, V. A., and Wilson, K. V., 1983, Flood characteristics of urban watersheds in the United States: U.S. Geological Survey Water-Supply Paper 2207.

14

Planetary analogs for geomorphic features produced by catastrophic flooding

Lisa A. Rossbacher and Dallas D. Rhodes

ABSTRACT

Terrestrial landforms created by catastrophic flooding can be used to interpret features on other planetary surfaces that may have been created by similar processes. The use of terrestrial analogs has limitations; for example, similar morphologies do not necessarily imply similar origins. The surfaces of planetary bodies have differing physical conditions, such as atmospheric density or gravitational acceleration, which also complicate the use of analogs. Use of analogs in geomorphology offers perspective on variables such as gravity that are typically considered fixed in terrestrial studies. Terrestrial analogs for Martian outflow channels have always emphasized floods associated with natural dam collapses, but the failure of artificial dams, such as the St. Francis Dam in California, also results in processes and landforms analogous to the Martian ones. The planetary perspective prepares us to consider the possibility of fluvial processes elsewhere in the Solar System, including liquid methane on Saturn's moon Titan and flowing nitrogen on Triton, Neptune's largest moon.

INTRODUCTION

Reasoning by analogy for Earth and Mars

Although Earth is the only planet in the Solar System where liquid water can exist under current temperatures and pressures, it is not the only planetary surface to exhibit landforms that appear to have been created by running water. The surface of Mars has a variety of channel forms, including some that are larger than any on Earth. The best terrestrial analogs for many of these Martian channels are features that were formed by catastrophic floods on Earth (Milton, 1973; Baker and Milton, 1974; Baker, 1982). The general consensus among planetary geomorphologists is that the large outflow channels on Mars were carved by great volumes of flowing water, although this interpretation is not unanimous. Many of the arguments for a flooding origin are based on the similarity of Martian channels to presumably analogous terrestrial features.

Geomorphic interpretation using analogies is a valuable tool, but too often it is the *only* tool that is available. More than 40,000 photographs of the Martian surface are available from NASA's Viking mission, and interpretation of these remotely sensed data is the primary approach available to study the surface of Mars, until manned exploration or more sophisticated unmanned missions are possible.

Using analogs to study planetary surfaces has some shortcomings. Mutch (1979) noted several of these; by using analogs, a researcher assumes that the landform reflects the process that created it, an assumption that is increasingly complicated by our growing understanding of polygenetic landforms. Interpretations and analogies are also necessarily limited by the experience of the individual doing the work. On Earth, geomorphic processes are also influenced by vegetation, soil type, and climatic conditions. These and other restrictions on the use of analogs in studying the surfaces of other planets have been discussed by Mutch (1979), Baker (1982), Schumm (1985), and Rhodes and Rossbacher (1985).

An additional assumption in the use of analogs in planetary geology is that the analogous geomorphic process is understood on Earth. In a sense, this approach to planetary geology is like applying the principle of uniformitarianism in historical geology; rather than assuming that the present is the key to the past, analogy in planetary science assumes that the Earth is the key to the Solar System. To apply this theory effectively, we must also assume that we understand the processes operating on Earth. In many situations, including a variety of periglacial processes, this assumption is not valid (Rhodes and Rossbacher, 1985). As a result, planetary studies have offered valuable feedback for terrestrial geology by pointing out aspects of our own planet that need more study.

All efforts to apply our understanding of terrestrial processes to other planetary surfaces must consider the differing physical conditions on other planets. Martian environmental conditions that are most significant for geomorphic processes are listed in Table 1. A better understanding of the role of such planetary conditions is one of the most valuable contributions of planetary geology to terrestrial studies. For example, a term for gravitational acceleration is hidden in most equations used in traditional geomorphology. Applying an understanding of terrestrial processes to another planet requires an appreciation of that gravity term. This has been done for a number of relationships dealing with fluid flow (Carr, 1979; Komar, 1979, 1980; Baker, 1982). The planetary perspective has already made important contributions to a better understanding of terrestrial geology (Sharp, 1980).

For large-scale flooding, the lower gravitational acceleration on Mars has both direct and indirect effects on sediment transport (Table 2). For example, the mean flow velocity required to move a particular sediment size is lower on Mars than on

Table 1. Comparison of physical conditions affecting geomorphic processes on Earth and Mars.

Parameter	Earth	Mars
Gravitational acceleration	978 cms^{-2}	371 cms^{-2}
Atmospheric density	1.29 kgm^{-3}	1.83 x 10-2 kgm^{-3}
Average wind speed (range)	0-60 ms^{-1}	0-150 ms^{-1}
Solar constant	1.94 cal cm^{-2}min^{-1}	0.86 cal cm^{-2}min^{-1}

Earth, but the sediments also weigh less, and so they can be transported more easily (Baker, 1982). The lower gravity also results in lower settling velocities; gravel and cobbles could move in suspension on Mars (Komar, 1980), and boulders measuring several meters across could be transported as bedload (Baker and Ritter, 1975). As shown in Table 2, most geomorphic parameters that affect sediment transport are 62% lower on Mars than they are on Earth, if everything but the gravity term is held constant.

Elsewhere in the Solar System

Although Earth and Mars are the only terrestrial planets that are known to exhibit fluvial landforms, Venus might also have experienced fluvial activity in its past. Studies of the deuterium:hydrogen ratio indicate that Venus may have had oceans of water, although only a trace is now measurable in the atmosphere (Donahue et al., 1982). The planned U.S. Magellan mission to Venus will use radar to look for geomorphic evidence for liquid water on that planet.

Titan, the largest moon of Saturn, is hidden by a nitrogen-rich atmosphere that is about 1.6 times as dense as Earth's. The Voyager spacecraft observations of the Saturnian system also revealed a variety of carbon-nitrogen compounds, including ethane, acetylene, ethylene, and hydrogen cyanide (Greeley, 1985). Titan's atmosphere was probably released from its interior during early differentiation, and photochemical reactions may have produced large quantities of methane. The temperature, around -180°C, is close to the triple point of methane and ethane, and some researchers have even speculated about the existence of a 1-km deep methane ocean, with islands of solid water ice rising above the waves (Sagan and Dermott, 1982). The next chance to explore the geomorphology of Titan is likely to be a NASA radar mission that is tentatively planned for the 1990's.

Table 2. Effect of different gravitional constant on sediment transport relations for Mars relative to Earth assuming all other factors constant. Martian differences are either 62% greater or 62% smaller than values on Earth.

Sediment Transport Parameter	Mars relative to Earth
Froude number	greater
Darcy-Weisbach friction factor	smaller
Chezy coefficient	smaller
Chezy velocity	samller
Shear velocity	smaller
Shear stress	smaller
Stream power	smaller
Stokes Law (for spheres)	smaller
Impact settling	smaller
Potential energy	smaller

The other candidate for possible catastrophic flooding in the outer Solar System is the larger moon of Neptune, Triton. With a diameter of 3500 km, this satellite would be a useful addition to the group of planetary bodies that exhibit fluvial activity. Based on spectroscopic evidence, Triton appears to have a tenuous atmosphere, and it may have seas of liquid nitrogen and land masses of solid methane (Cruikshank and Silvaggio, 1979; Cruikshank et al., 1983). If all goes well with the Voyager 2 spacecraft, our first close look at this smoggy body will come in August 1989.

The possibilities for liquid methane and nitrogen in the outer Solar System offer new perspectives for the geomorphic role of flowing liquids. Thus far, geomorphologists have been restricted to the processes accompanying water. The similarities and differences between water-carved landforms and those eroded by other liquids will be a dramatic expansion of our understanding of fluvial processes in the Solar System.

MARTIAN LANDFORMS AND THEIR TERRESTRIAL ANALOGS

Origin of the Martian channels

With the availability of Mariner 9 images of the Martian surface in the early 1970's, a wide range of possible origins were suggested for the channel forms. In addition to liquid water (McCauley et al., 1972; Masursky, 1973; Milton, 1973), other volatiles were proposed, including carbon dioxide hydrates (Milton, 1974) and liquid hydrocarbons (Yung and Pinto, 1978). Other processes of formation that have been suggested are lava flow (Carr, 1974; Schonfeld, 1977), faulting

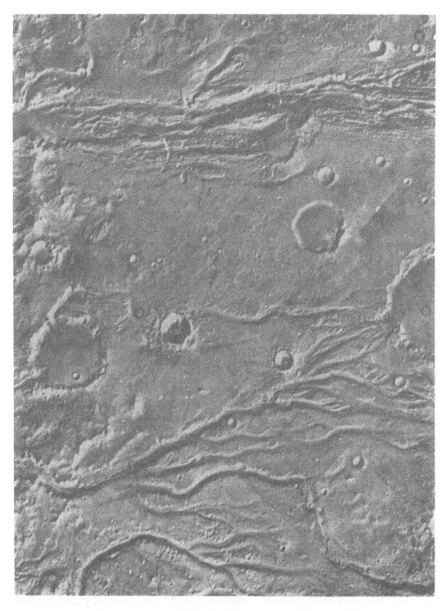

Figure 1. Runoff channels on Mars. The area shown is about 300 km across and is centered at 17° N, 55° W. Craters shown here are both older and younger than the flow. (Viking Orbiter 1 image P- 17698, frames 44A50-47A58, courtesy NASA/JPL)

Figure 2. Outflow channels on Mars. This area, near one of the potential Viking lander sites, is about 300 km across, with east toward the bottom of the picture. The 20 km wide channel emerges fully formed from a depression containing chaotic terrain; this type of feature may be formed by sudden release of subsurface groundwater or melted ground ice (Viking Orbiter 1 image P-16983, frames 14A67-14A69).

(Schumm, 1974), viscous mud flow (Nummedal, 1978), glacier movement (Lucchitta, Anderson, and Shoji, 1981), and wind (Cutts and Blasius, 1981). The most persuasive arguments support an origin by flowing water; these arguments are outlined in detail by Baker (1982).

Martian channels are generally grouped into runoff channels, outflow channels, and fretted channels (Sharp and Malin, 1975). The runoff channels are systems of narrow, sinuous valleys, usually less than 20 km wide, with a network of tributaries (Fig. 1). Despite their resemblance to terrestrial drainage networks, these valley networks are generally parallel rather than dendritic, and they probably formed by sapping and removal of subsurface fluid rather than by surface runoff (Pieri, 1976, 1980).

Outflow channels have relatively limited source areas, and they occur most frequently in the equatorial region. The source areas are topographically lower than the surrounding plains, and the floors of these depressions typically have a complex, irregular topography, with slump blocks and debris (Fig. 2). These channels arise fully developed from the chaotic source area; they may be 20-180 km in width and extend hundreds of kilometers before they dissipate in an ill-defined terminus (Sharp and Malin, 1975; Mars Channel Working Group, 1983). These outflow channels are the ones that most resemble catastrophic flood features on Earth.

The fretted channels are most extensively developed along the global boundary between the lower northern plains and the older cratered highlands that dominate the southern hemisphere. These channels are generally wide, flat-floored valleys, often without any tributaries. The channel floors do not have features that could be clearly identified as having been formed by running water. The origin of these features is likely to be sapping, involving removal of either subsurface fluid (Milton, 1973) or ground ice (Sharp, 1973).

The ages of the various Martian channels have not been accurately determined, but a few general relationships are known. The small runoff channels that occur on the ancient cratered terrain can be assigned a relative age by their degree of degradation. As a whole, the runoff channels appear to be older than the larger outflow channels, based on crater-counting techniques. Carr (1979) has outlined a geomorphic history for Mars in which the small runoff channels developed earlier, under a denser, warmer atmosphere. This was followed by a global cooling that created a thick permafrost layer which trapped groundwater. Periodic outbreaks, possibly initiated by local geothermal heating, could have created the catastrophic outflow channels. All of these channels are likely to have formed between 3.7 and 1 billion years ago (Baker, 1982), and most outflow channels have crater populations that indicate ages greater than 2.5 billion yrs. (Carr, 1981). Evidence of recent volcanism suggests the possibility of even more recent fluvial activity (Lucchitta, 1987).

Terrestrial analogs from the collapse of natural dams

Because of its gross morphological similarity, the Channeled Scabland of eastern Washington state is the most frequently cited terrestrial analog for the outflow channels on Mars (Baker and Milton, 1974; Baker, 1978; Komar, 1979; Baker, 1982). Features observed on both planets include streamlined obstacles and residual uplands, longitudinal grooves, inner channels, bar complexes downstream from expansion points, low sinuosity, and high width-to-depth ratios (Baker and Milton, 1974; Baker and Kochel, 1979; Baker, 1982).

In addition to the morphologic similarities to Martian features, the Channeled Scabland is an appealing analog for philosophical reasons. Baker (1978) has compared the development of the catastrophic flooding interpretation of the Channeled Scabland origin to the evolution of a similar theory for the outflow features on Mars. The efforts of Bretz (1923, 1969) to support his theory have made the work of planetary geologists significantly easier.

The similar morphologies only suggest that similar processes may have created these landforms. Baker (1982) has summarized the possible flow hydraulics under Martian conditions. In addition to considering the effect of Martian conditions on fluid flow, he also evaluated probable flow resistance, macroturbulence, cavitation, and ice-water interactions. He concludes that all of these factors could be operating together to generate the suite of landforms we observe on the Martian surface.

The Channeled Scabland analog depends on the number of geomorphic features similar to those on Mars and the relative similarity in their sizes. The two primary discrepancies between the Martian and Scabland features are scale and the underlying geology. Although the Channeled Scabland exhibits some of the largest fluvial features on Earth, the Martian landforms are consistently several times larger than the terrestrial analogs. Other catastrophic floods on Earth have been suggested as analogs for those that created the Martian features; all have some major point of resemblance, but no terrestrial fluvial features are the same size as the Martian ones. This problem with scale is not unique to fluvial features; nearly all Martian landforms are larger, often by an order of magnitude or more, than their terrestrial counterparts. In some cases, like volcanoes and landslides, these differences can be explained by the differing conditions at the planetary surfaces. For outflow channels and polygonally fractured ground, the explanation is more difficult.

A terrestrial analog for the Martian flood features that does offer a similarity in the type of bed material is in the Båldakatj area of Swedish Lapland (Rossbacher and Rhodes, 1984; Elfstrom and Rossbacher, 1985). The entire area is underlain by the Precambrian Arvidsjaur granite, with surficial deposits of large till and glaciofluvial material (Daniel, 1975). The till has been fluvially eroded into streamlined forms (Elfström, 1983). The large grain-size material approximates the inferred size of near-surface materials on Mars (Binder, et al., 1977).

The Båldakatj area is also characterized by a "boulder delta" that testifies to the power of the catastrophic floods that occurred during the last deglaciation of that region (Elfstrom, 1983). Streamlined islands of till on the Båldakatj delta offer a reasonable morphologic similarity to the Martian outflow features (Fig. 3) (Elfström and Rossbacher, 1985). The similarities in geological materials and landform morphology, if not scale, are the strongest aspects of the Båldakatj analog for Martian outflow channels.

Figure 3. View across Baldakatj boulder delta, Swedish Lapland, looking west (upstream). This area seems to provide a good analog for Martian catastrophic flood features because of the similarity in size of the near-surface material.

Figure 4. Tabular slabs of granitic bedrock on Baldakatj boulder delta, Swedish Lapland. The relationship between the large slabs and the underlying rounded cobbles may indicate the nature of transport during large floods in this region.

Reconstruction of the Båldakatj flood hydrology has been attempted by Williams (1983). The flows that carved the Båldakatj erosional remnants were probably confined by ice walls, and therefore Williams (1983) assumed that the water-surface width is represented by the upstream width of the delta lobe. The height of the streamlined summits of the Båldakatj remnants suggests the water depth may have reached 6 to 10 m, and Williams has estimated the minimum depth around 18 m. Other approaches to reconstructing the paleohydrology indicate depths as great as 200 m. Williams (1983) considers these estimates unrealistically high, but he also notes that Baker (1973) estimated a water depth up to 152 m in the Lake Missoula Flood. Estimated peak discharges for catastrophic floods are 1.3-1.8 x 10^4 m^3s^{-1} in Båldakatj (Williams, 1983), up to 2.1 x 10^7 m^3s^{-1} in the Channeled Scabland (Baker, 1971, 1973), and 8.6 x 10^7 m^3s^{-1} in the Ares Channel on Mars (Masursky et al., 1977). The difficulties in estimating peak discharges for floods in Båldakatj are similar to those in estimating flood discharges on Mars.

Field investigations at Båldakatj also offer some clues to mechanisms of sediment transport under catastrophic conditions. A significant feature at Baldakatj is the occurrence of slabs of granitic bedrock, 2-8 m in length, resting on small, rounded boulders (Fig. 4). The tabular blocks appear to be fracture-bounded pieces hydraulically quarried from outcrops that are located in the upstream paleoflow direction. Williams (1983) used these large slabs to estimate flow parameters, but he assumed the flood transported the blocks directly. The slabs may actually have moved over the rounded boulders that were rolling along the bed of the channel. This live boulder bed would have acted like ball bearings, decreasing the force required to move the slabs.

Other aspects of large terrestrial floods also suggest features that may be analogous to the outflow channels on Mars. For example, the Martian features may include boulder deltas. Masursky et al. (1977) have already inferred the presence of boulders up to 2 m in diameter on the channel floor of Ares, similar to the boulders observed at the Viking 1 lander site (Mutch et al., 1977). Large-scale bedforms on the Swedish boulder delta have been reported (Rossbacher and Rhodes, 1984), although Williams (1983) reported seeing no bedforms on the delta itself.

Terrestrial analogs from the collapse of artificial dams

Although catastrophic flooding from natural and constructed dam collapses have been linked in terrestrial studies, only natural-dam collapses have been explored as analogs for Martian channels. Of course, man-made dams do not exist on the surface of Mars, but the collapse of constructed dams on Earth may offer some valuable insights into the processes that operated in the creation of Martian outflow channels.

Significant dam failures number in the hundreds (Ward, 1978), however few post-disaster investigations have devoted much attention to the downstream geomorphologic effects. The lack of information about downstream morphologic changes caused by catastrophic floods may be one reason why artificial-dam collapses have not been used before as terrestrial analogs for Martian features. Historically, attention after dam failures has concentrated on the damage caused by the flood, and pressures to "clean up" the downstream effects can be intense. Much of the geomorphological data may be lost within a very short period of time.

A well documented example of an artificial dam collapse is Hell's Hole Dam on the Rubicon River of California (Scott and Gravelee, 1968). This dam failed on December 23, 1964, as a result of five days of torrential rain during which more than 55 cm fell on the drainage basin above it. The flood surge resulting from the dam collapse produced a discharge that peaked at approximately 8,500 m^3s^{-1} and a wave that reached more than 13.7 m in height. The average velocity of the flow exceeded 6.7 ms^{-1}. Although the flood wave caused considerable erosion along the valley sides, the amount of actual landscape sculpture was relatively small. The major geomorphological change in the valley was alteration of its cross section from a V-shape to a broad U-shape as a result of aggradation. Most of the depositional forms in the valley were modified and reestablished.

Scott and Gravlee (1968) suggested the following as possibly unique aspects of catastrophic flooding. First, there is an extraordinary transportation of boulder-size material. Boulders in the most turbulent part of the flood may be suspended for some time and deposited well above the valley floor. Boulders may also move as subaqueous particle flows and in wave forms. Although the volume of material moved and the grain sizes are large, individual boulders are usually stranded a relatively short distance beyond their sources. Second, the floods are likely to trigger periods of greatly increased mass movements from the valley sides due to the oversteepening of slopes. Third, the supply of coarse material supplied to the stream will be greatly increased. All of these have applications to Martian catastrophic flooding.

Another significant example is the St. Francis Dam in California. This dam was designed and constructed (1924-1926) to be the major reservoir for water diverted to Los Angeles from the Owens Valley. The dam was located in the San Francisquito Canyon, about 73 km northwest of Los Angeles. At completion, the dam was 62.5 m high and impounded a reservoir containing 4.69 x 10^7 m^3 (38,000 acre-feet) of water. The reservoir was filled in early March, 1928. On the night of March 12, 1928, at about 11:57 p.m., the northern third of the structure collapsed catastrophically.

At the dam site, the initial flood surge was about 56 m high, and the force of the flow was sufficient to move pieces of the dam weighing 1800-2700 metric tons nearly a kilometer downstream (Fig. 5). At its maximum, the flood discharge was 17,000-22,600 m^3s^{-1} (Outland, 1977). During its first 2.4 km of travel, the wave moved at about 8 ms^{-1}. Between that point and the Santa Paula bridge, about 62 km downstream from the dam, the velocity averaged 5 ms^{-1}. The flood moved through the Santa Clara River valley to the Pacific Ocean (87 km from the dam) in about 5.5 hr, causing more than 400 deaths and in excess of $10 million in property damage.

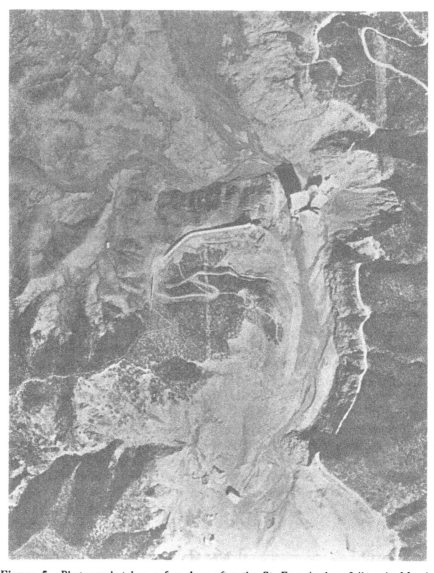

Figure 5. Photograph taken a few days after the St. Francis dam failure in March 1928. The photo shows the broken dam, large concrete fragments that have been deposited downstream, streamlined erosional forms at varying levels in the channel, and a narrow sapping channel above the dam site. Area shown is about 3 km across. (Photograph from the Fairchild Aerial Photography Collection at Whittier College)

Very little information is available about the downstream effects of historical dam failures, although in rare instances evidence may be preserved. Figure 5 is one of series of 340 aerial photographs that were taken within days of the St. Francis Dam failure; these are one of the only sources of data about the downstream fluvial changes caused by that catastrophic event.

A variety of morphologic features shown in Figure 5 and other photographs taken at the same time are similar to landforms associated with the Martian outflow channels. These include streamlined depositional areas, deeper inner channels, low sinuosity, and bar complexes downstream of expansion points. The valley extending diagonally up and to the left (northeast) of the dam resembles many of the sapping channels on Mars, which probably formed by removal of subsurface fluid as the local groundwater level fell rapidly. The large blocks of the dam visible downstream of the dam in this photograph testify to the power of the flood.

These similarities between features associated with the St. Francis Dam collapse and Martian outflow channels support the use of artificial-dam collapses as an analog for the Martian forms. Data are difficult to collect because of pressures to repair the damage quickly, but the morphologic similarities support the value of pursuing this terrestrial analog. This analogy also suggests that a number of Martian landforms associated with outflow channels, such as small sapping channels in the source area of the flood and landslides along the flanks of the valley, may exist on Mars, but below the resolution of the Viking images.

CONCLUSIONS

Investigation of geomorphic features on other planets offers valuable perspective on terrestrial landforms. When our knowledge is extended beyond terrestrial conditions, we are forced to consider the variable nature of factors that are fixed on Earth. Gravitational acceleration, atmospheric density, and temperature all become variables when geomorphic studies are expanded to include other planetary surfaces. The Martian surface is unique in the rest of the Solar System in providing illustrations of fluvial landforms, including ones that were created catastrophically, that developed under different environmental conditions.

Terrestrial analogs are a valuable tool for interpreting catastrophic flood features on Mars, but they are not a perfect tool. Available analogs generally offer similar processes, but details like the magnitude of flow, near-surface materials, gravitational acceleration, and other factors make the analogs less than totally applicable. The discrepancy between the analogs emphasizes the unique historical aspects of geomorphology. The analogs cannot be perfect, because geologic events are individual and non-reproducible. The variations between landforms illustrate some of the important differences in geologic processes, history, and resulting landforms between the planets.

Although terrestrial analogs for Martian channels have always concentrated on floods associated with the failure of natural dams, artificial-dam collapses also result in similar processes and landforms. Post-failure data are usually difficult to obtain, but study of man-made dam collapses, like the St. Francis Dam in California, can provide valuable information about catastrophic flooding.

Investigation of the catastrophic outflow features on Mars has better prepared us to interpret images of other planetary surfaces that may have been influenced by flowing volatiles. To date, only Earth and Mars show evidence of fluvial processes, but further exploration of the Solar System may provide new examples.

ACKNOWLEDGMENTS

Field work in the Båldakatj area of Swedish Lapland was done while both authors were visiting researchers at the Department of Physical Geography, University of Uppsala. Special thanks go to Professor Åke Sundborg for making that visit possible, and to Åsa Elfstrom for useful discussions in the field. Research support was provided by the Faculty of Mathematics and Natural Sciences of the University of Uppsala (to D.D.R.) and the NASA Planetary Geology and Geophysics Program (Grants NAGW-517 and 715 to L.A.R.).

REFERENCES CITED

Baker, V. R., 1971, Paleohydrology of catastrophic Pleistocene flooding in eastern Washington: Geological Society of America Abstract with Programs, v. 3, p. 497.

Baker, V. R., 1973, Paleohydrology and sedimentology of Lake Missoula flooding in eastern Washington: Geological Society of America Special Paper 144, 79 p.

Baker, V. R., 1978, The Spokane Flood controversy and the Martian outflow channels: Science, v. 202, p. 1249-1256.

Baker, V. R., 1979, Erosional processes in channelized water flows on Mars: Journal of Geophysical Research, v. 84, p. 7985-7993.

Baker, V. R., 1982, The channels of Mars: Austin, University of Texas Press, 198 p.

Baker, V. R., and Kochel, R. C., 1979, Martian channel morphology: Maja and Kasei Valles: Journal of Geophysical Research, v. 84, p. 7961-7983.

Baker, V. R., and D. J. Milton, 1974, Erosion by catastrophic floods on Mars and Earth: Icarus, v. 23, p. 27-41.

Baker, V. R., and Ritter, D. F., 1975, Competence of rivers to transport course bedload material: Geological Society of America Bulletin, v. 86, p. 975-978.

Binder, A. B., and others, 1977, The geology of the Viking Lander 1 site: Journal of Geophysical Research, v. 82, p. 4439-4451.

Bretz, J. H., 1923, The Channeled Scablands of the Columbia Plateau: Journal of Geology, v. 31, p. 617-649.

Bretz, J. H., 1969, The Lake Missoula floods and the Channeled Scabland: Journal of Geology, v. 77, p. 505-543.

Carr, M. H., 1974, The role of lava erosion in the formation of lunar rilles and martian channels: Icarus, v. 22, p. 1-23.

Carr. M. H., 1979, Formation of Martian flood features by release of water from confined aquifers: Journal of Geophysical Research, v. 84, p. 2995-3007.

Carr, M. H., 1981, The surface of Mars: New Haven, Yale University Press, 232 p. Cruikshank, D. P., and Silvaggio, P., 1979, Triton: A satellite with an atmosphere: Astrophysical Journal, v. 233, p. 1016.

Cruikshank, D. P., Brown, R., and Clark, R., 1983, Nitrogen on Triton: Bulletin of the American Astronomical Society, v. 15, p. 857.

Cutts, J. A., and Blasius, K. R., 1981, Origin of martian outflow channels: The eolian hypothesis: Journal of Geophysical Research, v. 86, p. 5075-5102.

Daniel, E., 1975, Glacialgeologi inom kartbladet Moskosel in mellersta Lappland: Sveriges Geologiska Undersokning, Series Ba, 121 p.

Donahue, T. M., Hoffman, J. H., Hodges, R. R., Jr., and Watson, A. J., 1982, Venus was wet: A measurement of the ratio of deuterium to hydrogen: Science, v. 216, p. 630-633.

Elfström, Å., 1983, The Båldakatj boulder delta, Lapland, northern Sweden: Geografiska Annaler, v. 65A, p. 201-225.

Elfström, Å., and Rossbacher, L. A., 1985, Erosional remnants in the Båldakatj area, Lapland, northern Sweden: Geografiska Annaler, v. 67A, p. 167-176.

Greeley, R., 1985, Planetary landscapes: Boston, Allen & Unwin, 265 p.

Komar, P. D., 1979, Comparisons of the hydraulics of water flows in Martian outflow channels with flows of similar scale on Earth: Icarus, v. 37, p. 156-181.

Komar, P. D., 1980, Modes of sediment transport in channelized water flows with ramifications to the erosion of the Martian outflow channels: Icarus, v. 43, p. 317-329.

Lucchitta, B. K., 1987, Recent mafic volcanism on Mars: Science, v. 235, p. 565-567.

Lucchitta, B. K., Anderson, D. M., and Shoji, H., 1981, Did ice streams carve martian outflow channels?: Nature, v. 290, p. 759- 763.

Mars Channel Working Group, 1983, Channels and valleys on Mars: Geological Society of America Bulletin, v. 94, p. 1035-1054.

Masursky, H., 1973, An overview of geological results from Mariner 9: Journal of Geophysical Research, v. 78, p. 4009-4030.

Masursky, H., Boyce, J. M., Dial, A. L., Schaber, G. G., and Strobell, M. E., 1977, Classification and time of formation of Martian channels based on Viking data: Journal of Geophysical research, v. 82, p. 4016-4038.

McCauley, J. F., Carr, M. H., Cutts, J. A., Hartmann, W. K., Masursky, H. and others , 1972, Preliminary Mariner 9 report on the geology of Mars: Icarus, v. 17, p. 289-327.

Milton, D. J., 1973, Water and processes of degradation in the martian landscape: Journal of Geophysical research, v. 78, p. 4037-4047.

Milton, D. J., 1974, Carbon dioxide hydrate and floods on Mars: Science, v. 183, p. 654-656.

Mutch, T. A., 1979, Planetary surfaces: Reviews of Geophysics and Space Physics, v. 17, p. 1694-1722.

Mutch, T. A., and others, 1977, The geology of the Viking Lander 2 site: Journal of Geophysical Research, v. 82, p. 4452-4467.

Nummedal, D., 1978, The role of liquefaction in channel development on Mars: Reports of Planetary Geology Program - 1977- 1978, NASA Technical Memorandum 79729, p. 257-259.

Outland, C. F., 1977, Man-made disaster: The story of the St. Francis Dan (2nd ed.): Glendale, California, The Arthur J. Clark Company, 275 p.

Pieri. D. C., 1976, Martian channels: Distribution of small channels on the Martian surface: Icarus, v. 27, p. 25-50.

Pieri, D. C., 1980, Martian valleys: Morphology, distribution, age, and origin: Science, v. 210, p. 895-897.

Rhodes, D. D., and Rossbacher, L. A., 1985, The surface of Mars, in Himmel och jord: Ymer, v. 105, p. 7-29.

Rossbacher, L. A., and Rhodes, D. D., 1984, Catastrophic flood features in Swedish Lapland as a terrestrial analog for Martian channel forms: NASA Technical Memorandum 87563, p. 319-321.

Sagan, C., and Dermott, S. F., 1982, The tide in the seas of Titan: Nature, v. 300, p. 731-733.

Schonfeld, D., 1977, Martian volcanism: Lunar Science, v. 8, p. 843- 845.

Schumm, S. A., 1974, Structural origin of large Martian channels: Icarus, v. 22, p. 371-384.

Schumm, S. A., 1985, Explanation and extrapolation in geomorphology: Seven reasons for geologic uncertainty: Transactions, Japanese Geomorphological Union, v. 6-1, p. 1-18.

Scott, K. M., and Gravlee, G. C., Jr., 1968, Flood surge on the Rubicon River, California - Hydrology, hydraulics and boulder transport: U.S. Geological Survey Professional Paper 422-M, 40 p.

Sharp, R. P., 1973, Mars: Fretted and chaotic terrains: Journal of Geophysical Research, v. 78, p. 4073-4083.

Sharp, R. P., 1980, Geomorphological processes on terrestrial planetary surfaces: Annual Review of Earth and Planetary Sciences, v. 18, p. 231-261.

Sharp, R. P., and Malin, M. C., 1975, Channels on Mars: Geological Society of America Bulletin, v. 86, p. 593-609.

Ward, R., 1978, Floods: A geographical perspective: New York, John Wiley and Sons, Inc., 244 p.

Williams, G. P., 1983, Paleohydrological methods and some examples from Swedish fluvial environments, I - Cobble and boulder deposits: Geografiska Annaler, v. 65A, p. 227-243.

Yung, Y. L., and Pinto, J. P., 1978, Primitive atmosphere and implications for the formation of channels on Mars: Nature, v. 273, p. 730-732.

15

Dynamics of a Missoula Flood

Richard G. Craig

ABSTRACT

Dynamic behavior of the Missoula Floods, which inundated the Scablands of Washington about 15,000 years ago, cannot be understood by field observations alone. It is necessary to link such observations to a model of the physics of fluid motion. Such models can direct investigations and formulation of hypotheses.

Controls upon flood dynamics that must be understood include: ice sheet configuration and dynamic behavior, topographic boundary conditions and the physics of ice dam breakup. In the study reported here, a digital elevation model (DEM) at a spacing of 2 arc minutes of latitude and longitude is used to represent the topographic controls. Volume-depth-area relations, ice sheet configuration and fluid flow are all constrained by this DEM. A simple kinematic model of ice sheet dynamics is used to represent the control of ice margin configuration upon lake volume, ice dam elevation and flood flow paths. Global controls upon the growth of the ice sheet are represented with an empirical global ice volume model coupled to insolation variations due to the changing orbit of the earth.

This model predicts a cycle of at least 35 floods during the Fraser glaciation; even if the discharge of the Clark Fork River was no different than today. A doubling of its discharge as suggested by Waitt (1980) would produce 67 floods, about the number that Atwater's (1986) data suggest. Floods were first very small, then increased, reached a peak magnitude and fell off in size more quickly than the rise. The largest floods within the Pasco Basin lasted about 11 days and were never close to an equilibrium condition during hydraulic ponding.

INTRODUCTION

Floods that burst from Lake Missoula and carved the loess and basalt of the Columbia Plateau into Scablands about 15,000 years ago were among the most cataclysmic the earth has experienced. Lake Missoula reached depths of 610 meters. When the glacial dam impounding it failed, the currents produced as the lake drained created some ripple trains with wave lengths of 150 m and amplitudes of 15 m (Pardee, 1942, p. 1587). The largest floods released over 2,500 km^3 of water within an 11 day period at velocities exceeding 30 ms^{-1}. Both sediment transport and bedrock scour dwarfed any historic floods. Boulders over 10 m in median diameter were carried more than 3.5 km (Baker, 1973). These floods challenge our abilities to *reconstruct* them, both in the field and with models capable of describing the attendant physics. Various methods of representing physical controls on the flood dymamics are reported here.

J Harlan Bretz first recognized evidence of the 'Spokane Flood' as he christened it (Bretz, 1925). The history of his careful field investigations and numerous contributions to understanding the flood has been recounted by Baker and Bunker (1985). Bretz clearly showed that the flood (in his early studies only one was recognized) reached velocities sufficient to entrain enormous boulders and scour great chunks of basalt from the bedrock exposures. He carefully reconstructed high water levels reached by the floodwaters using scour marks and sediment deposits (Bretz, 1969). In the course of a long career he traced the path of the flood and identified all of the major channels it occupied (Bretz, 1923; 1959). In later studies, Bretz (1969) recognized that more than one such flood has occurred; sparking a controversy that still reverbrates (Baker and Bunker, 1985; Waitt, 1985).

Next to the work of Bretz, the contributions of Baker have most increased our understanding of the hydraulics of these floods. Baker's (1973) computations of hydraulic parameters and statistical analyses of the flood evidence established the magnitude of such floods. Later, Baker combined new aerial photographs and satellite imagery to more clearly demonstrate the dimensions of these floods (Baker, 1978).

The idea that the floods were the result of failure of an ice dam that had impounded glacial Lake Missoula was an idea which germinated slowly (Pardee, 1910; Bretz, 1930). Recently, Clarke et al. (1984) have provided an analysis of the one-dimensional dynamics of the flood behavior at the ice dam during the failure. Their data suggest that velocities of water reached at least 26 ms^{-1} and discharges may have exceeded 13.7×10^6 m^{3}s^{-1}.

The most recent debate about the floods centers around the question of the number of floods during the last glacial maximum (the Fraser Glaciation in the Pacific Northwest). Chambers (1971; Alt and Chambers, 1970; see also Curry, 1977) describes glacial Lake Missoula deposits suggesting 37 or more cycles of filling and emptying of the lake during the last glaciation, each cycle requiring 30 to 60 years. Waitt's (1980; 1984) analyses in the distal portions of the floodways have identified deposits of forty to eighty floods that occurred between 17,500 and 13,500 years ago. Waitt estimated the time between floods to be 40 to 60 years. Based on deposits from glacial Lake Columbia, Atwater (1986) identified at least 89 floods (most from Lake Missoula) that invaded the Sanpoil Valley; each separated by 35 to 55 years of quiet sedimentation within the lake. Baker (1973) offered the suggestion that the sedimentary deposits can be explained with fewer floods. Each flood was assumed to form multiple layers of deposits in back valleys due to surging within the basin. There is much to be learned from the field evidence that can help us to improve the models that are available. However, complete pictures of the floods will be difficult to reconstruct without more clear notions of the physical behavior that can be expected for flows of these magnitudes. Such a modelling exercise is introduced here.

Few two-dimensional models of fluid dynamics are presently available for describing floods (see, for example: Lee and Froehlich, 1985). The nonlinear second-order unsteady-flow Navier-Stokes equations that must be solved are complex and their solution is rarely warrented for normal flood problems where one- dimensional approximations are satisfactory. The two-dimensional models that do exist are difficult to apply to the Missoula problem because they describe channels in which water flows prior to the incursion of the flood wave. This is not the case for the Missoula Floods where the Scablands were dry prior to the flood. Models that are available were generated for more `normal' flood flows where river channel gradients are gentle and commonly vary gradually in width. Assumptions such as these cannot be justified for the Missoula Floods. For these reasons, a new model is presented here specifically to study the two-dimensional dynamics of these unique events.

Two investigations of the floods are described here. The first involves the dynamics of floods at the dam failure point during the early moments of the floods when velocities and depths were the greatest. The second analysis focuses upon the hydraulic ponding that occurred in the Pasco Basin later in the flood. This requires the synthesis of a discharge hydrograph for inflow and outflow from the Pasco Basin for use as boundary conditions of the solution. Creation of that discharge hydrograph presents interesting problems in its own right and is discussed in some detail.

BOUNDARY CONDITIONS AT THE DAM FAILURE POINT

The immediate physical conditions leading to the floods was the creation of an ice dam, blocking the flow of the Clark Fork River in the area of modern day Lake Pend Orielle. A lobe of ice -- the Lake Pend Orielle lobe -- at the southern margin of the Cordilleran Ice Sheet passed down the Purcell Trench until it collided with the Bitterroot Mountains (Fig. 1). Unable to breach that wall of ice, the Clark Fork River began to back up in its basin. In at least one of the floods it ultimately reached a depth of 610 meters (Pardee, 1942, p.176) before it exerted sufficient pressure upon the ice to finally float it and break it up. The actual mechanism of ice dam failure in modern examples is described by Nye (1976). The depth of water, and so the volume and areal extent of the lake, depended upon the thickness of

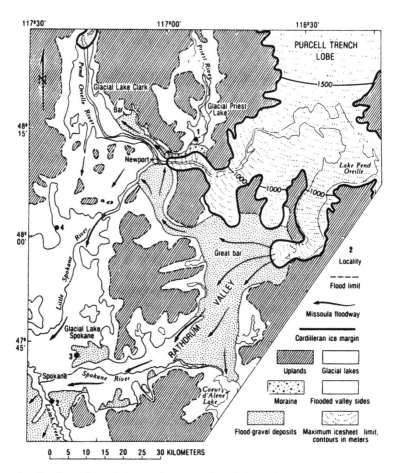

Figure 1. Configuration of the southern margin of the Lake Pend Orielle lobe of the Cordilleran Ice Sheet in the vicinity of the ice dam which created Lake Missoula. From Waitt (1984).

the ice lobe at the dam site.

The ice margin influenced the flood behavior in other ways. The Flathead lobe in Montana advanced into the basin of the Clark Fork River far enough to displace lake waters. Several configurations of the ice sheet are known (Fig. 2), including the Kalispell, Polson and Mission moraines (Alden, 1953). At some of these configurations less water would be held in the lake than its theoretical maximum capacity. In the Scablands, the Okanogan Lobe blocked the flow of the Columbia River creating glacial Lake Columbia. The existence of this lake, and the lobe of ice itself, probably modified the path that flood waters took. For these reasons, the configuration of the southern margin of the Cordilleran Ice Sheet provides impor-

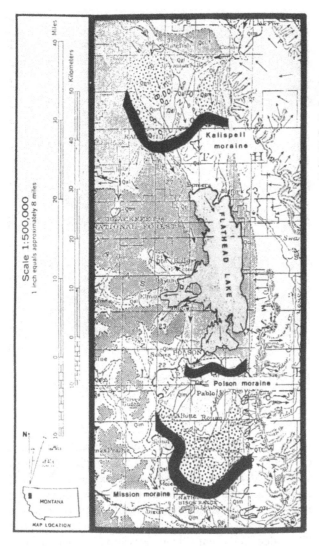

Figure 2. Moraines of the Flathead lobe at the southern margin of the Cordilleran Ice Sheet during the last glaciation. Adapted from Alden (1953).

tant boundary conditions for the flood model. Dynamic specification of that margin is incorporated into the model used here.

Solution of the two-dimensional dynamics of the flood also requires precise descriptions of the lake. Both the spatial and vertical distribution of the water supply determine flood behavior. Pardee (1942) has shown that high velocity currents developed in certain portions of the basin during its drainage when lake levels lowered to the levels of certain divides or where flow was constricted in narrow valleys. The multifaceted form of the basin (Fig. 3) introduces special constraints upon the rate at which water can be supplied to the flood.

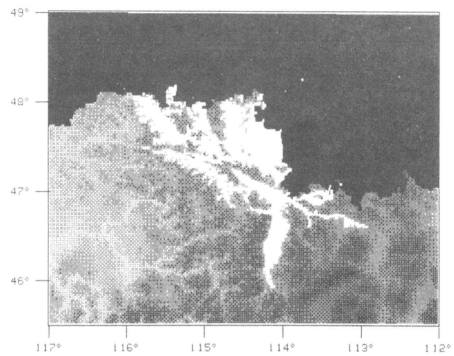

Figure 3. Configuration of Lake Missoula at its highest elevation (4150 feet) during the last glaciation. This configuration is solved from a digital elevation model.

To represent the complex basin geometry, a grid of elevation data, spaced at 2 minutes of latitude and longitude (averaging 9.21 km^2) was employed. This allows definition of water depth at each point and the drainage paths available as water is removed. For example, certain paths represent divide crossings that are occupied only when the water level exceeds the elevation of that divide. The same digital elevation model (DEM) is extended on the west to describe the topography in the path of the flood. The values represent average elevation of data taken at a spacing of 30 seconds of arc within each cell.

A simple kinematic model of ice flow is used to describe the motion of the southern margin of the ice sheet as the ice dam forms. Flow of ice is assumed to be proportional to the slope of the ice surface and is controlled by a fixed elevation of the ice sheet at the 49th parallel. This elevation is defined by the elevation of nunataks as reported by Richmond et al. (1965). With this method, all controls that the ice will exert upon the flood are represented.

A sequence from the solution of the ice flow equations is depicted in Figure 4. The first ice lobes course down the widest and deepest valleys (Fig. 4a). Following this initial spread, the lobes thicken and widen, beginning to occupy the higher and

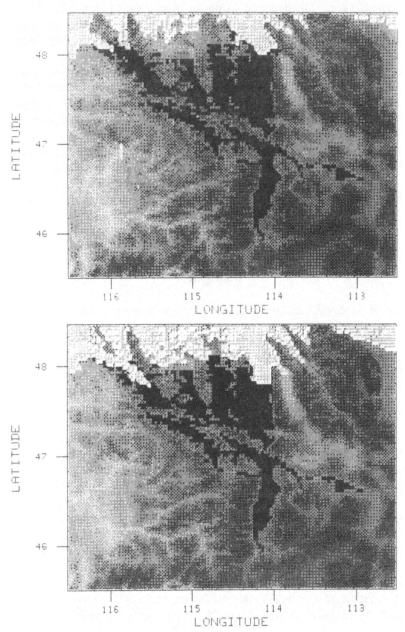

Figure 4. Simulated advance of the southwestern margin of the Cordilleran Ice Sheet during the last glaciation. (a) Early stage flow occupies the deepest valleys, (b) in later stages the ice sheet thickens and covers the higher elevations.

narrower valleys (Fig. 4b). In the east, the broad ice lobes spread out over the area of Flathead Lake. The dynamics of the ice lobe at the ice dam location are controlled by topography (Fig. 5a). The lobe moves south, spreading out slowly. It splits around areas of higher relief in the valley (Fig. 5b); one lobe flows east, up the Clark Fork River valley. Other lobes move south to cover the area of Lake Pend Orielle, and west towards the area of Sand Point. The computer code used includes a monitor of the elevation of the ice surface in the region of the ice dam that created Lake Missoula. It is found that the ice dam first becomes thick enough to create the deepest lake known in the Missoula valley before the ice lobe has reached the mapped ice margins southwest of Lake Pend Orielle. Thus, it is inferred that this margin must represent the edge of the ice at some later flood in the sequence and that the margin continued to advance as the sequence of floods progressed.

One difficulty was encountered in the ice dam modelling exercise that has led to a clearer understanding of the dynamics of a jökulhlaup system. The margins of most terrestrial ice sheets in equilibrium approximate a parabolic profile (Weertman, 1963; Figure 6a).

$$h = k \sqrt{d}$$

where: h = ice thickness (meters), d = distance from ice margin (meters), and k = a constant.

Maps identifying nunataks in northern Montana during the last glaciation (Richmond, et al., 1965) were combined with an iterative solution of the ice sheet surface elevation to determine the best fit estimate of k (over the range 1.0 to 4.5 in increments of 0.1). For this configuration of the Cordilleran Ice Sheet, k=2.2 provides the best fit.

This ideal parabolic shape does not explain the formation of a lake dammed by the ice lobe. As the ice advances over an existing river, as happened with the Clark Fork River, the river will be cutting a relatively thin portion of the ice. No dam exists until the ice collides with the opposing mountain range. If the ice was so thin, how did a 610 m deep lake form? The answer is, it did not. At least not before the first flood. That first flood is quite small, hardly worthy of the appellation "Missoula" since it would not have been deep enough to inundate the site of that city. The effect of the flood is hardly significant anywhere except at the toe of the ice dam. There it has a profound effect. The flood will erode the ice lobe at this toe (both mechanically and thermally), so that the remaining ice has a truncated form (Fig. 6b).

It is suggested here that the internal dynamics of the ice sheet are not in equilibrium with this form. The ice will continue to flow forward under the force resulting from the gradient of the ice surface. When the ice margin again butts up against the Bitterroot Mountains, it will be thicker, the lake formed behind the ice dam will be larger before it fails, and the erosive impact of the flood upon the ice margin shape will be greater. This sequence: ice flow - dam formation - lake

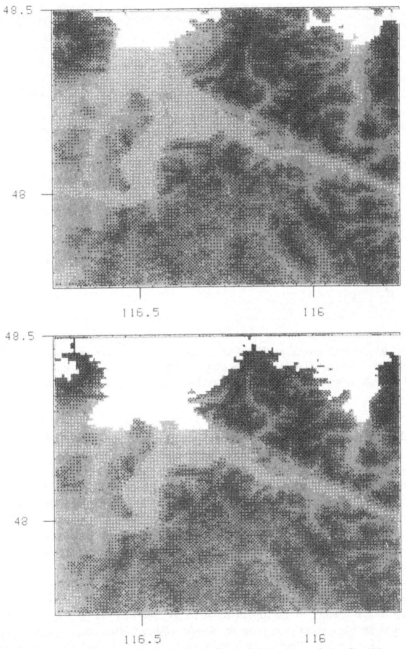

Figure 5. Simulated growth of the Lake Pend Orielle lobe of the Cordilleran Ice Sheet during the last glaciation. (a) early stages of the advance into northern Idaho, flow is confined to the lowest elevations. (b) lobe splits around topographic high and spreads east and west.

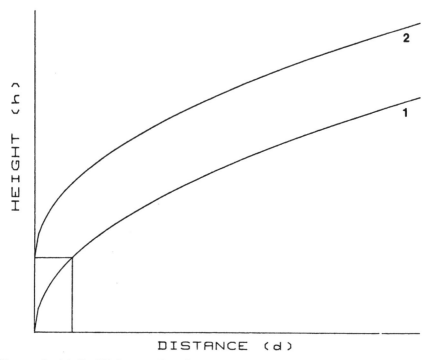

HEIGHT (h)

DISTANCE (d)

Figure 6. (a) Equilibrium profile of the margin of an ice sheet as solved using the equation of Weertman (1963). (b) Truncation of the "toe" of the ice sheet margin by flood erosion, resulting in a non-equilibrium form and a thicker margin.

buildup - failure of dam - erosion of toe of ice lobe, will continue. Each time the lake will be larger and the flood more catastrophic. The varve counts of Atwater (1986, his Figure 17) at the Manilla Creek section suggest a systematic variation in flood timing that, to a first degree, fits the model forecast. Evidence of less than catastrophic draining of Lake Missoula has been cited by both Pardee (1942) and by Richmond (Richmond et al., 1965). Weber (1972, p. 39) shows that in the higher elevations of the Bitterroot Valley, only nine high levels of the lake are recorded. This could be explained if only a small number of maximum discharge floods occurred.

It is difficult to compute a theoretical estimate of the number of floods that could be expected to occur during a glacial cycle. Deposits of the most recent glaciation, which reached its maximum about 15,000 years ago in the Pacific Northwest, provide abundant evidence of the number of floods that must have occurred (Waitt, 1985; Atwater, 1986). Atwater's estimate (89) was inferred from deposits in proglacial Lake Columbia. Waitt used field evidence from the distal area of the floodways to estimate that at least 40 floods occurred. From this he concluded that the discharge into Lake Missoula must have been at least double the modern discharge of the Clark Fork River at St. Regis, Montana (Waitt, 1980, p. 675). Waitt (1985) suggests that a larger area was tributary to the basin of Lake Missoula during these events. This would especially include contributions from the basin of the Kootanai River.

An independent estimate of the number of floods that could have occurred during the last glaciation can be made by considering the global controls on ice sheet advance and lake formation. The period of expansion of the ice sheet can be separately estimated from the Imbrie and Imbrie (1980) model of global ice sheet volume changes (Fig. 7). Knowing the time of initiation of ice advance, the rate of ice advance and the distance between the center of advance and the ice dam location allows us to determine the time at which the ice dam will be in place. This curve also allows an estimate of times when the ice margin will retreat; so that the total period of potential flooding can be determined. When combined with an estimate of the rate of infilling of the lake and a hypsometric curve for the lake, we can determine at what time the lake will reach an elevation sufficient to lead to break up of the ice dam -- since its elevation is also known from the ice model.

These steps have been followed to compute the number of floods that would be expected under conditions that prevailed at the last glaciation (Table 1). Under an assumption of no change in the rate of infilling of the lake, a mean rate of advance of the ice front of $50myr^{-1}$ and a mean rate of retreat of $54myr^{-1}$, the model predicts a sequence of 35 floods of increasing, then stable, then decreasing magnitude. Only 10 floods exceeding 1000 km^3 of water released will occur; these are in the middle of this sequence. These results generally agree with the field evidence reported by Waitt (1980, 1985). Chambers (1971) also recognized 37 fillings and emptyings of Lake Missoula. It would seem likely that the smaller lakes would not be recorded, or the record of these would be removed by subsequent floods. Atwater (1984, 1986) identified 89 floods -- many of them originating from Lake Missoula. An doubling of discharge of the Clark Fork River as suggested by Waitt (1980) leads to 67 floods in the model.

An interesting application of the modelling effort is to provide better estimates of times to fill the lake using the varve records. All available direct estimates of times required to fill Lake Missoula are based upon such varve deposits. The model allows us to recognize that at any given site (i.e. elevation) within a lake, a number of years of lake filling will occur before the lake reaches a given altitude. We can use a volume-depth relation (Fig. 8), coupled with an assumed discharge to compute this time. This allows an estimate to be made of the "missing time" not recorded in the varves.

The estimate of time to fill Lake Missoula provided by Waitt (1985) is based upon the duration of the period that floods were active and requires a doubling of the discharge of the Clark Fork River in order to fill the lake completely in the average time he computed. The computations reported here imply that -- based upon more regional controls -- the number of floods could have been as suggested in the literature and the average discharge may also have been as at present; however, the flood record can be accommodated even more easily if the rate of filling of the lake

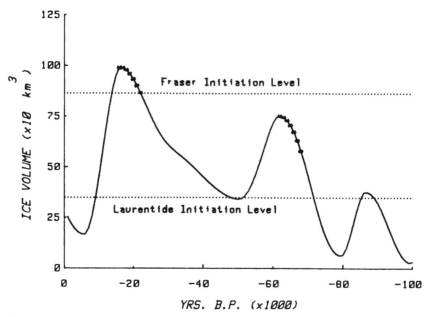

Figure 7. Period of advance of the Cordilleran Ice Sheet (marked by asterisks) during the last glaciation as inferred from the model of Imbrie and Imbrie (1980).

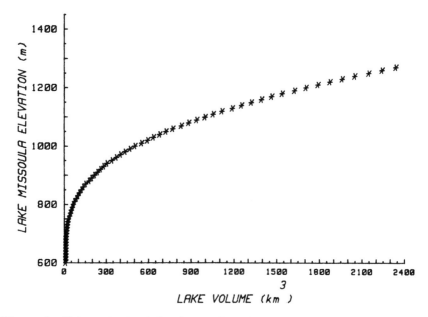

Figure 8. Volume-depth relation for the Lake Missoula basin at the last glaciation when the Flathead lobe is at the Mission moraine. Solved with a digital elevation model at a spacing of 2 minutes of arc, latitude and longitude.

changed by about as much as Waitt suggested. In making these estimates, it is important to recognize that not all floods were maximum size; many were considerably smaller and involved much less time for the lake to fill.

What controlled the size of the floods? The control must have been given by the ice dynamics since there was no natural outlet that was occupied which would provide a limit to the size of the lake. Pardee (1942, p. 1570) gives the elevation of the highest strandline as 4150 ft (1265 m). Other than the Clark Fork Valley, the lowest divide that would be available during a glaciation is Lookout Pass in the Bitterroot Mountains with an elevation of 4750 ft (1447.8 m). Control by ice dynamics explains why there was not one lake level that was most frequently occupied; no single strandline is especially notable. Instead, we observe a continuous sequence of strandlines, each at a different elevation, each adjacent pair of strandlines appears to be separated by about the same distance.

Probably only the strandlines which record the declining stages of Lake Missoula are well-preserved. In general, the strandlines are indistinct depositional forms; few strandlines eroded into bedrock are known. The soft sediments of the earlier beaches would probably have been eroded by the rising waters of the large lakes in the middle of the flood sequence. Beach ridges of the descending sequence of lake sizes would have been preserved since lake surfaces did not reach their elevations again. If this idea is correct, there would be some value in actually recording the number and elevation of preserved strandlines; an observation that has not yet been reported.

There are at least three ways the ice dynamics could have controlled the lake depth and so the size of the floods. First, the ice surface may have reached an equilibrium state with the flood erosion. If the profile of the ice sheet were nearly in equilibrium at the ice dam area, then each lake, and so each flood will be of nearly the same size. The nearly uniform thicknesses of the flood deposits at the Burlingame Canyon section (Waitt, 1980) suggests that each flood that reached that site must have been of approximately the same magnitude.

A second way the ice sheet could control lake depth is through a continuous dynamic variation in ice dam elevation in response to the external controls upon mass balance of the ice sheet. In this view, the size of the largest floods was coincidentally set by these external dynamics; a longer glacial cycle would have produced even larger floods. The data of Atwater (1986) seem to imply that there was a maximum size that was commonly reached and that there would not be an increase of flood size over a longer time of glaciation.

A third control upon the lake size is conceivable. It depends upon the formation of the lake itself. The existence of the lake could provide a lubricating bed of saturated, coarse-grained sediments -- late stage deposits of the previous flood -- which could lower the basal sliding friction of the ice, leading to a surge. Once the sequence of floods had advanced to the state that the ice lobe size made a surge possible, subsequent surges would be expected each time the lake formed. A surge could be the mechanism by which the lake finally disrupted the ice dam, producing the next flood. This mechanism might produce a sequence of floods of about the same magnitude and so could also be consistent with the record of Atwater (1986).

The dynamics of each flood will differ and will depend upon the volume of water that is released from Lake Missoula. This volume was different in different floods (Atwater, 1985, p. 345) so it is necessary to estimate this amount from the available information in the model. The principal control on lake size, besides the height of the ice dam itself, is the configuration (location and profile) of the ice margin along its contact with the lake. A lesser factor is the amount of sedimentation in the lake. Up to 10 m of sedimentation occurred in the lower reaches of the valleys during the last glacial event (Chambers, 1971, p. 30). This could change the volume of water held by the lake by about 5%. Both of these factors can be accounted for easily in the two-dimensional representation available in the model; the method is described next.

To determine the volume of water held in the lake in any flood, the points that can be occupied by the lake are identified by computing the drainage system from the DEM. The computer code used recognizes when the ice dam is in place, fills the lake in the natural sequence and removes the ice dam when the waters have reached the appropriate depth. There are certain places within the lake basin at which the narrow constrictions in the river cannot be represented in the DEM. Because of this, closed basins appear in the DEM drainage system that are not present in the field. To improve the representation of the drainage system -- which is essential to allow the lake to correctly drain -- modifications in the DEM were made. Three schemes have been implemented to correct the connectivity of the drainage network.

Starting from an assumed zero ice configuration, the ice advance computations are performed until the Lake Pend Orielle ice lobe blocks the flow of the Clark Fork River. A matrix of points is created to list the elevation of each point in the Clark Fork drainage system below the elevation at the top of the ice dam. These are sorted according to elevation, from lowest to highest. The height of the ice dam (at the point of failure) is multiplied by a failure factor (usually about 0.9, the density of ice) to obtain an estimate of the elevation of the water surface at the time of failure. Sediment deposition as the lake is filling is computed from the average sedimentation rate and time to filling of the lake. From this and the DEM, the actual depth of water at each point can be determined. Each point is tested for the presence of ice as solved by the ice advance code. The thickness of the ice at the point is also known. If that thickness is less than 90% of the depth of the water, it is assumed that the ice is floating on the water, or is calved off the glacier front. Ice thicker than this is assumed to remain to block the growth of the lake. Once all points that are within the lake have been determined, the volume of the lake is estimated as the product of the depth of the water in each cell times the assumed surface area of the grid cell. Because of the latitude-longitude grid used, surface area is a function of latitude and is solved directly for each latitude. This forms the basis of the estimated volume of each of the 35 last glacial maximum floods listed in Table 1.

Table 1. Predicted depth of Lake Missoula, surface area of lake and volume of water released in each flood during the Fraser glaciation.

	Depth (m)	Surface Area (km^2)	Volume (km^2)
1	880	1519	136
2	880	1519	136
3	880	1519	136
4	880	1519	136
5	910	2011	191
6	910	2011	191
7	940	2723	270
8	940	2723	270
9	960	2968	329
10	990	3839	431
11	1020	4200	553
12	1060	5274	745
13	1110	6210	1035
14	1170	7930	1460
15	1240	9780	2075
16	1340	12146	3191
17	1340	12146	3191
18	1270	10405	2381
19	1210	8589	1793
20	1150	7084	1308
21	1110	6210	1035
22	1070	5367	799
23	1030	4379	579
24	1010	4041	511
25	980	3199	392
26	960	2968	329
27	940	2723	270
28	940	2723	270
29	910	2011	191
30	910	2011	191
31	880	1519	136
32	880	1519	136
33	880	1519	136
34	880	1519	136
35	880	1519	136

MATHEMATICAL MODEL

Saint-Venant type equations are employed (maintaining continuity of mass) in a two-dimensional form to simulate flood behavior. A one-dimensional form is not adequate for an understanding of flood dynamics because the flows bifurcate and rejoin at numerous locations. Even where flows do not anastomose, channel width variations and secondary currents are significant components of the flows.

The differential equations of continuity of momentum used in this model are:

$$\frac{\partial u}{\partial t} + \frac{u \partial u}{\partial x} + \frac{v \partial u}{\partial y} + g\frac{\partial h}{\partial x} + g\frac{\partial h}{\partial y} + H_1 = 0$$

where: $\partial u/\partial t$ = local acceleration,

$u\partial u/\partial x + v\partial u/\partial y$ = convective acceleration,

$g\partial h/\partial x + g\partial h/\partial y$ = pressure and elevation head, and

H_1 = friction.

In difference form, these become,

$$\frac{\Delta u}{\Delta t} = -\left[\frac{u \Delta u}{\Delta x} + \frac{v \Delta u}{\Delta y} + g\frac{\Delta (d+z)}{\Delta x} + g\frac{\Delta (d+z)}{\Delta y} + H_1 \right]$$

and, expanding these with centered difference approximation yields:

$$u_{j,t+1} = u_{i,j,t} - \Delta t \left[\frac{u_{i,j,t}(u_{i+1,j,t} - u_{i-1,j,t})}{2\Delta x} + \frac{v_{i,j,t}(u_{i,j+1,t} - u_{i,j-1,t})}{2\Delta y} \right.$$

$$+ \frac{g(d_{i+1,j,t} - d_{i-1,j,t})}{2\Delta x} + \frac{g(d_{i,j+1,t} - d_{i,j-1,t})}{2\Delta y} + \frac{g(z_{i+1,j,t} - z_{i-1,j,t})}{2\Delta x}$$

$$\left. + \frac{g(z_{i,j+1,t} - z_{i,j-1,t})}{2\Delta y} + H_1 \right]$$

where H_1 is estimated using the Manning equation.

In the case of water, it is reasonable to assume that the fluid is incompressible so that density will not change over space or time. Compressible flow would occur when velocities exceed the speed of sound in water. By making the incompressibility assumption, we are able to simplify the equation of continuity of mass resulting in:

$$\rho \frac{\partial u}{\partial x} + \rho \frac{\partial v}{\partial y} = \frac{\partial h}{\partial t}$$

For the model we are using, the grid cells are only "square" in the sense that they represent equal distances of latitude and longitude. In particular, the data are spaced two minutes apart in latitude and longitude. This distance roughly corresponds to 3 km in both directions. However, the distance is not constant in the east-west direction at different latitudes, but changes significantly between the northern edge of the study area and the southern edge. Moreover, the distance in the north-south direction is nowhere the same as the distance in the east-west direction. The required equation to describe this slight complexity is:

$$\frac{\partial (hu)}{\partial x} + \frac{\partial (hv)}{\partial y} = \frac{\partial h}{\partial t}$$

A centered difference in space and a forward difference in time is used for a numerical representation.

$$\frac{h_{i,j,t}(u_{i,j+1,t} - u_{i,j-1,t})}{2\Delta x} + \frac{u_{i,j,t}(h_{i,j+1,t} - h_{i,j-1,t})}{2\Delta x} + \frac{h_{i,j,t}(v_{i+1,j,t} - v_{i-1,j,t})}{2\Delta y}$$
$$+ \frac{v_{i,j,t}(h_{i+1,j,t} - h_{i-1,j,t})}{2\Delta y} = \frac{(h_{i,j,t+1} - h_{i,j,t})}{\Delta t}$$

A method that has been devised to enhance the stability of the computational procedures is the diffusing difference approximation (see, for example: Viessman, Harbaugh and Knapp 1972, p. 198). The diffusing difference approximation makes use of average neighbor terms to estimate the value of the dependent variable at the

point of interest. The form of the diffusing difference approximation that is being
used in the two- dimensional model described here is:

$$\frac{\partial v}{\partial t} = \left[v_{i,j,t} - \frac{1}{4} \left(v_{i+1,j,t} + v_{i-1,j,t} + v_{i,j+1,t} + v_{i,j-1,t} \right) \right] / \Delta t$$

An analogous form is used to approximate $\partial u/\partial t$ and $\partial h/\partial t$. The diffusing difference approximation is used at several points in the code. In particular, it is required
in estimating the depth at a point for the continuity of mass equation and for use in
Manning's equation for friction.

As would be expected intuitively, it is not possible to make an arbitrary choice
of time steps. If such were the case we would simply choose a time step that
would carry us from the immediate flow initiation to flow at the point when the
flood waters reach a point of concern. In this case a minimum number of time
steps would be required. Unfortunately, such a time step is far from being practical. When time steps are too large, the approximations used become unstable.
Specific stability criteria for the solution of these equations in the one-dimensional
case have been described and are commonly known. The criterion was developed by
Courant, and is:

$$\Delta t \leq \Delta x / \omega + c$$

where Δx = the space step size, in our case ~3200 m, Δt = the requisite time step,
c = celerity of the flood wave, and ω = velocity of the water in the flocd.

The Courant condition provides a theoretical maximum time step that can be
used for stability. This time step is solved by assuming the celerity of the flood
wave is zero with respect to the flow. For given values of flood velocity and the
spacing of the grids in the x (and y) directions, we can solve for the longest time
span that will allow stable solutions. The stability criterion only describes the
characteristics of the finite difference approximation of the differential equations. It
does not include stability problems that might arise due to rounding errors in the
computer code. Practically speaking, it is usual that a time step length of no more
than 20% of the stable time step length is used.

A problem in applying this stability criterion is determining which velocity
would be used in that equation. In effect, one might apply values of velocity at
every point within the flood and compute the time step that would be acceptable for
each of those cases. Ideally then we would use the shortest time step, and this
would provide for stability at every grid point for the next set of solutions. The
equivalent of this is simply to find the maximum velocity that occurs within the
flood at a given time step and use that in the Courant equation to find what time
step is stable for the next set of computations. We have incorporated this methodology in the computer code presently available.

As can be seen, it will be expected that, because flood velocities will change during the evolution of the flood; and because the maximum flood velocities that will be encountered will change also, the desirable time step will not be the same during each iteration of the computations. Two alternatives are possible, we could go through and consider the worse case time step that would be usable. This time step would then be applied at *every* iteration of the computations. On the other hand, a more reasonable approach computationally is to use the longest time step that will provide stable solutions for each iteration. In this way certain iterations may cover longer time spans than other iterations. We use this variable time step methodology to obtain our solutions in this code.

A final consideration for the choice of the time step evolves from the fact that we have used the diffusing-difference approximation of certain terms in the equations. This approximation carries with it its own constraints concerning time step. Usually, the diffusing-difference approximation will allow greater stability in the computations and therefore use of a longer time step. The diffusing-difference approximation allows more reasonable computational times for solution of a flood event. The stability criterion applied for the diffusing- difference approximation is listed below.

$$\Delta t \le R^{4/3} / (g\, n^2\, |\omega|)$$

where Δt = time step, R = hydraulic radius, g = gravitational acceleration, n = Manning's roughness coefficient, and ω = velocity.

In practice, we compute the needed time step for each of these two stability criteria. From these, the minimum of the two is chosen as the time step length to be used in the next iteration. This will usually result in a choice of a longer time step than would be chosen simply by using a time length equal to 20% of that provided by the Courant condition. For the early stages of the flood, time step lengths are typically on the order of 5 - 10 seconds. Thus, a very large number of iterations are required in order to reach a final solution of the flood characteristics through its entire evolution since this would commonly take on the order of two weeks of real time. We have not made many improvements to the code which provide for rapid computational solutions. Therefore, at present the code runs very slow even on a relatively fast dedicated computer. Great improvements to run time can be made.

The expanding wavefront of the Scablands Floods is a special problem. The wave covers ground that is dry so that the normal assumptions used for flood problems cannot be applied. Stability in the solutions is maintained by assuming that 1 m of water covers the land surface and it has a velocity of 0.1 ms^{-1}. This 'film' of water has a minimal effect upon solutions with water at least several meters deep and moving in excess of 5 ms^{-1}. Special tests are made to recognize where the flood wave is expanding. Because the drainage of the lake limits the water available to the expanding flood, points where water level is lowered and/or that are removed from the lake area are also identified. Over 14,000 points occur in the grid and computations take several days on a 32 bit computer.

DYNAMICS OF THE FLOOD

Within three minutes of the release of the wall of water at the ice dam, velocities have reached 34 ms^{-1}. They continue to rise through the earliest hour of the flood. Velocities are so high throughout the Rathdrum Prairie that no unconsolidated sediments could remain. The flood wave passes the location of Spokane within twenty minutes. Several channels become occupied and waters also course up the valley of Lake Coeur D'Lane. Inertia carries breaking waves far up the side of ridges along the Rathdrum Prairie. The highest velocities occur at the constriction of Eddy Narrows. These maxima are within the limit set by simple gravity wave theory (77 ms^{-1}). Within the Eddy Narrows, heads drop within minutes of the complete dam failure. A maelstrom of turbulence forms at the entrance to that gorge. No secondary surges of waves into side valleys are observed in these areas.

Downstream in the Pasco Basin, a secondary lake was created during at least some of the Scablands Floods (Bretz, 1930). Because there are multiple inflow and outflow points for the Pasco Basin, it is difficult to establish the hydrograph at each of these locations. Such hydrographs are needed in order to solve the dynamics of a flood within the Basin. The following analysis describes the solution of the hydrographs for the simplest type of flood which could have occurred, one confined to the channel of the Columbia River.

The flood in this area is studied with a separate set of boundary conditions established at the inflow at Sentinel Gap and the outflow at Wallula Gap. Stage-discharge hydrographs can be established there using one-dimensional models. Multiple analyses with HEC-2 (COE, 1976) was made to determine the relation between stage and discharge. The relations were first established at Wallula Gap. This gap was represented as a set of five cross sections; six more cross sections were used upstream from the gap. The precise areas of backwater were estimated from the maps of Grolier and Bingham (1971). Discharges were varied from 1.0 x 10^6 m^{3}s^{-1} to 14.2 x 10^6 m^{3}s^{-1} (Fig. 9). Using these values, water surface gradients at the upstream end of the Wallula Gap cross sections were projected 64.4 km to Gable Mountain at the downstream end of a set of cross sections that extended to Sentinel Gap (Fig. 10). Similar flow analyses were then made for the Sentinel Gap cross sections. With this approach, discharges could be matched between the two gaps.

The results suggest that the high water marks reported at Wallula Gap (1150 ft, Bretz, 1969) involved a discharge of 12.5 x 10^6 m^{3}s^{-1}, whereas those at Sentinel Gap did not exceed 9.5 x 10^6 m^{3}s^{-1} and must have been produced during hydraulic ponding at Wallula Gap as a backwater effect. These figures, as would be expected, are somewhat lower than peak discharges of 13.7 million m^{3}s^{-1} at the ice dam failure location in the Rathdrum Prairie.

To synthesize a discharge hydrograph for the Basin from these stage-discharge relations, the hydrograph form had to be fixed. To produce a nearly symmetric hydrograph which corresponds approximately with that observed from modern jokulh-

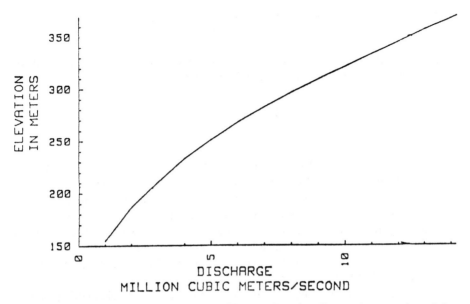

Figure 9. Stage-discharge relation at Wallula Gap for flow volumes released by floods from Lake Missoula.

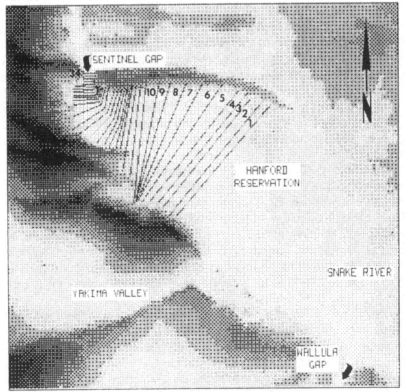

Figure 10. Locations of cross sections used in the computation of stage-discharge relations for Sentinel Gap in the Pasco Basin.

laups, and which attenuates with time and in the downstream direction, it was assumed that the variation of discharge follows the form:

$$Q_t = Q_{total} \, e^{-\lambda} \, \lambda^t / \, t!$$

where t= time (s), Q_t = discharge (million cms at time t), Q_{total}= total discharge during one flood (miilion cms), and λ = an adjustable rate parameter.

Time steps of six hours were chosen for solution. The equation was iteratively scaled (Fig. 11a) to the maximum discharge at Sentinel Gap in a single time step as obtained from the HEC-2 analysis. When the parameter value, λ, was fixed at Sentinel Gap, iterations were made for the value at Wallula Gap that would result in cumulative ponding in the Pasco Basin matching that corresponding to a high water mark of 1150 ft (Fig. 11b).

The final hydrograph improves upon the computations of Baker (1973) who assumed a constant discharge at Wallula Gap to estimate a flood duration of 7-14 days. The hydrograph method implies that the flood stayed at least 1.05 times normal modern flow at Sentinel Gap for nine days. At Wallula Gap, the flood lasted for eleven days. The best fit value of λ at Sentinel Gap (if only integer values of λ are considered) implies the high water mark was formed 3.75 days after the initial surge into the Basin. The two hydrographs suggest that a lag of approximately one day occurred between the high water level at Sentinel Gap and that at Wallula Gap. The maximum velocities computed with HEC-2 imply about 20 hours would be required for a parcel of water to pass through the Basin. This illustrates the problems of correlating high water mark evidence within a single flood. Such evidence may not be usable for estimation of water surface slopes; a point which will be discussed more later.

The DEM for the Pasco Basin also was used to compute the volume-depth relations there (Fig. 12). This allows us to estimate the elevation of the water surface at each stage of the flood (Fig. 13a) since the hydrograph procedure described above yields an estimate of volume of water ponded in the Basin at each time step. The stage-discharge solutions at the two gaps using HEC-2 also allow estimates of the velocity of water at each stage as a function of water surface elevation. Combining the water surface elevation information with the stage-discharge results provides us with an estimate of the velocity of the flood waters at each stage of the flood (Fig. 13b).

With the availability of the velocity and depth curves at Wallula and Sentinel Gaps, the same two-dimensional model used for the dam failure study can be applied in the Pasco Basin. Current results of that exercise suggest a very complex pattern of flow within the Pasco Basin that changes radically during the flood. Changing water elevations appear to be the most important factor in these dynamics and once again, surging does not appear to be important.

LIMITATIONS OF THE GEOLOGICAL RECORD

Testing of dynamic models of floods of this magnitude are still underway and introduce special problems that were not apparent when analyses were limited to equilibrium computations. The evidence available consists of median diameters of the largest boulders moved, ripple train data, high water marks, divide crossings and inferences about water slopes. Each type of evidence yields distinct information about the floods. There is some difficulty in assigning each piece of evidence to a particular flood. For example, high water marks at Wallula Gap only tell us about the water profile at the time of maximum discharge from that gap. At other times, water profiles are little constrained or are completely unknown from field evidence.

Even if the evidence could be assigned to a specific flood, the time-transgressive nature of the flood wave and sedimentary and erosional evidence that results makes it difficult to perform detailed tests with that data. As Baker (1978, p.66) has said, "Time variant aspects of the flood surges cannot be quantitatively deduced from the existing field evidence". For example, although boulder sizes allow an estimate of velocity of transport, it is not correct to assume that the size of the largest boulder can define the maximum velocity. Flood velocities may have greatly exceeded that required to move any available boulders. All we can say is that velocities must have been at least high enough to entrain that boulder. Boulder size sets a minimum value for the maximum velocity.

Similar arguments apply to the water surface slope data. These values provide an estimate of the water surface slope (and by inference velocity) at the time that the water was at its highest levels. That does not tell us the maximum velocity during the flood. Such maxima may have occurred before or after the water reached its highest levels. Within the Pasco Basin, velocities were almost surely lower at the highest water levels than earlier and later in the flood -- when hydraulic ponding was not as effective. Because passage of the flood wave required finite time (it took about a day to pass through the Pasco Basin) water surface slopes inferred at different locations in the Scablands almost surely represent characteristics of different times in the flood and, of course, may represent different floods.

Further difficulties arise because many floods have occurred. Evidence about a flood in one place may not represent the same flood as in another place. Thus, we might estimate the maximum transport capacity of one flood from sediment size data. Estimates of water surface slope from high water marks may represent a different flood. In most cases we cannot know which evidence to attribute to which flood. If we synthesize a picture of 'the flood' by combining evidence of extremes from many different floods, we may create a "monster flood" which at no time did occur. It is because of this that the impression of Missoula floods gleaned from the field evidence may help us little in understanding the characteristics of a single flood.

Mathematical procedures are available to help to solve some of the outstanding problems of the Missoula Floods. To apply the available field evidence in recon-

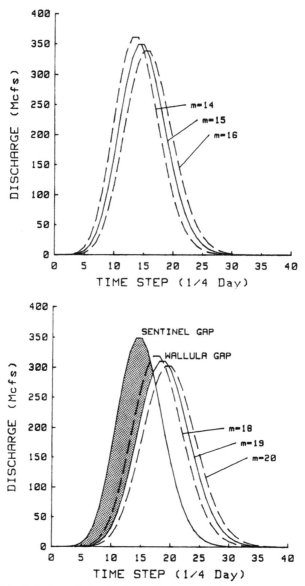

Figure 11. Solution of the discharge hydrograph for the Pasco Basin from the stage-discharge relations at Sentinel and Wallula Gaps. (a) Iterative determination of the value of λ for Sentinel Gap to match the maximum discharge at that gap as solved from the HEC-2 analysis. (b) Iterative determination of the value of λ at Wallula Gap which is capable of yielding a cumulative storage in the Pasco Basin matching the high water mark there. Storage in the final model used is indicated by shaded area.

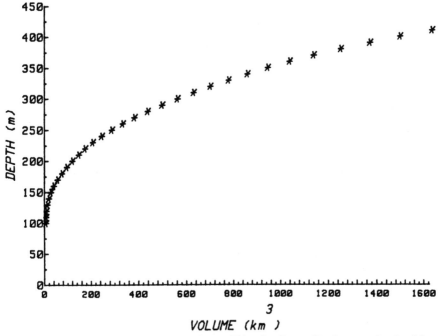

Figure 12. Volume-depth-area relation within the Pasco Basin as solved with a digital elevation model at a spacing of 2 minutes of arc, latitude and longitude.

structing the characteristics of individual floods will require the interaction of modelers and field researchers. The modelers can provide specific hypotheses that will motivate tests in the field, and may direct attention to areas that require investigation. This area of study also holds promise of providing constraints on models for simpler problems of flood dynamics encountered on more typical scales. Much work on modern examples of such floods -- such as the outburst of Lake Russell through the Hubbard Glacier on October 7, 1986 -- may allow validation of the models we would like to apply to interpretation of the paleoflood record.

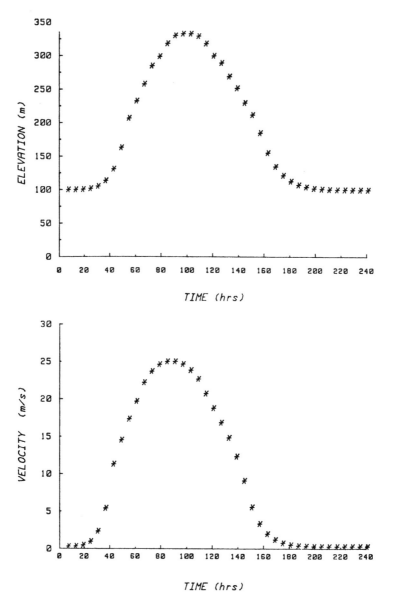

Figure 13. (a) Elevation of the water surface at each time step of a Missoula flood in the Pasco Basin if water arrives only through Sentinel Gap and hydraulic ponding reaches the recorded high water mark at Wallula Gap (same scenario as Figure 11).(b) Velocity of flood waters at Sentinel Gap during each stage.

REFERENCES

Alden, W. C., 1953, Physiography and Glacial Geology of Western Montana and Adjacent Areas, U. S. Geological Survey Professional Paper, 231, 200 p.

Alt, D., and Chambers, R. L., 1970, Repetition of the Spokane Flood: American Quaternary Association, Abstracts of the First Biennial Meeting, v. 1, Bozeman, Montana.

Atwater, B. F., 1983, Jökulhlaps Into the Sanpoil Arm of Glacial Lake Columbia Northeastern Washington: U. S. Geological Survey Open-File Report, 83-456.

Atwater, B. F., 1984, Periodic Floods from Glacial Lake Missoula Into the Sanpoil Arm of Glacial Lake Columbia, Northeastern Washington,: Geology, v. 12, p. 464-467.

Atwater, B. F., 1986, Pleistocene Glacial-Lake Deposits of the Sanpoil River Valley, Northeastern Washington, U. S. Geological Survey Bulletin 1661, 39 p.

Baker, V. R., 1973, Paleohydrology and Sedimentology of Lake Missoula Flooding in Eastern Washington, Geological Society of America Special Paper, 144, 73 p.

Baker, V. R., 1978, Paleohydraulics and Hydrodynamics of Scabland Floods, in Baker, V. R., and Numedal, D., editors, The Channeled Scabland, National Aeronautics and Space Administration, Washington, D.C., p. 59-79.

Baker, V.R. and Bunker, R. C., 1985, Cataclysmic Late Pleistocene Flooding from Glacial Lake Missoula: A Review: Quaternary Science Reviews, v. 4, p. 1-41.

Bretz, J H., 1923, The Channeled Scablands of the Columbia Plateau: Journal of Geology, v. 31, p. 617-649.

Bretz, J H., 1925, The Spokane Flood Beyond the Channeled Scablands I: Journal of Geology, v. 33, p. 97-115.

Bretz, J H., 1929, Valley Deposits Immediately East of the Channeled Scablands of Washington I and II: Journal of Geology, v. 37, p. 393-427.

Bretz, J H., 1930, Lake Missoula and the Spokane Flood: Geological Society of America Bulletin, v. 41, p. 461- 468.

Bretz, J H., 1959, Washingtons Channeled Scablands, State of Washington: Division of Mines and Geology Bulletin, v. 45, 57 p.

Bretz, J H., 1969, The Lake Missoula Floods and the Channeled Scabland: Journal of Geology, v. 77, p. 505-543.

Chambers, R. L., 1971, Sedimentation in Glacial Lake Missoula: Unpublished M.S. Thesis, Univ. of Montana.

Clarke, G. K. C., Mathews, W. H., and Pack, R. T., 1984, Outburst Floods from Glacial Lake Missoula: Quaternary Research, v. 22, p. 289-299.

COE, (Corps of Engineers), 1976, Water Surface Profiles: HEC-2- Users Manual, U. S. Army Hydrologic Engineering Center, 290 p.

Curry, R. R., 1977, Discussion in Curry, R. R., Lister, J. C., and Stoffel, K., editors, Glacial History of Flathead Valley and Missoula Floods: Geological Society of America Rocky - Mountain Section Meeting, Field Guide No. 4, Department of Geology, University of Montana, Missoula.

Grolier, M.J , and Bingham, J. W., 1971, Geologic Map and Sections of Parts of Grant, Adams and Franklin Counties, Washington: U. S. Geological Survey Miscellaneous Geologic Investigations Map I-589.

Imbrie, J., and Imbrie, J. Z., 1980, Modelling the Climatic Response to Orbital Variations: Science, v. 207, p. 943-953.

Lee, J. K., and Froehlich, D. C., 1985, Review of Literature on the Finite-Element Solution of the Equations of Two-Dimensional Surface-Water Flow in the Horizontal Plane: U. S. Geological Survey Open-File Report, 85-96, 137 p.

Nye, J. F., 1976, Water Flow in Glaciers: Jökulhlaups, Tunnels and Veins: Journal of Glaciology, v. 17, pp. 181-207.

Pardee, J. T., 1910, The Glacial Lake Missoula: Journal of Geology, Vol. 18, no. 4, pp. 376-386.

Pardee, J. T., 1942, Unusual Currents in Glacial Lake Missoula: Geological Society of America Bulletin, Vol. 53, pp. 1570-1599.

Richmond, G. M., Fryxell, R., Neff, G. E., and Weis, P. L., 1965, The Cordilleran Ice Sheet of the Northern Rocky Mountains, and Related Quaternary History of the Columbia Plateau, in Wright, H. E., Jr., and Frey, D. G., editors, The Quaternary of the United States: Princeton University Press, Princeton, New Jersey, p. 231-242.

Viessman, W., Jr., Harbaugh, T. E., and Knapp, J. W., 1972, Introduction to Hydrology, Intext Educational Publishers, New York.

Waitt, R. B., Jr., 1980, About forty Last-Glacial Lake Missoula jökulhlaups through southern Washington: Journal of Geology, v. 88, p. 653-679.

Waitt, R. B., Jr., 1984, Periodic jökulhlaups from Pleistocene Glacial Lake Missoula - new evidence from varved sediment in northern Idaho and Washington: Quaternary Research, v. 22, p. 46-58.

Waitt, R. B., Jr., 1985, Case for perodic, colossal jökulhlaups from Pleistocene Glacial Lake Missoula: Geological Society of America Bulletin, v. 96, p. 1271-1286.

Weber, W. M., 1972, Correlation of Pleistocene Glaciation in the Bitterroot Range, Montana, with fluctuations of Glacial Lake Missoula: Montana Bureau of Mines and Geology, 42 p.

Weertman, J., 1963, Rate of growth and shrinkage of non-equilibrium ice sheets: Journal of Glaciology, v. 5, p. 145-159.

16

Effects of a high magnitude flood in a Mediterranean climate: A case study in the Jordan River basin

Moshe Inbar

ABSTRACT

Rare events in Meditteranean environments are a major factor in channel and floodplain modifications. The catastrophic flood of 19-23 January 1969 in the Upper Jordan River was caused by a rare climatic event and intensified by the combined effect of human intervention. The season rainfall amount reached 180% to 250% of the annual average and was the highest amount measured on the whole region for a 150 year period. Daily maximum precipitation measured values were 170 mm and 190 mm. The flood was 10^8 m^3 in volume with peak discharge and velocity of 215 m^{3}s^{-1} and 6 ms^{-1} respectively. Lake Kinneret-base level of the Upper Jordan River basin reached the highest level in a 100-year period.

Suspended sediment from the river basin and the removal of the lower floodplain soil cover - about 2 X 10^6 m^3 - were transported to the lake and formed a new delta at the Jordan's outlet; since its inception there has been a permanent growth in its size by consolidation of sediments, vegetation and a continuous supply of sediment of the river.

Bedload is composed of basaltic boulders. D_{50} values are between 100 and 500 mm with a maximum size of 2000 mm b-axis. The boulders are derived from rotational bank failures and landslides along the main river basaltic canyon and during the flood were deposited in a braided 0.02 slope floodplain.

The flood resulted in a sporadic influx of large quantities of eroded material, about 20 times the annual discharge values. The annual average bedload discharge is about 0.1% of the total floodplain sediment storage.

The movement of coarse material in the braided system correlates with power expenditure of the flowing water as suggested by Bagnold. Morphological structures in the channel can be explained by changes in the unit stream power values. The flood generated four major large scale elements of bed relief; megabars, channels, erosional scarps and terracetes, and rectilinear harrows. The distribution of these macroforms, and their alignement and orientation with respect to the mean flood propagation axis allow a tentative reconstruction of the kinematic structure of the flood, in terms of secondary currents operating within the main flow. Megabars did not change since deposition during the flood.

The high magnitude-low frequency flood created a new adjusted morphology of channel and floodplain. The established relationship is reflected in planimetric braided system values, channel macroforms and channel hydraulic geometry.

INTRODUCTION

In Mediterranean climates, catastrophic or exceptional floods, play an essential role in modifying the fluvial environment. The Wolman and Miller (1960) concept is valid for the geomorphic work expressed in total load transport or basin denudation rates. However, in Mediterranean and semi-arid climates, rare events are basic features in the shaping of the acting fluvial system (Starkel, 1983).

In channels where bedload is composed mainly of gravel and boulders, a shear high threshold is needed for initiate sediment transport (Baker and Ritter, 1975). Coarse channel bed enviroments require high energy in order to cause major changes in the hydraulic geometry. Vegetation development on the fluvial terraces, as well as the accumulation of sediments during the period between major floods, change the adjustment. Floods determine a new adjustment of the fluvial system, based on the approach that long-term equilibrium conditions develop between flow parameters, channel hydraulic geometry and planimetric channel pattern (Schumm, 1977). Therefore, a general relationship can be established between them.

The "catastrophe theory" (Graf, 1983) provides a useful framework for an integrated theory about the development of a "arroyos" in the western United States. In a general model for flood plain formation, the shaping of very coarse-grained alluvium flood plains is attributed to catastrophic events (Nanson, 1986). In this study, the term "catastrophic flood" is used to denote an event characterized by low frequency and high magnitude exceeding the equilibrium threshold of the fluvial system (Schumm, 1973). Catastrophic events, as major factors in fluvial system adjustment, has received wide attention and stress (Baker, 1977; Graf, 1983; Rapp, 1963; Shea, 1982; Thornes, 1980).

Nevertheless, there are few studies on the present effect of catastrophic floods in a boulder bedload environment (Baker, 1984). Direct and indirect instrumental measurements during extreme events are lacking because of the difficulties in collecting such data; it is important to study a variety of environments in order to understand the impact of high magnitude floods on fluvial systems.

This paper deals with the effect of the January 1969 catastrophic flood in the Upper Jordan River; it will consider the climatic, basin and channel factors as well as their influence in the shaping of the fluvial environment.

THE PHYSICAL SETTING

The Jordan basin covers an area of 1590 km^2 and is an elongated valley, 90 km in length, and 10 to 30 km in width; it is situated between Latitude 32°48'-33°29' North and Longitude 25°35'-35°53' East (Fig. 1). The basin's general outline is determined by the tectonic features of the Jordan Rift Valley which is part of the Dead Sea Transform. Elevations range from 2,814 m at the summit of Mt. Hermon to -210 m at the level of Lake Kinneret, an intermediate lake which drains the Jordan River basin toward the Dean Sea.

Lowlands are intensively cultivated under irrigation. The main area is mountainous and, due to thousands of years of overgrazing and deforestation, the natural mediterranean forest is sparse. Mean annual precipitation is 800 mm, varying from 1600 mm in the upper area to 400 mm in the southern lake region. The entire amount falls as winter rainfall, with about 70% in the December-February period. The runoff/rainfall ratio is 40% and one-third of the total runoff is in the form of floods caused by rainstorms in the northern part of the catchment area and the Hula Valley. Summer baseflow is low, about 5 ms^{-1}, somewhat higher during the winter with an average of three to five floods having a peak discharge of 50 m^{3}sec^{-1} to 150 m^{3}s^{-1} (Hydrological Service, 1971). Mean annual sediment discharge is approximately 70,000 tons or 44 tons km^{-2}yr^{-1} (Inbar, 1982).

METEOROLOGICAL SETTING

In general, cold fronts associated with depressions coming through the Eastern Mediterranean, are the main synoptic systems producing rainfall in the December-February period. The fronts cause rain upon reaching the Lebanon and Israel land area. The time interval between successive troughs indicates an apparent seasonable cycle of about a week; a Northern European (Scandinavian) cold stream meeting a Mediterranean low strengthens the cyclonic circulation and causes high amounts of rain in the region. During December 1968 and January 1969, the cyclonic circulation persisted, and the period was characterized by exceptional precipitation in Northern Israel and Lebanon (Blanchet, 1972). The high amounts of precipitation in the storm under study were preceded by heavy rains throughout the months of December and January. In fact, the entire rainfall season of 1968-69 was exceptional in the North of Israel and throughout the Jordan basin. Seasonal rainfall totals reached 180% to 250% of the annual average (Fig. 2). (Israel Meteorological Service, 1969)

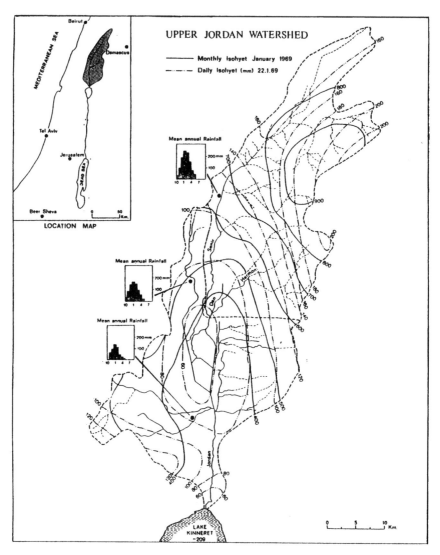

Figure 1. Upper Jordan watershed and location map. January 1969 isohyet and daily isohyet for 22 January 1969.

A look at the meteorological data collected at various stations illustrates the extent of the record-breaking rainfall that season. In Beirut, average rainfall for the December-January period is 385 mm; the annual mean amount is 895 mm. And yet, the Beirut station, breaking a 95 year record, registered a total rainfall for December-January 1968/69 of 1,104 mm -- an increase of 289%. In northern Israel, annual precipitation values exceeded all previous records in stations monitoring weather conditions and rainfall levels in the region for more than 50 years.

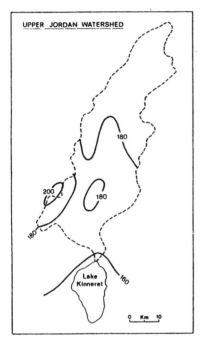

Figure 2. Annual rainfall for 1968-69 in percent of mean rainfall for 1931-1960.

The cold fronts, usually associated with depressions that move in an easterly direction through the Mediterranean, and which are the common synoptic situation on rainy days, brought uncommon rainfall that December and January. The major storm of the period occurred between January 15th and 25th. Favorable storm conditions started to develop early in January; from January 3rd to 8th, a ridge of high pressure extended from Europe over the Mediterranean. Cold air advected toward the Eastern Mediterranean. During January 7th and 8th, large amounts of rain and snow fell in the upper mountains. A Mediterranean Low developed, fed by a Central Russian High. Snow fell in areas, above 900-1000 m a.s.l., in the Upper Galilee in Israel and in Lebanon. The Low persisted until the end of the month and caused cold air advection into the Mediterranean; cyclonic circulation continued for the entire period. At the end of January, a Scandinavian cold stream reached the Eastern Mediterranean, strengthening the cyclonic circulation; snow and rain fell during most of the two months.

The isohyetal map presented in Figure 1 illustrates total storm rainfall as well as the rainfall recorded on January 22nd, which was the day of maximum rainfall. On the same day, maximum runoff discharge values were recorded in the lower order streams of the basin. The general spatial distribution of the storm values is similar to the annual distribution of precipitation in the basin. Daily values for the period under examination were between 40-50 mm and 170 mm at an elevation of 1000 m; at the higher elevations, the values reached up to 200 mm.

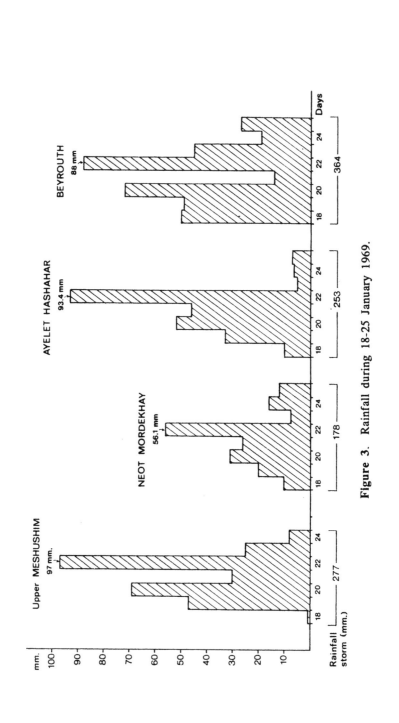

Figure 3. Rainfall during 18-25 January 1969.

Rain continued to fall intermittently for 80 hours in the central part of the basin. Such duration is quite infrequent in the region. Rainfall storm values reached amounts of 120 mm near the Kinneret Lake, 250 mm in the Central Golan Heights, and between 350 and 500 mm in the mountainous areas, 1000 to 2000 m high (Fig. 3). Rainfall intensities were high: 15 mmhr^{-1} in Amudia (400 m a.s.l.) and 6 mmhr^{-1} in Dafna (Hula Valley at 400 m a.s.l.). In Beirut, on the rainiest day, a total of 139 mm of maximum intensities were recorded, with 14 mm in 5 minutes. The duration of the storm and the total amount of rainfall - and not rainfall intensity - were therefore, the main meteorological factors affecting the flood.

Precipitation storm mean values for the entire basin were 300 mm, and it seems that these are the maximum values under present climatic conditions for exceptional rains. Mean annual maximum values for storm periods are 100 to 150 mm. The January 1969 storm, therefore, exceeded the mean intensity by a factor of 3. The conversion to flood runoff is complex; different tributaries had different runoff values for the storm, but all of them exceeded the recorded and observed records - giving greater credence to the conclusions that the floods resulted from an extremely rare event.

THE FLOOD OF JANUARY 20-31, 1969

Stream gage and discharge data were collected from 12 gaging hydrometric stations of the Israel Hydrological Service in the Upper Jordan River basin. The major tributaries of the Jordan River and the main Upper Jordan Southern Station were included in the monitoring procedure. Added for hydrological analysis were two nearby sites from the Meshushim River which included two small experimental basins (5 and 15 km^2.) with gaging stations. Lake Kinneret, which drains the Upper Jordan River, reached a maximum known level of -208.30 m, the highest level in a 100 year history of record-keeping.

The flood in the Upper Jordan started on January 20th, and continued until January 31st. The total flow reached 156 x 10^6 m^3 with a peak of 214 m^{3}s^{-1} on January 23rd. Flow above 200 m^{3}s^{-1} persisted for more than 24 hours. It is assumed that flood peak velocities, based on hydraulic geometry considerations, reached 6 - 7 ms^{-1}. During maximum annual flows of 100 m^{3}s^{-1} measured velocities reached 4 and 4.5 ms^{-1}.

The passage of the flood crest was recorded at the Jordan Southern Station at 21.00, on January 23rd. The lag time between the center of the mass for the rainfall distribution and the peak hydrograph was 48 hours. Lag time for main tributaries was 24 hours, with 4 to 6 hours in the small basins.

Flood wave movement is affected by the pondage of the Hula Valley. Since 1957, following the drainage of the lake and swamps which previously covered the area, the attenuation effect of the valley in the downstream translation of flood waves has been considerably reduced. For example, the time delay between peaks

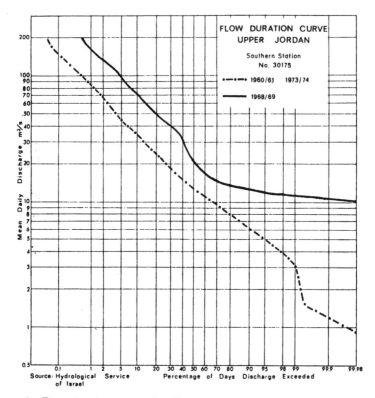

Figure 4. Flow duration curves for Upper Jordan river.

of the January 1969 flood moving between the Northern and Southern stations, at the inlet and outlet of the Hula Valley, was 20 hours. Prior to drainage, the interval between peaks over the same area would have taken several days.

During the entire twelve day period under examination, discharge reached more than 100 m^3s^{-1}, which is the average annual peak discharge and usually lasts between 12 - 24 hours (Fig. 4). Peak discharge of most of the tributaries exceeded the known peaks for the monitored period.

Daily water volumes were 18 x 10^6 m^3 during January 23 and 24 (as opposed to the average daily discharge of 1.7 x 10^6 m^3 during the rainy season). The specific discharge was relatively low: 0.14 $m^3s^{-1}km^{-2}$. For smaller watersheds in the Jordan basin and in adjacent areas, values were higher -- Meshushim (160 km^2); 2 $m^3s^{-1}km^{-2}$; Orvim (30 km^2): 1 $m^3s^{-1}km^{-2}$. The increase of the basin area, which has a distinctive karstic physiography, explains the low specific discharge.

Recurrence intervals for peak floods were computed at all gaging stations using a log Pearson type III distribution. The data used covers the 28 year period following drainage of the Hula Lake (Fig. 5). Peaks measured during the 20 years prior to drainage were much lower. The obtained recurrence interval of 1:120 to 1:200 for maximum discharge seems to be in keeping with meteorological rainfall data collected over 60 years of measurement. An analysis of a synthetic series of yearly volumes (Simon and Yatir, 1981) based on rainfall data and runoff/rainfall relationship, gives the same order of recurrence interval. The assumed recurrence interval for the seasonal rains in Beirut over a 2000 year period (Blanchet, 1972) seems speculative, to us; the statistical frame is built on the extrapolation of more than three times the measured period of rainfall or runoff (Dunne and Leopold, 1978).

The 1969 flood peak discharge is 2.25 times the mean annual peak discharge and 2.1 times the 2.3 recurrence interval per year value. These values are much lower than those suggested by Reich (1973): the 2.33 recurrence interval should be multiplied by 3.5 and 3.8 to get the 100- and 200-year return period, respectively.

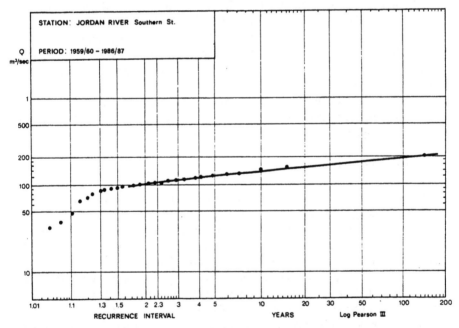

Figure 5. Recurrence interval of annual peak flows for the Jordan River.

MORPHOLOGICAL EFFECTS OF THE FLOOD

The main morphological effects of the flood on the fluvial system depended upon: 1) the local physiographic conditions, and 2) the particle size involved in the sediment transport. Two main areas affected by the flood were: a) the Jordan's outlet to the lake where a new delta was formed; b) the braided floodplain composed, for the most part, of boulders and coarse material.

The Delta Area

The delta acted as a sedimentary environment receiving fine materials - mostly fine sand and silt, transported by the fluvial system. The annual sediment yield from the Jordan basin is relatively low and reaches mean values of 44 tonskm^{-2}yr^{-1} (Inbar, 1982). Since the drainage of the intermediate Lake Hula, total sediment yields entering the Kinneret increased (Inbar, 1982). During the 1968-69 season, large quantities of sediments were transported to the lake. One million tons, or about 20 times the total annual average, is the rough estimate of sediment transported by the flood.

Concentration of suspended sediment during floods reaches values of 1000 - 7000 ppm. Estimated mean concentration for the January 1969 flood is 5000 ppm. A layer of fine sediments was washed from the alluvial braided area, with an assumed volume of about 0.5x 10^6 m^3. The Jordan stream carried part of the fine material into the lake for a distance of several kilometers from its mouth.

The formed delta did not exist prior to 1969. As indicated by early aerial photos as well as the evidence available from examination of the terrain, it is assumed that a delta did not exist in the most recently past centuries, nor even in earlier historical periods (Inbar, 1974).

The present delta itself acts as a morphological factor promoting the accumulation of sediments; it prevents rapid dispersion of sediments by local currents and serves as a self-perpetuating sediment trap (Fig. 6).

The Braided Floodplain

Before entering Lake Kinneret, the Jordan channel is deeply incised by a basaltic saddle forming a 13 km long canyon with steep basaltic slopes which provide boulders and coarse particles for the channel bedload. Between the canyon and the lake area, the channel diverses into a number of permanent and ephemeral channels, forming a braided floodplain about 2 km long and 300 m wide.

During the January 1969 event, the entire valley floor was reworked, as has been documented by detailed aerial photographs, photometric surveys, and field work conducted before and after the flood. This evidence gives increased credence to the importance of catastrophic floods (Lewin, 1983). For paleohydraulic assumption, it shows that a floodplain of the Jordan type may be reworked during a single event lasting only about two days.

Vegetation cover is confined to the banks and higher bars and therefore offers only limited protection; furthermore, during high floods, the weak cohesiveness of the material promotes erosion of the vegetation cover. During the January 1969 flood, large trees were transported as debris by the river. The examination of detailed aerial photos, field observations before the flood, as well as the study of remnant areas after the flood, make it clear that a shallow vegetated soil layer covered the braided area. Following the flood, the soil and vegetation cover were eroded, and a new layer of coarse material was deposited (Fig. 7).

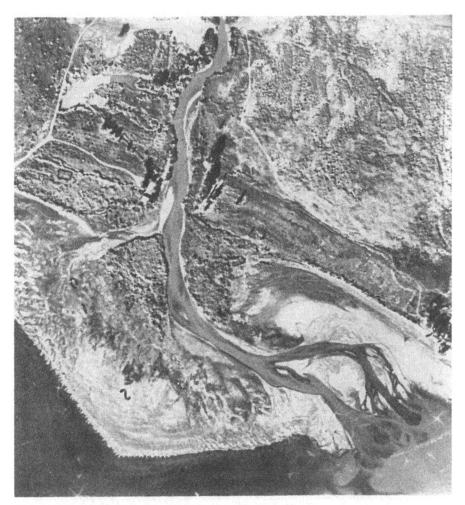

Figure 6. Airphoto of delta developed at river mouth.

For the entire canyon reach, the channel slope is 0.02; values are 0.04 in the riffles for 100 m length of channel reach. In the lower section of the braided area, the channel slope is 0.01.

The bedload is composed, mainly, of basaltic boulders (Fig. 8), which are derived from rotational bank failures and landslides along the basaltic slopes in the canyon between the Hula outlet and the alluvial section. No other source for coarse particles is possible, as the Hula valley, upstream of the canyon section, is 100% efficient in trapping coarse material. Thus, the traveling distance for such material is short, with the largest boulders having a travel distance of only a few hundred meters.

Figure 7. Oblique aiphoto of the Jordan River channel, flowing from the canyon (in upper part) through a braided reach to the delta and into Lake Kinneret (upper part).

In the 4 km channel reach between the canyon and the delta, there is a distal decrease of sediments. In the upper area, the median size of the paticles is 300 mm, with the largest boulders measuring 2000 mm. In the central region, bedload is composed of boulders and cobbles with a median size of 100 mm; the largest particles here, measure 500 mm. No boulders reach the lower channel and delta area, which is mostly sand and silt.

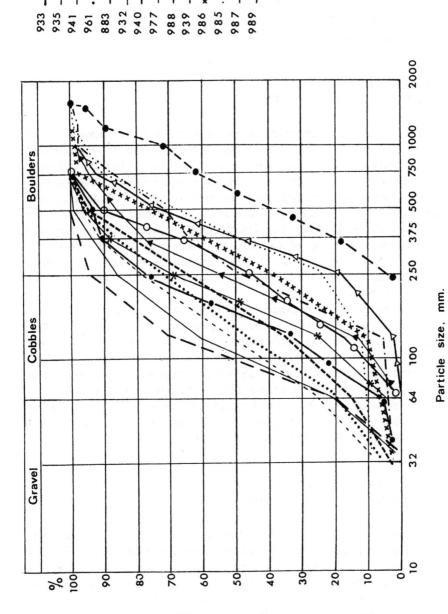

Particle size, mm.

Figure 8. Size distribution of bed size material.

During the flood, large boulders moved tens of meters, while fine material and gravels were transported several kilometers; such movement is indicative of a discontinuous transport in the system.

In the central section of the flood plain, the depth of bedload, measured in bank exposures and by geoseismic survey, reaches 3 m to 5 m. The estimated mass of sediment deposited downstream of the reference section, near the outlet of the canyon, is 3.6×10^6 m^3. The annual bedload discharge is about 0.1% of the total floodplain sediment storage.

BEDLOAD TRANSPORT
AND HIGH STREAM POWER

The very coarse bed material transported during the flood was deposited in the braided area. Deposition volumetrics and particle size distributions were measured after the January 1969 flood event; measurements continued to be taken following posterior events until 1986. Particle motion was monitored by the examination of painted rocks and repeated photographs of the region (Fig. 9). The bedload actually transported during the flood is represented by the bed material measured after the flow (Inbar and Schick, 1979). Boulders measuring between 1500 mm and 2000 mm, the largest rock size in bed material, moved only in the January 1969 event. Mean annual peak flows of about 100 m^{3}s^{-1}, result in the movement of approximately 6% of the boulders in the 500-1000 mm size class and 12% of the boulders in the 250-500 mm size class.

The movement of coarse material in the braided system correlates with the power expenditure of the flowing water, as suggested by Bagnold (1973). The data plotted in Figure 10 illustrates bedload-transport rate per meter of width (i_b) as a function of unit stream power (ω); both values are in units of kg m^{-1}s^{-1}. Data collected from field measurements of the rate of the movement of boulders larger than 500 mm agree with the calculated values obtained from the bedload transport rate/ unit stream power ratio established by Leopold and Emmett (1977) and verified by particles 0.5-2 mm in size. The data collected in this study show that the unit stream power ratio is valid for boulder size bedload movement.

CHANNEL AND FLOODPLAIN
ADJUSTMENT TO THE FLOOD

The morphological adjustment to the flood by the braided area below the canyon outlet resulted in a new valley floor, with an increased planimetric braided drainage pattern. The new valley floor and planimetric pattern evolved in 24 to 48 hours, the time lapse of the flood at high magnitude. The timescale is similar to flood channel formation in other braided systems with very coarse bed material (Baumgart-Kotarba, 1985). The floodplain widened due to lateral erosion. The val-

Figure 9. Painted rocks in the monitored longituinal bar.

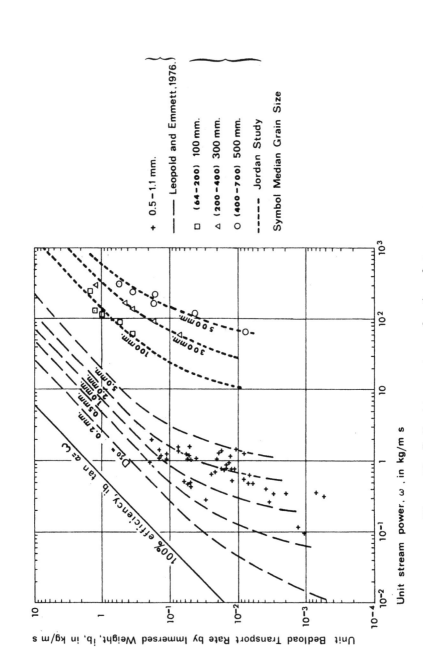

Figure 10. Bedload-transport as a function of stream power

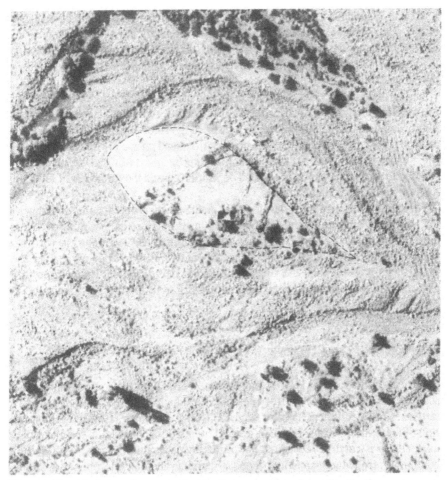

Figure 11. Airphoto of remnant bar in the area stripped by the flood.

ley floor was not completely reworked and remnant bars, now about one meter higher than the new valley floor, indicate that sediment stripping and scouring occurred during the flood (Fig. 11). The stripping affected the whole width of the flood plain, across a 100 m to 400 m wide area along a length of approximately 2000 m of flood plain. Similar episodic flood plain erosion events were described by Nanson (1986).

From 1969 until 1987, there have been no major changes in the planimetric pattern; megabars, which occupy the largest part of the flood plain, were only slightly eroded in the active banks. There was no overbank deposition on them.

The macroform sedimentary accumulations (Church and Jones, 1982) which determine the channel morphology were completely changed by the flood. Four unit bar forms are recognized in the braided reach: longitudinal, transverse or megabars, point and diagonal bars. Longitudinal bars are typical of the main channel and built by very coarse material. Transverse or megabars are most frequent in the braided reach and include fine material and boulders. Point bars are found close to the inner convex bank and are covered during floods. Diagonal bars act as riffles in the active channel.

During the January 1969 flood, the development stages advanced, as follows: (1) In the rising limb of the flood, the fine-sized layer of sediments, covering the flood plain, was stripped; (2) during the high stage of the flood particles of all sizes were reworked; sediment load moved as a boulder front, and coarse particles settled, forming typical rhomboidal transverse bars; (3) during the falling stage, streamflow scoured the main channels and bedload deposition caused the formation of longitudinal inner, point, and diagonal bars.

Weathering and vegetation development on the megabars are the major changes to have occurred since the flood. A monitored inner longitudinal bar was covered by vegetation and, for 16 years, served as a sediment trap. In 1985, the bar was completely eroded during a 1:20 year recurrence interval flood. The main pattern and major features have persisted since the January 1969 flood until the present; it is assumed that only another major flood will be able to modify the adjusted flood plain morphology.

BED RELIEF ELEMENTS

The 1969 flood generated distinct morphological structures. Four major large-scale elements of bed relief were generated: megabars, channels, erosion scarps and terracetes, and rectilinear harrows (Fig. 12) (Karcz and Inbar, 1978). The distribution of these macroforms, and their general orientation in respect to the mean flood axis, allows a tentative reconstruction of the flood structure. The non-steady flow conditions resulted in intensified secondary currents and motions, oblique and transverse to the main flow. Other factors that intensified secondary motions included: the expanding stream geometry from the river canyon, the changing channel cross section, flood dynamics during rising and receding stages, and local depositional bar formations.

CONCLUSIONS

The January 1969 flood in the Northern Jordan Valley was a catastrophic event caused by heavy and prolonged rainfall. Precipitation resulted from extended cold front conditions in the Eastern Mediterranean, strengthened by Northern European cold streams. The 1969 storm established precipitation records for the month of January and the 1968-69 season for most of the Northern Israel and Lebanon monitoring stations.

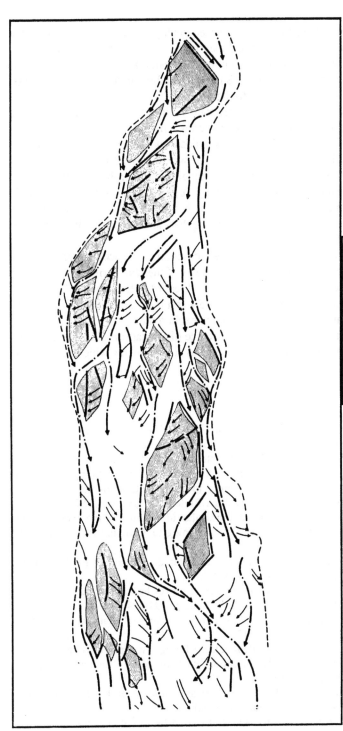

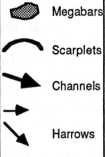

Megabars

Scarplets

Channels

Harrows

Figure 12. Reconstruction of flow pattern.

Peak discharges at the Jordan River exceeded all previously known levels; the obtained recurrence interval of the flood measured between 1:120 years and 1:200 years. The length of the flood was exceptional; it lasted for 12 days as against an average annual peak of 12 to 24 hours. Suspended sediment discharges were large, about twenty times the mean annual discharge.

The main geomorphic effects were pronounced in the depositional areas of the basin: at the outlet of the canyon in the braided reach, and in the new delta area of the river. In the braided reach, the high magnitude flow and associated stream power caused the mobilization of the inner bars as wel as a front wave movement of coarse material. All bedload sizes moved, changing the morphology of the floodplain planimetric features. Movement of very large boulders -- more than 1 m in diameter -- was recorded in other events with higher frequency in the active channel, however, no major changes occurred in the magabars and major floodplain features in the 18-year post-flood period.

The new delta developed because of the promotion of the accumulation of sediments and the prevention of the dispersion effects of the lake currents. Changes in the morphological features and sedimentological record of the coarse alluvial environment of the Jordan braided flood plain are a result of rare catastrophic events; there is no need to assume climatic changes to explain them.

The given example illustrates the importance of the catastrophic events in promoting changes and causing new adjustments to environments which require high threshold stresses.

REFERENCES

Bagnold, R. A., 1973, The nature of saltation and of "bedload" transport in water. Proceedings of the Royal Society of London; Series A, v. 332, p. 473-504.

Baker, V. R. and Ritter, D. F., 1975, Competence of rivers to transport coarse bedload material: Geological Society of America Bulletin, V. 86, pp. 975-978.

Baker, V. R., 1977, Stream channel response to floods with examples from central Texas. Geological Society of America Bulletin, v. 88, pp. 1057-1071.

Baumgart-Kotarba, M., 1985, Different timescales of examining the river bed and valley floor evolution: Studia Geomorphologica Carpatho-Balcanica, v. 19, p. 61-76.

Blanchet, G., 1972, Les precipitations exceptionelles de Decembre 1968 et Janvier 1969 au Liban. Hannon, v. 7, p. 3-27.

Church, M. and Jones, D., 1982, Channel bars in gravel bed rivers, in Hey, R. D. et al. Gravel bed rivers: John Wiley & Sons, p. 291-324.

Dunne, T., and Leopold, L. B., 1978, Water in environmental planning: W.H. Freeman and Co., San Francisco, 818 p.

Graf, W. L., 1983, The arroyo problem-paleohydrology and paleohydraulics in the short term, in Gregory, K. J., editor, Background to Paleohydrology, John Wiley & Sons, p. 280-302.

Hydrologic Service of Israel, 1971, Hydrological Yearbook of Israel: Ministry of Agriculture, Water Commission, Jerusalem.

Inbar, M., 1974, River delta on Lake Kinneret caused by recent changes in the drainage basin, *in* Geomorphologische Prozesse und Prozesskombinationen in der Gegenwart unter verschiedenen Klimabedinungen: Akad. de Wissenschaften in Gottingen, No. 29, p. 197-207.

Inbar, M. and Schick, A. P., 1979, Bedload transport associated with high stream power, Jordan River, Israel. Proceedings of the National Academy of Sciences, USA, v. 76, No. 6, p. 2525-2517.

Inbar, M., 1982, Spatial and temporal aspects of Man-induced changes in the hydrological and sedimentological regime of the Upper Jordan River. Israel Journal of Earth Sciences, v. 31, pp. 53-66.

Inbar, M., 1982, Measurement of fluvial sediment transport compared with lacustrine sedimentation rates: The flow of the River Jordan into Lake Kinneret. Hydrological Sciences Journal, Dec. 1982, No. 4, p. 439-449.

Israel Meteorological Service, 1969, Monthly weather report - January 1969, Bet Dagan, Israel.

Karcz, I. and Inbar, M., 1978, Flood-generated streambed relief and reconstruction of the flow pattern, Upper Jordan River, Israel, *in* Friedman, G. M., editor, Tenth International Congress on Sedimentology Abstracts Jerusalem, v. 1, p. 347-348.

Leopold, L. B. and Emmett, W. W., 1976, Bedload measurements, East Fort River, Wyoming: Proceedings of the National Academy of Sciences, USA, v. 73, No. 4, p. 1000-1004.

Lewin, J., 1983, Changes of channel patterns and floodplains, *in*: Gregory, K. J., editor, Background to Paleohydrology: John Wiley & Sons, p. 303-319.

Nanson, G. C., 1986, Episodes of vertical accretion and catastrophic stripping: A model of disequilibrium flood-plain development, Geological Society of America Bulletin, v. 97, p. 1467-1475.

Rapp, A., 1963, The debris sides at Ulvadal, Western Norway. An example of catastrophic slope processes in Scandinavia: Nachrichten der Akademie Wissenschaften Gottingen, Math. - Phys. Kl., No. 13, p. 195-210.

Reich, B. M., 1973, Log-Pearson Type III and Gumbel analysis of floods, *in* Schulz, E. F. et al., editors, Floods and Droughts: Water Resources Publ. Fort Collings, Colorado, pp.291-303.

Schumm, S. A., 1973, Geomorphic thresholds and complex response of drainage systems, in: Morisawa, M., editor, Fluvial Geomorphology: New York State Univ. Pubs. Geomorphology, p. 299-310.

Schumm, S. A., 1977, The Fluvial System: Wiley-Interscience Pub. New York, 338 p.

Shea, J. H., 1982, Twelve fallacies of uniformitarianism. Geology, v. 10, p. 455-460.

Simon, E. and Yatir, Y., 1981, Synthetic series of flows in the Upper Jordan catchment: Tahal Report 01/81/102, Tel Aviv, 43 p. (Hebrew).

Starkel, L., 1983, The reflection of hydrologic changes in the fluvial environment of the temperate zone during the last 1500 years, *in* Gregory, K. J., editor, Background to Paleohydrology: John Wiley & Sons, p. 213-235.

Thornes, J. B., 1980, Structural instability and ephemeral channel behaviour, Zeitschrift fur Geomrophologie (Supplementbank), v. 36, p. 233-244.

Wolman, M. G. and Miller, J. P., 1960, Magnitude and frequency of forces in geomorphic processes: Journal of Geology, v. 68, p. 54-74.

17

Storm-induced catastrophic flooding in Virginia and West Virginia, November, 1985

G. Michael Clark, R. B. Jacobson,
J. Steven Kite, R. C. Linton

ABSTRACT

The November 1985 floods in the Central Appalachians resulted from prolonged and locally intense precipitation exceeding 250 mm in 3 days in the upper
Greenbrier, James, Monongahela (Cheat), Potomac and Roanoke drainage basins.
Discharges exceeded the one hundred year event at many gaging stations. Discharges exceeding the five hundred year event occurred on the Cheat and South Branch
Potomac rivers in West Virginia. Fluvial alterations resulting from flooding in
these two basins include truncation of debris fans and of valley wall colluvium, intense scour of the floodplains, lateral retreat of cutbanks, and enlargement and subsequent partial filling of channels. Depositional forms resulting from the flooding
include localized scour- and gravel-splays, large-scale point bars, slack-water sediments, large dune- and bar-shaped deposits of alluvium and debris, and widespread
overbank deposits. Mass movement, although locally significant, is described here
for only one drainage basin. Mass movement is possibly related to more intense
precipitation cells or to the effects of local geology, soils, vegetation, and land use.
The impact of the flooding in the narrow, bedrock-defended valleys typical of the
Appalachian Plateau and Valley and Ridge provinces was more severe than would
be predicted for the currently defined one hundred to five hundred year events. We
consider this paper to be a preliminary report on features that we have selected for
initial study; we hope it will stimulate further investigations.

INTRODUCTION

This paper reports the meteorological events producing the November 1985 floods and the impact of these floods on the geomorphology of the Cheat and Potomac River basins. The November 1985 floods in the Central Appalachians resulted from prolonged and locally intense precipitation exceeding 250 mm in three days in the Cheat (upper Monongahela) and upper-Potomac, Greenbrier, James, and Roanoke river basins (Fig. 1). Lescinsky (1986) estimates discharges equal to or exceeding the one hundred year event at more than fifty gaging stations.

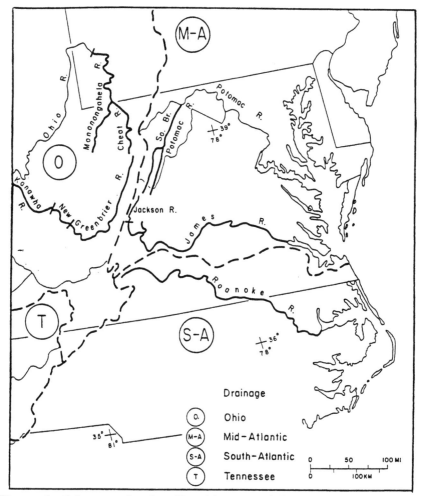

Figure 1. Index map of region affected by storm of November 1985. Major drainage divides shown by dashed lines; rivers that experienced major flooding in their headwaters shown in heavy solid lines.

Our study has concentrated on the Cheat and South Branch Potomac rivers located in the Appalachian Plateau and Valley and Ridge province respectively (Thornbury, 1965). In these two drainage basins, the 1985 flood discharge is estimated to have locally exceeded the five hundred year event. Regional reconnaissance found widespread erosional and depositional effects of the November storm (Kite, 1986). Evidence of mass movement was concentrated mainly in Pendleton County, West Virginia (Jacobson and others, 1987) although isolated landslides occurred elsewhere. Approximately $750,000,000 of property damage and the loss of nearly seventy lives resulted from the flooding.

The Cheat river drainage pattern is dendritic (Fig. 1). Major headwater tributaries to Cheat River are Shavers Fork, Glady Fork, Laurel Fork, Dry Fork, and Blackwater River (U. S. Geological Survey, 1966). From Parsons, Tucker County, West Virginia, Cheat River flows northward through an increasingly narrow, entrenched valley, joining Monongahela River in southernmost Pennsylvania (Fig. 1). The drainage basin is underlain by Upper Paleozoic sandstones, siltstones, shales and conglomerates which are gently folded into broad anticlines and synclines (Cardwell and others, 1968).

Both South Branch Potomac River and North Fork South Branch Potomac River rise in Highland County, Virginia, and flow northeastward through Pendleton County, West Virginia (Fig. 1). In their headwaters, both rivers are classic examples of trellis drainage. The course of North Fork South Branch Potomac River parallels the strike of relatively unresistant, steeply-dipping, Silurian and Devonian limestones and shales between the northwest limb of the Wills Mountain anticlinorium and the base of the Allegheny front. At North Fork Gap, the course of the river shifts eastward along the Petersburg Lineament (Sites, 1978). From Upper Tract, West Virginia, South Branch Potomac River enters the gorge of Smoke Hole Canyon and crosses numerous resistant rock units in a tortuous, entrenched valley approximately 40 km long. Emerging from Smoke Hole Canyon, it joins North Fork South Branch Potomac River 7 km west of Petersburg. Two major constrictions, Petersburg Gap and the Trough (Tewalt, 1977) caused high 1985 flood levels in the towns of Petersburg and Moorefield respectively.

Mean annual precipitation in the Potomac River basin of West Virginia ranges between 800-1200 mm (Hobba and others, 1972). There are pronounced local variations of as much as 250 mm in mountainous areas and in valleys in rain shadows. The mean annual precipitation within the Cheat River basin is estimated by Baloch and others (1973) to be 1150 mm and probably varies over a range similar to that of the Potomac River basin.

Previous large floods in the upper Potomac River drainage in West Virginia include the flood of March, 1936 (Hobba and others, 1972) and a local but severe flood at, and upstream from, Petersburg in June, 1949 (Stringfield and Smith, 1956). The largest recorded pre-1985 floods in the Cheat River basin occurred in 1959, 1888, and 1844 (Baloch and others, 1973).

METEOROLOGICAL SETTING

Antecedent Conditions

October, 1985 was abnormally wet in some areas of north-central West Virginia and northern Virginia. Precipitation of 100-150 mm occurred over a broad band during that month (Fig. 2). This rainfall exceeds the normal rainfall by up to 220% (average based on N.O.A.A. records for the years 1941-1973). The great topographic relief of the area undoubtedly caused large, unrecorded, local variations in the rainfall.

Remnants from tropical cyclone Juan (October 26, 1985 - November 1, 1985), which made final landfall just west of Pensacola Florida October 31, 1985, were moved eastward by a low-level jet stream. Low-pressure cells over northeastern Tennessee, the Atlantic Ocean east of the Carolinas, and the Gulf of Mexico November 3, 1985 provided additional moisture and formed a low-pressure trough that

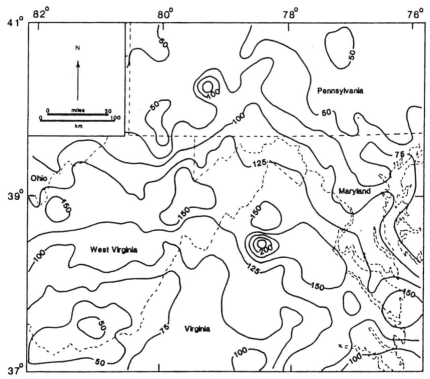

Figure 2. Map of total recorded precipitation for October 1985. Contour interval 25 mm. Data from N O.A.A. (1985c).

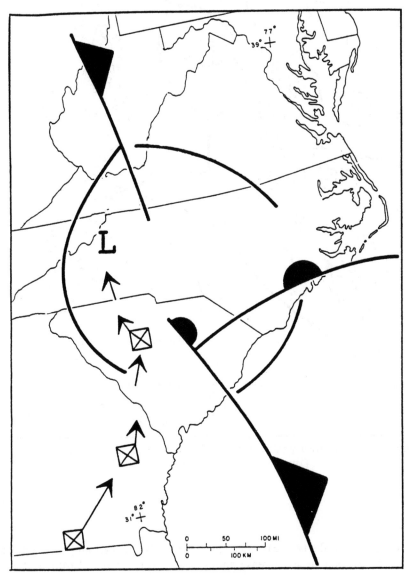

Figure 3. Simplified weather map for 7:00 a.m., E.S. T., 4 November 1985. Boxed X's indicate previous positions of center of low-pressure cell at 6, 12 and 18 h. Attaches arrows indicate direction of cell movement. Data from N.O.A.A., (1985b).

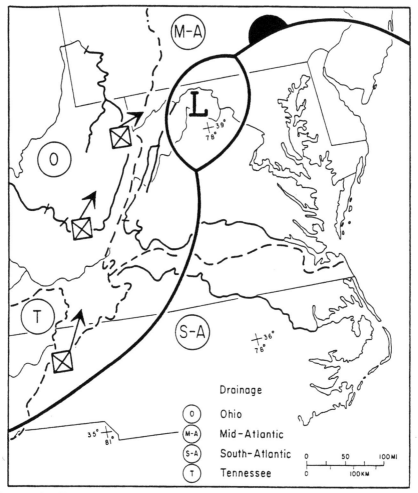

Figure 4. Simplified weather map for 7:00 a.m., E. S. T., 5 November 1985. For symbols see Fig. 3. Data from N.O.A.A., (1985b).

moved slowly to the northeast (Lescinsky, 1986). By the morning of November 4, 1985, a large single and intense anticyclone existed in the Carolinas (Fig. 3), bringing moist air up against a high pressure ridge which lay to the northeast off the New England Coast. A series of meteorological events was thus brought together which individually might have had little or no capability to cause catastrophic flooding. Collectively, however, they provided abundant precipitation.

Description of the Storm

The consolidated low-pressure cell that formed over the southeastern states 3-5 November moved north-northeastward into eastern West Virginia and northern Virginia (Fig. 4). Most of the precipitation in the Cheat and upper Potomac occurred November 4, 1985 (Fig. 5). Rainfall was concentrated along the Allegheny Front in Highland County, Virginia and in Pendleton County, West Virginia. The loci of maximum precipitation continued northeastward during the night of 4-5 November, producing intense rain along the Blue Ridge in Virginia during November 5, 1985. The total precipitation for the period 3-5 November is shown in Figure 6.

Except for a few stations, measured precipitation intensities during the storm were not extremely high. The most intense rainfall within the Cheat and upper Potomac River basins, excluding the cells along the Blue Ridge, was 38.3 mm/hour at Hot Springs, Virginia. Rainfall of this intensity is not rare in this area, recurring on the average of two to five years (Hershfield, 1961). The duration of precip-

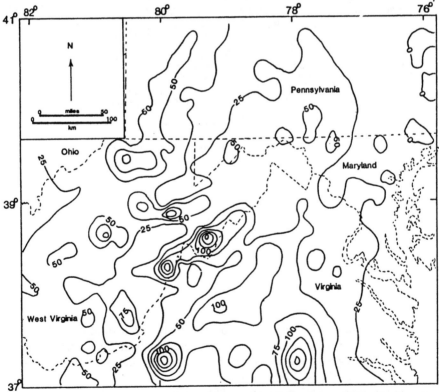

Figure 5. Map of total recorded precipitation for 4 November 1985. Contour interval 25mm. Data from N.O.A.A., (1985c).

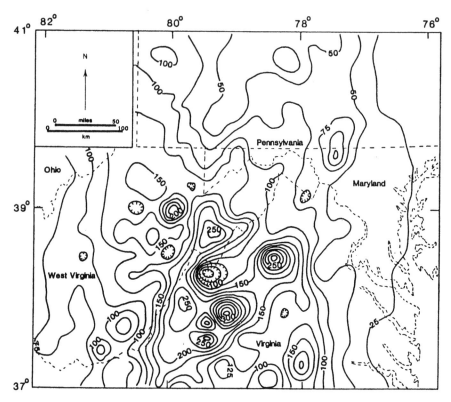

Figure 6. Map of total recorded precipitation for the time interval 3-5 November 1985. Contour interval 25 mm. Data from N.O.A.A., (1985c).

itation of specific lower intensities is not commonly summarized in the literature but may have been a factor in producing the observed geomorphic effects. Many stations in the area recorded steady rainfall of about 10 mm/hour for eight hours or more during November 4, 1985.

The precipitation for November 4, 1985 was in excess of 50 mm over much of the region, ranging up to 175 mm at Franklin, West Virginia. Table 1 summarizes the recurrence intervals (from Hershfield, 1961) for this precipitation.

Table 1. Precipitation recurrence intervals applicable to the study area (Hershfield, 1961)

24-hrainfall,(mm)	Recurrence interval, yrs
50	<1
100	2-6
150	10-70
200	>100

Table 2. Selected peak discharges in West Virginia, November 1985 (Lescinsky, 1986)

Stream: Station	Discharge cfs (m^3s^{-1})	Area mi^2 (km^2)	Discharge/Area cfs/mi^2 $(m^3s^{-1}km^{-2})$
POTOMAC RIVER BASIN:			
North Fork South Branch	90,000	314	287
Potomac River: Cabins	(2550)	(813)	(3.14)
South Fork South Branch	110,000	283	389
Potomac River: Moorefield	(3100)	(733)	(4.23)
South Branch Potomac	130,000	642	202
River: Springfield	(3680)	(1663)	(2.21)
South Branch Potomac	240,000	1471	163
River: Springfield	(6800)	(3810)	(1.78)
MONONGAHELA (CHEAT) RIVER BASIN:			
Dry Fork: Hendricks	100,000 (2830)	345 (894)	290 (3.16)
Cheat River: Parsons	200,000 (5660)	718 (1860)	279 (3.04)
Cheat River: Rowlesburg	230,000 (6500)	972 (2510)	237 (2.59)

THE FLOODS OF NOVEMBER, 1985

Introduction

Widespread overbank flow equalling or exceeding the hundred year event occurred in the headward areas of the drainage basins under the storm path (Fig. 3 and 4). The basins are typically 130-4000 km^2 in area. The paucity of gaging stations in the flooded areas makes it difficulty to determine discharges. In addition to flooding in the upper Cheat and Potomac rivers, the hundred year event was exceeded in the basins of the upper Roanoke, James (including the Jackson), Greenbrier, and Shenandoah River valleys. Although we have not closely examined the latter basins, regional reconnaissance indicates that flood effects in these more southerly valleys were also severe.

Floods on Upper Cheat and Potomac Rivers

Comparison between precipitation maps (Figs. 5 and 6) and discharge records (Table 2) reveals much about the causes of the flooding. The accuracy with which these causal relationships can be reconstructed, however, is limited by the paucity of rainfall data in mountainous areas and the lack of discharge records for small streams in the flood-affected areas. Two storm cells dropped more than 350 mm of rain between November 3 and 5 but they are defined by a few high-elevation precipitation stations in the northern Blue Ridge. Although some locally significant flooding was associated with these cells, the discharge data suggests that the precipitation projected for the cells is an overestimation. A preliminary bucket survey suggests that comparable precipitation cells occurred in the mountains of West Virginia (Purkey, unpublished data), but the precipitation data are not sufficient to define the influence of topography on rainfall intensity or distribution during the flood.

The discharge data do suggest that the heart-shaped isohyet cell shown in the center of Figure 6 was the most significant flood producer. Most of the headwaters of Cheat River and South Branch Potomac River were under this cell. Many streams within these basins had discharges estimated to exceed the five hundred year event. Parts of the Shenandoah, Greenbrier, and Jackson Rivers were also under this cell and had floods in excess of the one hundred year event. Streams under the cell had runoffs per area in excess of 2 $m^3s^{-1}km^{-2}$ in drainage areas less than 2500 km^2.

The greatest runoffs per area recorded for the November 1985 flooding were about 10.4 $m^3s^{-1}km^{-2}$ for small streams in the upper James River basin (Lescinsky, 1986). Comparable runoff probably occurred in the Cheat and South Branch Potomac basins but there are no gages on streams with drainage areas of less than 223 km^2. Most of the smaller streams in the Cheat River basin showed relatively little modification by the 1985 flood. With some exceptions in the South Branch Potomac River basin mentioned in the discussion of mass movement, the floodwaters appear to have had the most pronounced geomorphic impact in basins with drainage areas from 130 to 4000 km^2. Smaller streams may have been able to accommodate the moderately high precipitation because comparable rainfall intensities occur during conventional thunderstorms. Larger basins include streams which received significantly less precipitation so did not experience such dramatic geomorphic change.

When compared with two widely studied floods in the Appalachians, the November 1985 floods show significantly higher runoffs per area for drainage basins between 650 km^2 and 3900 km^2. Runoff per area for the Moorefield, Springfield, Parsons, and Rowlesburg stations (Table 2) are at least 25% greater than values reported for similar sized basins during Hurricane Camille floods in Virginia during 1969 (Prugh and others, 1986) and Hurricane Agnes floods during 1972 (Bailey and others, 1975).

Calculation of the recurrence interval for extreme floods is speculative at best. Gumbell-distribution plots of discharge on the South Branch Potomac and Cheat rivers (Figs. 7a-7d), however, show the relationship between the November 1985 floods and previous floods. If flood recurrence is determined by the log Pearson type III method, peak 1985 discharge was 2.1 times the one hundred year flood at the Petersburg station and 3.5 times the one hundred year flood at Parsons (U. S. Geological Survey, 1985). Although these flood data could be used to question whether standard flood-frequency calculation methods are appropriate for extremely high magnitude floods in the Appalachians, the data clearly show the unprecedented nature of the 1985 flooding.

One question associated with the November 1985 flooding is why five hundred year recurrence flooding occurred in response to what was, at most stations, a one hundred year (or less) 24-hour rainfall event. Clearly the antecedent late-October and 1-2 November rains brought soil moisture, water table levels, and stream discharge to relatively high levels before the 4-5 November storm. Evapotranspiration rates and interception storage by foliage were low because the deciduous trees had dropped their leaves. As severe as the November 1985 flooding was, it could have been much worse if precipitation had been greater and if the major precipitation cells had been located completely within one drainage basin instead of along the drainage divides of at least five different basins. In addition, if precipitation intensities had been higher, even locally, widespread debris slides and flows could have been triggered, impeding or damming flood water upstream. The breaching of such dams could have produced debris-laden flood surges, further exacerbating the devastation of the flood.

SELECTED GEOMORPHIC EFFECTS

Introduction

Because of the size and number of drainage basins affected by theNovember 1985 storm (Fig. 1) and the inaccessibility of many of areas within them, it has not been possible to complete a detailed reconnaissance of all of the flood affected areas. The effects of flooding in the upper Greenbrier, Shenandoah, James, and Roanoke rivers need to be studied. Landslides reported in the Roanoke valley also need investigation. High resolution remote sensing imagery may provide a means to further inventory the effects of the storm. When completed, the investigations of the headwater areas of South Branch Potomac River and Cheat River described below will serve as a control for much-needed models of overall storm effects.

Hillslopes

Considering the record flood effects in at least five drainage basins, it is surprising that regional reconnaissance following the November storm identified few areas affected by mass movement even in the Northern Blue Ridge area where precipitation in excess of 350 mm fell in three days (Fig. 6).

Most of the investigated mass movement (Spring, 1987) was concentrated in Pendleton County, West Virginia, along the upper South Branch Potomac River (Fig. 1) (U. S. Geological Survey Circleville and Onego 7.5' quadrangle maps) and adjacent areas where the 3-5 November rainfall was 200-250 mm. Within this area, soils on the Reedsville Shale were most susceptible to failure (Fig. 8). Many of the shallow slumps and planar slides mobilized into mud flows and delivered sediment directly to the streams at their base (Jacobson and others, 1987). Although it is unclear what proportion of total sediment yield resulted from these slope failures, field evidence strongly suggests that greater channel and channel-margin erosion occurred downstream from them. This suggests that sediment input by slope failure was instrumental in mobilizing additional sediment, probably by increasing the effective density of the flow and consequently increasing the basal shear stress and erosive capacity of the stream.

Fewer failures occurred on slopes underlain by limestones, siltstones, or sandstones in the same area of Pendleton County. Five large debris slides and flows were important anomalies but were scattered so widely as to be relatively unimportant in contributing sediment to the flood. It is significant that these large slides (slide scars areas in excess of 100 m^2 and flow tracks up to 2 km long) were triggered by the storm on dip slopes (Jacobson and others, 1987; Clark 1987). By contrast, in the much more local and intense storm of June 17-18, 1949 (Stringfield and Smith, 1956), nearly all major hollows on dip-, strike-, and obsequent-slope faces failed around the northern perimeter of North Fork Mountain (Clark, 1987, Fig. 3). Therefore, the November 1985 event may shed light on threshold antecedent soil moisture conditions and precipitation intensities necessary to trigger large mass movements under post-leaf-fall conditions.

Slope failures in upper South Branch Potomac River drainage were rare outside the area covered by the U.S.G.S. Circleville and Onego 7.5' quadrangle maps in Pendleton County, West Virginia. Along the Allegheny Front, surficial slumps and planar failures were observed in the valley of Brushy Run, tributary to Seneca Creek. Most of these failures were in colluvial soils derived from shales and silt-

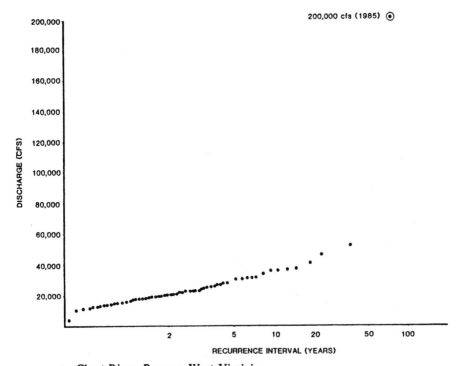

a. Cheat River, Parsons, West Virginia.

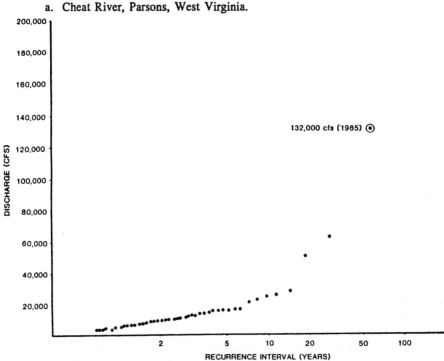

b. South Branch Potomac River, Petersburg, West Virginia.

Figure 7. Gumbell plots showing relationship of previously-recorded discharges (cfs) with discharge for the November 1985 floods.

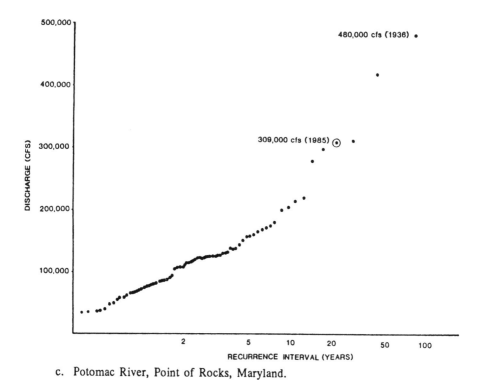

c. Potomac River, Point of Rocks, Maryland.

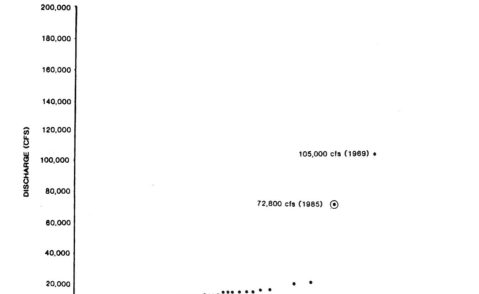

d. Maury River, Buena Vista, Virginia.

Figure 8. Shallow debris slides, skin slides, water blowouts and slumps along a tributary of Judy Run, North Fork Mountain, Pendleton County, West Virginia. Location is in central 1/9th of Circleville, West Virginia, 7.5' U. S. Geological Survey quadrangle. Bedrock is Reedsville Shale of Ordovician age.

stones of the Mauch Chunk Group. Few of the failures had runouts sufficiently long to reach flood waters. Similarly, few failures were observed in the Little River basin, Virginia, an area extensively effected by debris slide, debris flows, and widespread deposition of debris fans during the June 17-18, 1949 storm (Hack and Goodlett, 1960).

Perhaps the most historically-instructive slope feature brought to light by the 1985 storm are dramatic cross-sectional exposures of pre-existing debris fans. Many of the newly exposed cuts in debris fans reveal a complex stratigraphy produced by multiple episodes of deposition, weathering, erosion, and winnowing. Most of these deposits are diamictons. commonly greater than 4 m thick and frequently composed of massive to extremely poorly bedded, pebble or cobble block and boulder sediments (Kite, 1987). At some locations, a thick pebble diamicton, commonly exhibiting pedogenesis, is overlain by a thick cobble-block and boulder diamicton, suggesting the possibility of two or more discrete episodes of fan development. On both fans and aprons, sequences of relatively thin (2-3 m), normally graded diamicton beds are common and are underlain by cobble-block and boulder diamictons which fine upsection (Kite, 1987). Many diamicton sequences are capped by relatively matrix-free block and boulder accumulations at the present fan

Figure 9. Older debris fan showing compelx stratigraphy (Kite, 1987) exposed by truncation during the November 1985 flood. North Fork South Branch Potomac River in North Fork Gap, Grant County, West Virginia. Site is in northeast 1/9th of Hopeville, West Virginia, 7.5' U.S. Geological Survey quadrangle. Most of fan is composed of diamictons, but note essentially horizontal alluvium in fan near right side of photograph.

and apron surfaces. Tabular beds and lenses of alluvial sand, gravel, and boulder-cobble gravel commonly intercalate with diamictons in fan exposures (Kite, 1987) and demonstrate the polygenetic nature of the deposits underlying these landforms.

Stratigraphic and pedologic study of a large, statistically significant sample of fan sections may yield valuable data on past episodes of debris fan deposition and erosion, and would shed light on the relative importance of the processes involved in their formation and modification by subsequent geomorphic events. We warn, however, against the premature development of a chronology of catastrophic events based on fan deposition and erosion observed on a too local or otherwise unrepresentative sampling of the fan population. Even in the central areas of intense debris fan producing storms (see Clark, 1987), not all debris fans have been observed to mobilize, only certain areas of individual fans may do so. Careful soil-stratigraphic field work, and the examination of large numbers of fan exposures will be necessary in order to utilize this potentially valuable sedimentary record.

Figure 10. Example of bedrock-defended water gap where natural constriction of floodwaters typically produced intense upriver ponding and slack-water deposits, and downstream   scour and splay deposition.   South Branch Potomac River at Eagle Rock, Pendleton County, West Virginia. Location is in central 1/9th of Upper Tract, West Virginia, 7.5' U. S. Geological Survey quadrangle. Bedrock in ledge and high cliff is Oriskany Sandstone of Devonian age.

Floodplains

Geologic Setting. The November 1985 flood in upper Cheat and Potomac Rivers highlighted the effects of pre-existing controls on floodwaters, fluvial processes, erosion and sedimentation, and landforms. In both drainage basins, the floodplain geometry is dramatically responsive to differences in the resistance of the bedrock to weathering and erosion. Floodplains are widest upstream from bedrock-defended water gaps (Fig. 10), particularly in the vicinity of major stream confluences. For example, the Cheat River confluence at Parsons, West Virginia is near the head of a floodplain nearly 1 km wide that extends downstream for nearly 18 km. Large flood plains in the upper Potomac basin are at Petersburg where a 4 km wide plain extends upstream 7 km from Petersburg Gap, and over 10 km both northeast and southeast of Moorefield where floodplain widths are commonly 2 km upstream from the Trough (Tewalt, 1977). All of these floodplains were inundated up to 3-4 m during the 1985 flood.

In sharp contrast with broad alluvial plains are bedrock-bordered narrows having no floodplain. In the upper Potomac basin, excellent examples of such constrictions are the Trough and Petersburg Gap (Tewalt, 1977) and Smoke Hole where Sites (1972) finds narrow and discontinuous alluvial deposits. During the 1985 flooding, these bedrock narrows dammed waters upstream to heighten flood crests and funneled the discharge through the gaps at increased velocities to produce scours and splays downstream.

Between the extremes of well-developed flood plains and rock-walled gorges are the remainder of floodplain reaches. These reaches are characterized by narrow stretches of usually-thin alluvial fills alternating with narrower stretches in which only discontinuous floodplain sediments are present.

General Characteristics of Floodplain Erosion and Deposition. Erosional landforms resulting from the 1985 flood were most abundant and best developed in valley constrictions and in areas without trees. Intense scour occurred within and just downstream from bridge abutments (compare Ritter and Blakely, 1986) and bedrock-defended water gaps.

Away from valley narrows, floodplain erosion was concentrated in treeless areas such as fallow fields and unpaved roads. Cutbank retreats of more than 10 m were measured at several locations although smaller lateral incisions were more common. Abandoned channel scars left by avulsion were another landform developed during flooding. Avulsions and near avulsions most commonly occurred along the insides of meander bends and sloughs.

The volume of sediment entrained, transported, and redeposited in the affected drainage basins constitutes the most striking geomorphic and sedimentological effects of the November 1985 floods (Kite, 1986). Most of this sediment was derived from preexisting, lower and upper floodplain deposits and was redeposited in downstream reaches. Additional sediment was supplied by erosion of colluvium and debris fans on valley sides by flood-waters that reached depths of 3 to 4 m, completely inundating the valley bottom. In the area of Pendleton County, West Virginia, sediment was also supplied by many small landslides that were transformed into mudflows and delivered sediment directly to the streams (Jacobson and others, 1987). The contribution of these slides and flows to the total sediment yield is as yet unknown. Their effect, however, was to increase the severity of erosion downstream by increasing floodwater viscosities and shear stresses and by providing large angular boulders and vegetation which increased turbulence and provided erosive tools.

Widespread cobble or cobble-boulder splays formed downstream from large bridge abutments (see Ritter and Blakely, 1986) and water gaps. Their distal splay ends thin to cobble veneers which cover pre-flood alluvium (Fig. 11). Costa (1974) notes that coarse-grained sediment lenses within fine-grained alluvial sequences may record overbank deposition during large floods. The thickest (up to 1.5 m) sand and gravel deposits on the floodplain occurred as dunes immediately downstream from sites of intense scour of the floodplain. The abundance of these

Figure 11. Floodplain of South Branch Potomac River between Smoke   Hole Campground, which was destroyed in the flood, and Eagle Rock, Pendleton County, West Virginia. Location is in northeast 1/9th of Upper Tract, West Virginia, 7.5' U. S. Geological Survey quadrangle. View is upstream to south; note extensive scour, overbank deposition of sand, gravel, cobbles and boulders, and effects on vegetation. Smoke Hole Road bank (reconstructed) at lower right; South Branch Potomac River is at base of left hillslope just below left center.

large dunes indicates widespread, sustained, high-velocity flows of considerable depth. Even larger dunes and bars (over 3 m high) are composed of trees, automobiles, and mobile homes. Many of these dunes and bars were removed during the cleanup after the flood.

Slack-water Deposits on Cheat River Floodplain. Our studies of slack-water sediments have concentrated on deposits along the Cheat River between Parsons and Albright, West Virginia. Some have been obliterated or much modified by the cleanup after the flood but many were left undisturbed and provide the basis for our ongoing study of the sedimentology and geomorphology of slack-water deposits and their use as a paleoflood stage indicator (Linton and Kite, 1987).

Slack-water deposits formed at the intersection of tributary streams and the Cheat river wherever a topographic high upstream from the tributary obstructed flow of the Cheat River, creating a quiet-water pool. Slack-water alluvial sequences contain up to four units: a lower basal gravel unit which fines upward into a medium-to-coarse sand unit, overlain by a silt unit and capped by a discontinuous fine-sand unit.

The basal gravel unit is composed of poorly sorted and imbricated cobbles, pebbles, and sands that thicken towards the tributary mouth. The imbricated gravels dip in an up-tributary direction, indicating this unit is derived from the tributary basin. The basal gravel unit is restricted to tributary channel bars and the tributary bank. The basal gravel unit does not underlie finer sand and silt units deposited on the narrow tributary floodplains further from the tributary channel. The basal gravel unit occurs in higher-gradient tributary stream but is absent in shallow-gradient tributary streams.

The gravel unit fines upward into a moderately sorted, medium-to-coarse sand unit which also thickens towards the tributary mouth. The sand unit, present at every deposit studied along the Cheat river, thins away from the tributary and, at one location, is in a lateral facies relationship with the gravel unit. The sand unit is closely related to the gravel unit and probably originates from the tributary. The sand unit is thicker adjacent to higher gradient tributary streams and has an abrupt contact with the overlying silt unit.

The silt thickens towards the tributary mouth. The silt unit has a relatively uniform thickness across the floodplain of the tributary. There does not seem to be a correlation between silt unit thickness and stream gradient. The silt unit appears to originate from the Cheat river. It contains stratified and unstratified gray and light red sediments. It is present at every slack-water deposit studied along the Cheat river and is capped with a thin, discontinuous, fine-sand unit.

The fine-sand unit may have been originally deposited as a discontinuous body by the flood or have been initially deposited as a continuous unit which was subsequently eroded. This unit is readily eroded and has washed into 1-2 cm wide desiccation cracks that developed in the upper silt unit. The fine-sand unit appears to originate in the tributary basin. The unit is thickest closest to the tributary and pinches out laterally away from it.

The Cheat river slack-water deposits present some disturbing implications for paleohydraulic reconstructions of large floods in the Appalachians. The upper surface of a recognizable slack-water deposit studied in the Cheat River basin lies 2-3 m below high water marks indicated by other evidence. Small isolated sand patches occur near the high water marks but these poorly developed deposits appear to be very ephemeral and difficult to distinguish from similar sand patches found further up the tributary stream. Our preliminary investigation shows that retrodiction of the November 1985 Cheat River flood, based on the top of the slack-water deposits, underestimates the discharge by at least 50%.

The presence or absence of similar flood deposits beneath the slack-water deposits left by the 1985 flood could be used to determine whether or not comparable, extremely large floods occurred in the Late Holocene. No sedimentary packages analogous to the 1985 deposits have yet been found. Unstratified sandy loams with weak soil profile development are found locally beneath the 1985 deposits but they lack discrete sand and clayey units common to all 1985 slack-water deposits. It is possible that the unstratified sandy loam represents bioturbated or otherwise disturbed, originally stratified, slack-water packages but it is more likely that it represents upper floodplain sediments accumulated in relatively small increments during several or many floods that were significantly smaller than the 1985 event. The absence of older slack-water sediment packages may suggest that the 1985 event was the "flood of the millennia". Alternatively, their absence may indicate that slack-water deposits are relatively ephemeral (removed in decades or centuries) through reworking by minor floods on tributary streams during the intervals between major floods on the Cheat River.

Single Geomorphic Events. The flood also caused geomorphic events not commonly observed during floods of one hundred to five hundred year frequency. Incipient takeover of South Branch of Potomac River began at the drainage divide between Eagle Rock and Upper Tract, Pendleton County, West Virginia. Floodwaters, hydraulically dammed behind a bridge, a large mill, and the narrow, bedrock-confined water gap, poured over the low drainage divide and into the headwaters of Stony Creek, a tributary of North Mill Creek. This temporary diversion destroyed a building and scoured the road-bed of US 220. This may represent an early stage in the diversion of the river from a 40 km reach which includes the scenic Smoke Hole area. Along its present course, the South Branch transects the Tuscarora Sandstone six times and the Oriskany Sandstone at least thirty times (Dunne and Gerritsen, unpublished data; Sites, 1972). The valley of Stony and North Mill Creeks is a strike valley crossing neither of these resistant, quartz-rich sandstones and is underlain by shales for most of its course to Petersburg, West Virginia.

At Lost River Sinks, Hardy County, West Virginia, the swallet of Lost River became temporarily plugged during or after the flood. Because the Lost River Sinks have been known for well over a hundred years, this event may be unique in this time span. Large magnitude floods on uncleared flood plains in early settlement history, however, would probably not have yielded the sediment load produced by the November 1985 flood. Therefore, even great floods that predated extensive deforestation may not have provided the types and volumes of sediment carried by the flood of November, 1985.

DISCUSSION

The November, 1985 storm in the Central Appalachians inundated entire floodplains in at least five headwater drainage basins between 130-400 km^2 and produced numerous hillslope failures in at least one basin. This one hundred year to greater than five hundred year event modified diverse hillslope soils (Jacobson and others, 1987) and floodplain sediments (Kite, 1986) and produced erosional and depositional landforms on a scale greater than that suggested by the recorded precipitation intensities, durations, and total amounts. Some of this incongruity may be explained by antecedent effects and some may be due to the paucity of rainfall gaging stations and the spatially variable intensity of precipitation. Pedological and vegetational factors in the drainage basins examined thus far, particularly soil texture and structure, slope class and slope position, and the type and density of vegetation are reflected in the location and nature of both hillslope failures and floodplain scour and deposition. The structural attitude of bedding on fold limbs and the relatively high resistance to weathering and erosion of the sandstones that wall the narrows and water gaps locally modified the effects of the flood on the floodplain and adjacent hillsides.

Recovery time is the time necessary for the river to reacquire its equilibrium geometry after a catastrophic flood. In the Cheat and Potomac as well as in other affected basins, the floodplains are inhabited and cultivated thus much of the "recovery" has been anthropogenic. In these areas, the resultant channel configurations bear little resemblance to those produced in the November 1985 flood. Follow-up research on the natural recovery time will be done in isolated, steep-walled, high-gradient reaches, several of which are present in both the Cheat and Potomac River basins. These isolated reaches, however, are not representative of the settled and cultivated reaches of the rivers.

It is interesting to speculate about the long-term effects of great floods in bedrock confined valleys in temperate, humid regions such as those affected by the November 1985 storm. Except in upper Potomac River drainage, widespread mass movement on hillslopes has not yet been reported and the movements that have been reported (Jacobson and others, 1987) are predominantly not large-scale catastrophic debris slides and flows but their aggregate effect may be great. The regional effects of this storm may be regarded as predominantly fluvial in terms of both down-basin sediment transport and changes in floodplain landforms. The widespread floodplain scour and deposition observed in the 1985 floods, often extending across entire valleys and reworking older floodplain sediments, underscores the importance of infrequent, high-magnitude floods.

Regardless of the magnitude and frequency of the causative geomorphic events, the pervasive reworking of floodplains in Late Quaternary time may be more widespread in the Appalachian region than previously appreciated. Jordan, Behling, and Kite (1987) report ^{14}C dates from alluvium and colluvium from the valley of Pendleton Creek, tributary to Blackwater and Cheat River drainage, Tucker County, West Virginia. They find the modern floodplain alluvium is generally no older than 10,000 ^{14}C years in this high elevation (about 950 m) headwater drainage. Alluvial deposits on the floodplain margins, however, contain wood fragments dating approximately 17,000 B.P. and the basal colluvium along valley sides yield wood fragments generally between 22,000-29,000 B.P. These dates suggest that such floodplains have been almost completely reworked at least once during Holocene time. In the study just cited, the Pendleton Creek record shows Early Holocene aggradation followed by lateral channel migration without incision or aggradation. Whether this pattern is regional cannot, of course, be determined without additional studies but the data add to the growing evidence that Appalachian fluvial landscapes are of multiple ages and origins.

Landforms and deposits produced by floods may, under certain circumstances, be useful as paleoflood indicators and be used for calculating accurate recurrence intervals for flooding. Baker (1986) has stressed the need for accurate paleoflood reconstructions to help estimate recurrence intervals. In addition to expected flood crests, more information is needed on the effects of local hydraulic and debris damming, channel constrictions, severe lateral and vertical erosion, massive overbank deposition, and avulsions. The November 1985 flood demonstrated the complex interaction among precipitation patterns, runoff, hillslope and floodplain configuration, and the effects of water gaps and other constrictions. The current recurrence interval data underestimated the catastrophic geomorphic changes and the loss of life and property caused by the flood. This may be due to inherent weaknesses in the way recurrence intervals are calculated combined with nature of precipitation and flood data combined with the complex interactions among processes, sediments, and landforms. Because this investigation finds similar effects of the flooding on floodplains in different basins, the generalizations drawn from this study may be valid for other areas within the Appalachian highlands.

Because of the paucity of precipitation and discharge data and the preliminary nature of our research on storm effects, the interpretations drawn in this paper are tentative. We hope this paper will stimulate further research on the November 1985 storm and its effects.

REFERENCES CITED

Bailey, J. F., Patterson, J.L., and Paulhus, J. L. H., 1975, Hurricane Agnes rainfall and floods, June-July 1972: U. S. Geological Survey, Professional Paper 924, 403p.

Baker, V. R., 1986, Natural hazards in the hydrologic cycle (abstract): Geological Society of America, Abstracts with Programs, v. 18, p. 533.

Baloch, M. S., Henry, E. N., and Burchinal, J. C., 1973, Comprehensive survey of the Monongahela River: Volume 1, Inventory: West Virginia Department of Natural Resources, 319p.

Cardwel, D. H., Erwin, R. B., and Woodward, H. P., compilers, 1968 (revised 1986), Geologic map of West Virginia: West Virginia Geological and Economic Survey, 2 sheets, scale 1:250,000.

Clark, G. M., 1987, Debris slide and debris flow historical events in the Appalachians south of the glacial border: Geological Society of America, Reviews in Engineering Geology, in press.

Costa, J. E., 1974, Stratigraphic, morphologic, and pedological evidence of large floods in humid environments: Geology, v. 2, p. 301-303.

Hack, J. T., and Goodlett, J. C., 1960, Geomorphology and forest ecology of a mountain region in the central Appalachians: U. S. Geological Survey, Professional Paper 347, 66p.

Hershfield, D. M., 1961, Rainfall atlas of the United States: Washington, D.C., Weather Bureau Technical Paper 40, 111p.

Hobba, W. A., Jr., Friel, E. A., and Chisholm, J. L., 1972, Water Resources of the Potomac River Basin, West Virginia: West Virginia Geological and Economic Survey, River Basin Bulletin 3, 110p.

Jacobson, R. B., Cron, E. D., and McGeehin, J. P., 1987, Preliminary results from a study of natural slope failures triggered by the storm of November 3-5, 1985, Germany Valley, West Virginia and Virginia: U. S. Geological Survey Circular 1008, p. 11-16.

Jordan, M. K., Behling, R. E., and Kite, J. S., 1987, Characterization and Late Quaternary history of the alluvial and colluvial deposits of the Pendleton Creek Valley, Tucker Co., WV (absract): Geological Society of America, Abstracts with Programs, v. 19, p. 92.

Kite, J. S., 1986, Erosion and deposition in the Monongahela and Potomac drainage basins during the floods of November 1985 (abs): Geological Society of America, Abstracts with Programs, v. 18, p. 313.

Kite, J. S., 1987, Colluvial diamictons in the Ridge and Valley province, West Virginia and Virginia: U. S. Geological Survey Circular 1008, p. 21-23.

Lescinsky, J. B., 1986, Flood of November 1985 in West Virginia, Pennsylvania, Maryland, and Virginia: U. S. Geological Survey, Open File Report 86-486, 33p.

Linton, R. C., and Kite, J. S., 1987, Slack-water deposits along the Cheat River in east central West Virginia (abstract): Geological Society of America, Abstracts With Programs, v. 19, p. 95.

N.O.A.A., 1973, Monthly normals of temperature, precipitation, and heating and cooling degree days 1941-1970: Climatology of the United States No. 81, West Virginia, Virginia;, and D. C.: Asheville, N. C., National Climatic Data Center.

N.O.A.A., 1985a, Daily Weather Maps, Weekly Series, October 28- November 3, 1985: Washington, D.C., U. S. Government Printing Office, 8 p.

N.O.A.A., 1985b, Daily Weather Maps, Weekly Series, November 4-10, 1985: Washington, D.C., U. S. Government Printing Office, 8p.

N.O.A.A., 1985c, Climatological Data - West Virginia, Maryland, Pennsylvania - October and November, 1985: Asheville, N. C., National Climatic Data Center, v. 93, No. 10-11.

Prugh, B. J., Easton, F. J., and Lynch, D. D., 1986, Water Resources Data: Virginia: Water Year 1985: U.S. Geological Survey, Water-Data Report VA-85-1, 398p.

Ritter, D. F., and Blakley, D.S., 1986, Localized catastrophic distruption of the Gasconde River flood plain during the December 1982 flood, southeast Missouri: Geology, v. 14, p. 472-476.

Sites, R. S., 1972, Geology of the Smoke Hole region in Grant and Pendleton Counties, West Virginia (M.S. thesis): Morgantown, West Virginia University, 106 p.

Sites, R. S., 1978, Structural analysis of the Petersburg lineament, Central Appalachians (Ph.D. thesis): Morgantown, West Virginia University, 171 p.

Stringfield, V. T., and Smith, R. C., 1956, Relation of geology to drainage, floods, and landslides in the Petersburg area, West Virginia: West Virginia Geological and Economic Sruvey, Report of Investigations 13, 19 p.

Tewalt, S. J., 1977, Fluvial terraces of a part of the South Branch of the Potomac

River (M.S. thesis): Morgantown, West Virginia University, 87 p.

Thornbury, W. D., 1965, Regional Geomorphology of the United States: New York, John Wiley and Sons, 609 p.

U. S. Geological Survey, 1966, Drainage map of West Virginia: U.S. Geological Survey, scale 1:500,000.

U. S. Geological Survey, 1985, Natural water conditions, November 1985, Stream-flow during November, 7p.

18

A high magnitude flood in the Sinai Desert

Asher P. Schick and Judith Lekach

ABSTRACT

Wadi Mikeimin, a 13 km^2 watershed in the hyper-arid steep crystalline mountains of SE Sinai, experienced a local high magnitude flood in January 1971. A detailed reconstruction of the event was undertaken on the basis of: (1) high water marks; (2) geomorphic evaluation; (3) assumptions based on data transferred from comparable instrumented watersheds; (4) considerations of competence; and (5) analysis of the total sediment load deposited as an alluvial fan at the junction of Wadi Mikeimin with the large Wadi Watir.

The flow in Wadi Mikeimin peaked to 68.5 m^{3}s^{-1} at the outlet and to 80 m^{3}s^{-1} for a 5 km^2 tributary within it. Total flow was in excess of 100,000 m^3. The flow deposited 6,200 m^3 of coarse stratified sediment in the form of an alluvial fan on the channel bed of Wadi Watir. The January 1971 flood did not activate Wadi Watir itself, and its channel remained dammed by the Mikeimin fan until November 1972, at which time it was breached by a major flood which did not activate Wadi Mikeimin. For the 21 months that elapsed between these two events, a lake up to 400 m long existed, fed by the perched water table of the nearby oasis of Ein Fortagha.

Breached remnants of similar sediment dams were found in many localities in the main wadis of the mountainous Sinai. Their role as obstructions at junctions, caused by the localized flood pattern characteristic of deserts, is important in understanding some aspects of floods, especially their fronts. In evaluating the probability of their formation, factors to be considered include the junction angle, the difference in size and in channel slope between the junction tributaries, and the temporal and spatial distribution of rainfall. Such obstructions may also operate for short periods within a single flood event wherever the size difference between tributary and main stream insures the required lag between the recession in the tributary and the onset of the floodwave in the main stream.

INTRODUCTION

An exceptionally large rainstorm occurred in January 1971 over the watershed of Wadi Mikeimin, one of the numerous tributaries of Wadi Watir in southeastern Sinai. The resulting flood created an alluvial fan at the Mikeimin-Watir confluence. This confluence is located in the area of the Ein Fortagha springs, about 20 km upstream of the terminal alluvial fan of the Watir at Noueiba, on the western coast of the Gulf of Aqaba (Fig. 1). The Mikeimin fan dammed some of the shallow groundwater which collected upstream of the obstruction, to form a lake 400 m long and covering the entire width of the Watir channel bed (Meshel, 1972). This obstruction existed as long as there was no flow; it was breached by the first subsequent flood in Wadi Watir which occurred in November 1972.

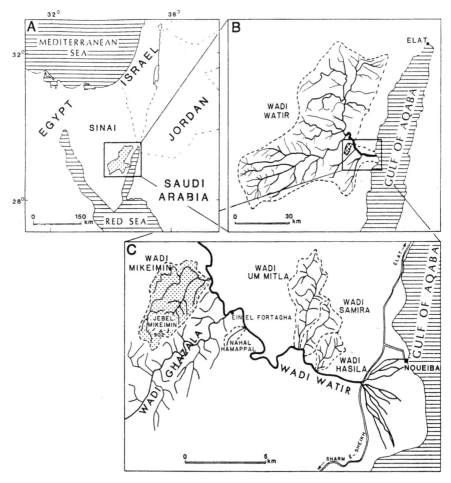

Figure 1. Location map: A - The Sinai Desert and environs; B - The Wadi Watir catchment; C - The lower Watir area.

The effect of damming or partial obstruction by sediment contributed by tributaries and subsequent breaching by flow in the main stream has been described by numerous authors. The heavy floods of March 1936 in the Connecticut River area produced tributary alluvial fans which covered large tracts of the floodplain (Jahns, 1946). Variations of this effect have been studied rather intensively in the Colorado River (Webb, 1987; Howard and Dolan, 1981; Cooley et al., 1977; Leopold, 1969; Graf, 1979), but are described from many other localities, especially in connection with debris flows (Benda, 1985; Benda, personal communication; Dietrich and Dunne, 1977; Gallino and Pierson, 1984; Suwa and Okuda, 1984), dam breaks (Jarrett and Costa, 1984), sedimentary processes (Blair, 1987), and man-made changes (Petts, 1984). A recent well documented study is by Harvey (1986).

Relatively less is known about the effect in arid regions. Channels of larger streams in Utah are sometimes dammed temporarily by coarse debris from tributaries (Woolley, 1946). Rio Puerco and Rio Salado occasionally build fans at their mouths which divert the Rio Grande to the opposite side of its valley. Depending on the flow of the main stem, removal of these features may take years (Wolman and Miller, 1960). Under arid to hyper-arid conditions, debris flow activity is very restricted and, in most cases, debris flows do not even reach talus footslopes (Gerson and Grossman, 1987). Still, damming of wadis by tributary fans occasionally occurs, reflecting the spatially and temporally erratic activity pattern of stream systems in arid regions (Gerson, 1982). Stratigraphically, it offers the possibility of interpreting fine fluvial material as having been deposited upstream of fan dams such as along Wadi Hanifa, in central Saudi Arabia (Hoetzl et al., 1976). Modern counterparts of this phenomenon are known to have occurred in the Minab Rud catchment of southeast Iran (Carls, personal communication).

This paper presents the results of a study of a present-day alluvial fan deposit formed by a tributary at its confluence with the main stream during a single flood event in a hyper-arid environment. The aim of the study is to reconstruct the storm, runoff, and erosion-deposition characteristics of the flood event. Using data from the Nahal Yael Research Watershed, situated 80 km to the north of Wadi Mikeimin, frequency considerations are established for the stochastic simulation of the temporal damming-breaching interplay under arid conditions.

THE PHYSICAL SETTING

Location

Wadi Mikeimin is one of the numerous tributaries of Wadi Watir -- a 7th order, 3,860 km^2 catchment draining a part of the mountainous desert of southeastern Sinai towards the Gulf of Aqaba (Fig. 1). In the last 30 km of its course, just west of the terminal alluvial fan of Noueiba, the Watir is deeply entrenched in a canyon-like channel many hundreds of meters lower than the adjoining mountain peaks. Wadi Mikeimin, a 3rd order, 12.9 km^2 catchment, joins the Watir in the upper part

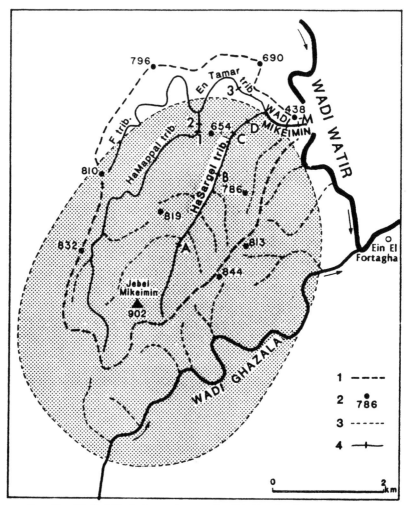

Figure 2.   The Wadi Mikeimin catchment: 1 - Divide; 2 - Elevation in meters;
3 - Hypothetical extent of rainstorm cell; 4 - Floodmarks section.

of this gorge, about 20 km upstream of the apex of the Noueiba fan and 4 km upstream of the main seepage area of the Ein Fortagha springs. The area of the Watir
catchment upstream of the Mikeimin confluence is 3,440 km². The springs,
mostly concentrated downstream of the confluence, yield small but semi-permanent
amounts of water. The continuous low flow over several kilometres along the
Watir channel results in a highly localized date palm oasis, in stark contrast to the
surrounding barren rocky desert.

Climate

Climatic conditions prevailing in the Wadi Mikeimin area can be estimated on the basis of the meteorological records at Elat, 80 km to the north. Summer temperatures exceed 40°C, with diurnal differences of up to 20° or more. Winter temperatures are 10-15° lower. The mean annual rainfall can be estimated at 30-40 mm, with high inter-annual variation. There are, as a rule, 5-10 rain days per year, and most of the amount is concentrated in a few hours.

The infrequent rainstorms are from two origins: Mediterranean weather systems with a deep southern trajectory; and southerly systems, mostly active in the transition seasons. Many storms are highly localized, though large scale regional rain events also occur.

Geology

Wadi Mikeimin is part of the strongly uplifted western margins of the great Aqaba rift. This geologic zone is the backbone of a continuous high mountain range of predominantly highly resistant basement rocks, towering many hundreds of meters above the nearby coastline to the east. Wadi Watir is one of the hydro-

Figure 3. Wadi Mikeimin. View is upstream. The surface of the fan deposited at the Watir confluence is in the foreground.

graphic outlets breached through this ridge along its 250 km from Sharm-el-Sheikh in the south to Elat in the north.

The Mikeimin catchment is entirely underlain by Precambrian basement rocks, especially granites and schists (Eyal et al., 1980), which are intruded by Tertiary basalt. Numerous faults trending NNW-SSE are present; one is clearly reflected by the 4 km long straight course of the HaSargel tributary (Fig. 2). Numerous other stream reaches in the area around the Mikeimin also exhibit this tendency.

Geomorphology

The catchment of Wadi Mikeimin has a steep mountainous topography (Fig. 3). Its divide is around 800 m and its mouth is 270 m a.s.l. The highest point in the catchment, Jebel Mikeimin, is 5 km southwest of the Mikeimin-Watir confluence (Fig. 2). Its elevation -- 902 m -- yields a relief ratio of over 12%.

Wadi Mikeimin unites three main tributaries, generally trending southwest to northeast, into a relatively short main reach. They have substantially different attributes. HaSargel tributary is the steepest -- more than 0.08 over its lower 4 km of channel. It has a gravelly bed with a median size of 53 mm on the active bars. Boulders 500 mm and larger are common, with sizes of up to 1,500 mm also present. Channel width is fairly constant around 35 m, without constrictions and with occasional widening to over 40 m.

En Tamar tributary (Fig. 2) unites two tributaries: HaMappal and F. HaMappal tributary has two dry waterfalls 15 and 12 m high. Polish by flood scour on the waterfall rocks makes the climb on them nearly impossible and ascent was by a roundabout path. Upstream of the waterfalls the channel slope is 0.03 and the channel alluvium is finer than below the waterfalls, though some very large boulders are present. The reach below the waterfalls is only 500 m up to the En Tamar confluence. It has a channel slope of 0.05, a median size on the active gravel bars of 29 mm, and lacks large boulders.

The F tributary is similar in most respects to HaSargel tributary. It is somewhat less steep and lacks very large boulders.

The slopes in the Mikeimin catchment are steep, with angles of 20° to 30° common. In some places there are cliffs descending nearly vertically into the basal channel (Fig. 3). HaSargel tributary is lined for several tens of meters by an alluvial terrace which forms one of the banks of the present channel. The rocky parts of the slopes are intensely jointed. Sediment available for transport is found on them in large quantities. We estimate that one half of the area of the Mikeimin slopes is bare rock and the other half is covered by debris.

METHODS

The study of floods in extremely arid environments is seriously hampered by lack of data, difficulties of forecasting storms, and the suddenness and fast transience of flood events (Schick, in press). Excepting unique instrumented catchments such as Nahal Yael, it is usual to study such floods post factum by evaluating floodmarks, geomorphic effects, and various indirect evidence.

The January 1971 event in Wadi Mikeimin was recognized as an exceptionally large flood only several months after its occurrence. The detailed study began only in August 1972. Though some of the floodmarks had been obscured by natural processes during the intervening 18 months, the hyper-arid climate guarantees good preservation. Fortunately, no flow occurred either in the Mikeimin or in the Watir during that period. Hence, the total sediment delivered to the Watir by the Mikeimin catchment during the event was preserved intact as an alluvial fan at the confluence. The reconstruction of the event is based on a correlation between the flood characteristics as deduced from floodmarks inside the catchment and the sediment deposited at its mouth. Use is made of detailed background information on the generation of desert floods collected over the last 20 years in and around the Nahal Yael Research Watershed, located in a similar setting 80 km north of Wadi Mikeimin (Schick, in press).

Peak Discharges and Reconstructed Hydrographs

Peak discharges were determined from floodmarks at eight sections within the Mikeimin catchment, corroborated by a slope-energy analysis done between two of the sections, which suggested a near equality between channel slope and energy line slope. Several factors contribute to the relatively low level of accuracy which can be claimed for the reconstruction: supercritical flow at and near the peak discharge at all sections, substitution of mean depth for hydraulic radius in the Manning formula, asymmetry and dubiousness of some of the floodmarks, and an above-average uncertainty in the determination of the friction coefficient.

Using time-to-peak and recession values typical for the southern Negev - southeastern Sinai environment, and taking into consideration differences in catchment size and morphology, hydrographs for HaSargel and En Tamar tributaries were reconstructed. Given the faster response of the former, a five minute lag difference betwen the two was arbitrarily assumed for the reconstruction of the unified Mikeimin hydrograph. Field evidence at the En Tamar - HaSargel confluence clearly showed that the flow in the latter was the first to cease. The flow in F tributary was insignificant.

Sediment

Bed material discharges were reconstructed for each of the eight sections based on the Einstein procedure, taking into account the varying flow geometry and mean particle size (Graf, 1971). Sediment size was determined for the gravel bars by the Leopold (1972) method, and the resulting mean values were arbitrarily reduced by two phi units in order to account for the finer in-channel material not sampled, as well as for the wash load. The proportion of sub-sand sizes in the Mikeimin 1971

Table 1. Estimated peak flow, Wadi Mikeimin catchment, event of January 1971.

Stn.	Flow width (m)	Mean flow depth (m)	Cross section area (m²)	Manning 'n'	Slope	Mean vel. (ms⁻¹)	Froude no.	Discharge (m³s⁻¹)	Data quality
1	18.8	0.45	8.5	0.030	0.051	4.5	2.1	38	fair
2	19.7	0.45	8.8	0.030	0.053	4.5	2.1	40	fair
3	45.0	0.30	13.0	0.035	0.075	3.5	2.0	47	fair
A	39.7	0.54	21.6	0.045	0.083	4.2	1.8	92	fair
B	20.0	0.67	13.4	0.030	0.087	7.4	2.9	100	poor
C	45.5	0.48	21.9	0.045	0.084	3.9	1.8	86	good
D	35.7	0.53	18.9	0.045	0.085	4.2	1.8	80	good
M	18.8	0.70*	12.3	0.040	0.087	5.8*	2.2		
		0.52@				4.7@		68.5	fair

Table headers use ms^{-1}, m^2, and $m^3 s^{-1}$ units.

* To the left of the central bar.
@ To the right of the central bar.

Table 2. Estimated time-to-peak and total flow volumes, selected stations, Mikeimin flood.

Stn.	Time-to-peak (min)	Total flow volume (m³)
D	6	74,400
1	4.5	38,900
M	11	106,200

Table 3. Estimates of catchment rainfall and runoff coefficients, Mikeimin flood.

Tributary	Drainage area Total (km²)	Drainage area Within cell (km²)	Peak discharge (m³s⁻¹)	Estimated rainfall (mm)	Estimated rainfall (m³)	Estimated runoff coefficient (%)
HaMappal	4.0	4.0	40	46	180,000	22
F	1.9	0.7	< 1			
En Tamar	6.9	5.0	47	43	300,000	
HaSargel	5.2	5.2	80	90	460,000	16
Mikcimin	12.9	10.7	68.5	50	650,000	16
M	4.9*	12@	68.5	5.3		

fan was very small, and no appreciable deposits of fines were found in the Watir channel downstream of the fan.

The reconstructed sediment discharge graphs were integrated to obtain total sediment volumes which were assumed to pass through each of the sections. This volume was calibrated for the lowest section M (Fig. 2) with the volume of the Mikeimin 1971 fan. No material moved down the Watir during the event beyond the clearly defined fan toe.

THE MIKEIMIN EVENT

Flood

Water discharge. En Tamar tributary contributed a peak discharge of 35-47 m^3s^{-1} and HaSargel tributary about 80 m^3s^{-1} (Table 1, Fig. 2). Compared to the sum of these peak flows -- 120 m^3s^{-1} -- the floodmark based peak flow estimate for the unified Mikeimin was only 60 per cent of this value, a discrepancy explainable by lag difference, floodwave attenuation, and the high peakedness of the hydrographs characteristic of this environment.

Table 2 lists the results of the reconstructed discharge computations from the onset of channel flow till the time of the peak for three selected stations in the Mikeimin catchment. The reconstructed hydrographs are shown in Figure 4. They reflect the unequal contribution of the tributaries also in terms of flood volume: the bulk of the floodwaters was contributed by HaSargel tributary -- more than twice the amount contributed by HaMappal tributary. Considerations of catchment relief, which favor more water losses in HaMappal and En Tamar sub-catchments, are insufficient to explain the discrepancy. It is, therefore, attributed to the assumed location of the rainstorm cell, which must have been centered over the HaSargel sub-catchment (Fig. 2; discussion follows).

Sediment. The Einstein method computations (Colby and Hubbell, 1981) of the sediment discharge at the mouths of the two main tributaries, based on their reconstructed hydrographs, are summarized in Table 3. These computations yielded a 'sediment rating curve' which was used to reconstruct the sediment discharge over the entire event (Fig. 5).

For HaSargel tributary the resulting total sediment yield for the event was about 8,000 tons, which were transported by a total volume of water 74,400 m^3, indicating a mean concentration of 11 per cent. En Tamar tributary yielded 3,400 tons of sediment, which were moved by a total flow of 38,900 m^3 -- a mean concentration of 9 per cent.

For similar and somewhat larger catchments off the Gulf of Aqaba coast in southeastern Sinai, the floods of November 1972 were estimated to contain 15 - 20 per cent sediment, mostly bed material (Floods, 1972).

Peak unit bed material transport rates ranged, according to one estimate, from 25 kg $m^{-1}s^{-1}$ at En Tamar tributary (site 1) through 58 kg $m^{-1}s^{-1}$ at HaSargel tributary (site D) to 92 kg m⁻¹s⁻¹ at the unified Mikeimin (section M) (Schick and Lekach, 1981).

Rainstorm

Location of the rain cell. No evidence of flow from slopes in the area of the Mikeimin mouth was identified (Fig. 2). In neighboring direct tributaries of Wadi Watir, both upstream and downstream of the Mikeimin, little or no flow occurred during the event; no untrenched fans from these tributaries were found. Further, tributary F, which drains the northwestern corner of the Mikeimin catchment, had a flow which peaked to less than 1 m³s⁻¹.

Thus, the main flow-generating zone of the rainstorm must have focused over the central and upper parts of the Mikeimin catchment. In general, peak discharges in the catchment diminish in a downstream direction (Table 1). The center of the storm cell must have coincided, more or less, with Jebel Mikeimin (Fig. 2). No indication that the storm was a moving one, such as the Ma'an flood (Schick, 1971a), was found or reported. The Mikeimin rainstorm must have been a highly localized, quasi-stationary and very intense cloudburst. It may have been associated with an infrequent southerly storm-producing synoptic configuration known as the Red Sea trough (Margalit and Sharon, in press), but there is no concrete evidence to support this assumption; the date of the event is not known.

Estimated rainfall amount. It is of interest to estimate the catchment rainfall which might have generated the Mikeimin flood. We used for the purpose a relation of catchment rainfall with peak discharge as a function of catchment area, established for the Ma'an 1966 flood (Schick, 1971, p. 123). Assuming that condi-

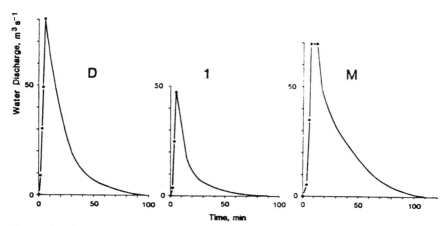

Figure 4. Reconstructed hydrographs at sections D (mouth of HaSargel tributary), 1 (mouth of HaMappal tributary), and M (outlet of Wadi Mikeimin). Location as in Figure 2.

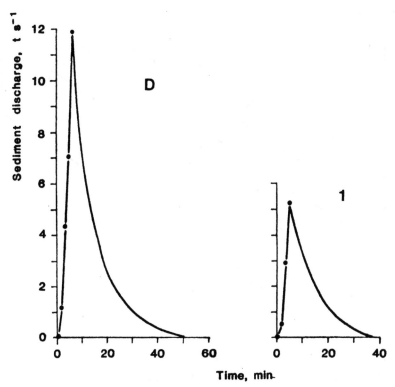

Figure 5.   Reconstructed sediment discharge curves at section D and 1.   Location as in Figure 2.

tions of both floods were similar, an estimate for the catchment rainfall for the Mikeimin flood can be obtained (Fig. 6).

The procedure yields values of catchment rainfall for HaMappal and En Tamar sub-catchments of 40-50 mm, and about double that value for HaSargel sub-catchment (Table 3). This agrees with the conclusion that the storm cell was centered on the upper and central part of HaSargel sub-catchment.

The resulting estimated total rainfall volumes (Table 3) are fairly consistent among themselves, though the sum of the sub-catchment rainfall volumes exceeds that of the entire Mikeimin catchment by 14%. The resulting runoff coefficients are likewise reasonable and lie around 20% -- a very high value for hyper-arid environments, where coefficients for medium size catchments seldom exceed 5%.

HaSargel tributary. The routing of the flood through HaSargel tributary deserves special attention. Unit peak discharges of over 25 $m^3s^{-1}km^{-2}$ are indicated for the two upper stations A and B (Table 4), though at both sites the interpretation of the

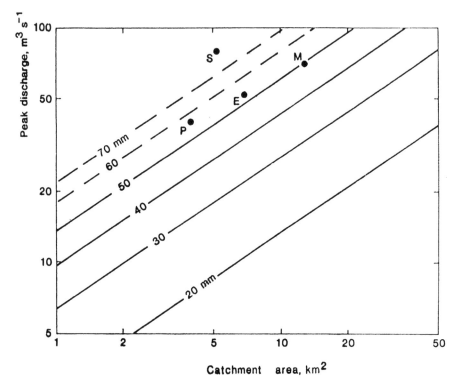

Figure 6.   Rainfall and peak discharge in selected catchments affected by the Ma'an flood event (after Schick, 1971a). Isolines indicate mean storm rainfall over catchment. Dots represent the Mikeimin 1971 flood event: E, P, and S = En Tamar, HaMappal, and HaSargel sub-catchment, respectively; M = Mikeimin catchment.

floodmarks is open to some doubt. However, we may use the Themed, Sinai record rainfall of 142-164 mm "in a few hours" (Ashbel, 1951, p. 124-125) on 18 November 1925 to investigate to what extent the upper HaSargel discharges are reasonable. To produce these discharges, and assuming a 100 per cent runoff coefficient for these small and rocky sub-catchments, an hourly intensity of about double the amount of the Themed value would be necessary. Under appropriate antecedent and catchment conditions, a runoff coefficient near 100 per cent is possible (Schick, 1971, Fig. 6, Event 7B). Event 17 in Nahal Yael had an intensity of 150 mm/h, but it lasted only 5 min (Greenbaum, 1987). Although exceptional, it is conceivable that similar intensities do last considerably longer, i.e., for periods exceeding the time of concentration of small catchments of at least up to 10 km² in area. It is, therefore, not completely impossible to assume that the Mikeimin rain cell may have reached at its center, in the uppermost part of HaSargel sub-catchment, a rain-

Table 4. Peak discharge along Hasargel tributary, Mikeimin flood

Station	Distance from divide (km)	Drainage area (km^2)	Peak discharge (m^{3}s^{-1})	Specific peak discharge (m^{3}s^{-1}km^{-2})
A	1.4	1.0	92	92
B	2.8	3.6	100	28
C	3.6	4.5	86	19
D	3.9	4.7	80	17
M	4.9*	12.9@	68.5	5.3

* Via HaSargel tributary; distance via longest tributary (HaMappal) is 8 km.

@ Includes entire catchment; if HaMappal tributary is excluded, drainage area effective for peak discharge is 6.2 km.

fall value two to four times higher than the mean value for the 5.15 km^2 sub-catchment. The very slow attenuation of the peak discharges from station A through D (Table 4), reminiscent of a dam breach effect, perhaps aided by fortuitously synchronized downstream cell movement, lends support to this conclusion (Sharon, personal communication; compare Costa, 1985b).

Among the maximum rainfall-runoff floods for small drainage areas in the United States analyzed by Costa (1985a), the July 20, 1971 event on Lahotan Reservoir tributary No. 3 near Silver Spring, Nevada (Moosburner, 1978) resembles in certain aspects the Mikeimin 1971 event as it occurred in the upper HaSargel tributary. At Lahotan -- an area with 114 mm mean annual precipitation and a 6-hour 100-year rainfall intensity of 43 mm, and with geometric and hydraulic characteristics similar to HaSargel tributary -- a drainage area of 0.57 km^2 yielded a unit peak discharge of 83.5 m^{3}s^{-1}km^{-2}.

The Mikeimin Fan and its Breaching

Description. From its formation in January 1971 until its breaching by the Watir flood of November 1972, the Mikeimin fan obstructed the entire width of the Watir channel (Figs. 7 and 8). The fan was nearly symmetrical, as expected from the 90 degree junction angle and the large difference in the channel slope of the Mikeimin (0.08-0.09) and the Watir (0.013). Minor irregularities, such as the narrow channel between the fan toe and the opposite valley wall of the Watir and the lobe which extends for 40 m downstream the Watir channel (Figs. 8 and 9), were caused by

Figure 7. Aerial photograph of the Mikeimin-Watir confluence area. Note the obstructing fan.

anastomosing recession flow on the newly formed fan surface. Some of these braiding afterflows deposited relatively fine material. The greatest part of the fan surface was, however, composed of gravel-bed channels with intervening gravel bars often including particles as large as cobbles (Fig. 10). Largest particle size at about one hundred grid intersections on the fan surface gave a mode of 150 mm (b-axis) and a maximum of 380 mm.

At its widest extent the fan was 120 m across. Further upstream the Mikeimin it was constricted to 65 m by remnants of larger and higher fans deposited in the past and subsequently trenched. The fan length from apex to toe was 80 m, and its

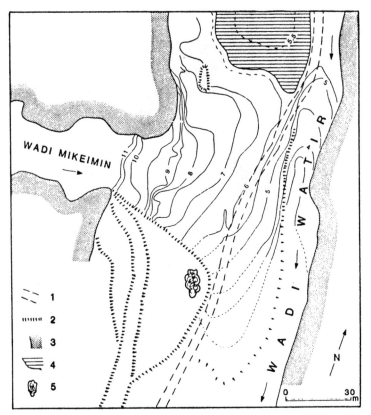

Figure 8. The Mikeimin fan. The contours, in meters, are from a plan table survey. 1 = dirt road; 2 = alluvial scarp; 3 = rocky sidewall; 4 = 'lake' shore; 5 = bush. 'Lake' bottom was at relative elevation ±5.0 m.

surface area was 7,000 m². The slope of the fan surface was generally 0.08-0.09, similar to the channel slope of the lower Mikeimin immediately upstream (Table 1, station M).

Sediment volume. The volume of the fan was estimated by relating surveyed cross sections to an assumed base representing the pre-1971 fan surface of the Mikeimin. This somewhat arbitrarily determined base relies at some points on the margins of the 1971 fan and takes into account evidence by local Bedouins as to the form and extent of the pre-1971 features. The breaching of the fan in 1972 exposed an up to 4 m high scarp which dissected the entire width of the fan at about mid-length (Figs. 11 and 12). The upper half of the cut scarp exposes material desposited in 1971, with a discontinuity at its base indicating the pre-1971 fan surface. The additional data obtained from this exposure confirmed the volumetric data determined before the breach.

Figure 9. The Mikeimin 1971 fan. A mosaic of oblique photographs taken from the top of the Watir valley wall opposite the junction. Note vehicle and goat herd for scale, dammed 'lake' (dark area) on right, and remnants of older fans at left. Dashed line indicates location of scarp formed by the 1972 breach (Figures 11-13).

Figure 10. The surface of the Mikeimin fan. Ruler is 15 cm long.

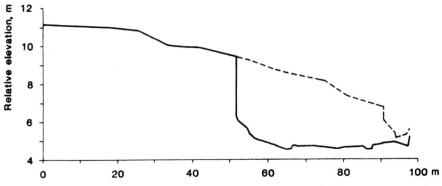

Figure 11. Longitudinal profile of the Mikeimin fan after the breaching. Dashed line is the original surface.

The cut face revealed that the central part of the fan was, on the average, 2 m thick; the left and right hand sectors were about 1 m thick. The mean depth for the entire fan, including its lateral and upper margins, was 0.86 m. The volume of the 1971 fan was determined to have been 6,200 m^3. Using a density of 1.6 tons m^{-3}, the fan accounts for 10,000 tons of sediment. This amount is 10% less than the combined estimated sediment volume yielded by HaSargel and En Tamar tributaries during the 1971 event. The balance may represent aggradation in the lower reaches of the Mikeimin, caused by the higher baselevel at the junction.

Stream power. Maximum unit stream power for the flood at HaSargel stations A, C, and D ranged between 1,560 and 1,880 $Nm^{-1}s^{-1}$. At the constricted section B it was 4,250 $Nm^{-1}s^{-1}$, and 3,110 $Nm^{-1}s^{-1}$ at the downstream Mikeimin station M. Assuming no infiltration at the instant of discharge into the Watir, the maximum unit stream power there was only 109 $Nm^{-1}s^{-1}$. The twelve largest rainfall-runoff floods in small basins analyzed by Costa (1985) range in maximum stream power between 212 and 8,131 $Nm^{-1}s^{-1}$. All of these were in semiarid or arid areas.

Earlier fans. The pre-1971 junction had a small fan which extended about halfway down the width of the Watir channel. It was probably deposited by one or more small events in the Mikeimin, during which the Watir remained inactive. The large difference in catchment size between the Watir and the Mikeimin explains the higher flow frequency in the latter (Schick, in press).

There is also evidence for larger fans at the junction in the form of two paired terrace-like relics which must have belonged, prior to their trenching, to fans much larger than the 1971 fan. Though baselevel changes in the bed of the Watir may be partly responsible, this is unlikely, as the Watir profile is controlled by bedrock several hundreds of meters downstream the Mikeimin junction. Hence the 1971 Mikeimin flood was not the largest in the recent geologic record. Possibly the damming of the Watir caused by each of these two fans was at least 2 m higher than that caused by the 1971 fan.

The breach. The breaching flood in the Watir lasted for three days and peaked to 320 m^3s^{-1} (Floods, 1972). Its many sharp peaks may reflect the non-synchronous input of tributaries as well as the breaching of tributary dams (Fig. 13). Most of these may have occurred within one and the same event, with the phasing caused by differential lag. The Mikeimin 1971 fan also acted as a temporary dam, as is suggested by the light-colored fine-grained Watir sediment found after the breach as a veneer on some upstream facing parts of the remaining fan surface (Fig. 12).

The breach took about 4,500 m^3 of Mikeimin sediment from the composite fan at the junction. About 2,500 m^3 of this amount belonged to the recent 1971 fan and the rest to older fans.

For the Watir the breaching flow was a medium to high magnitude event. In February 1975 a major regional rainstorm caused the Watir to peak to 1,170 m^3s^{-1} (Ben Zvi and Kornic, 1976).

Figure 12. The breached Mikeimin fan immediately after the November 1972 event. A mosaic of oblique photographs from the opposite valley wall. Note fine, light colored sediment veneer on upstream (Watir) side of fan.

Sediment Size

Median sediment size in the central part of the exposed fan face ranged from 29 to 40 mm, with D_{84} around 45 mm (Fig. 15). The material of the older fan below the 1971 fan base is generally slightly coarser and somewhat less well sorted. The median size of the material, interpreted as having been deposited during the flood recession, is considerably finer, 3-5 mm, and its D_{84} is around 20 mm. In all samples the fraction larger than 0.063 mm accounts for over 98% of the total material. An exception is the fine sand veneer deposited on the margins of the Mikeimin fan by the Watir flood which breached the fan. Its median size is 0.15 mm and it contained 15% silt and clay.

In total, the Mikeimin fan consisted of 53% pebbles and cobbles (>16 mm), 25% gravel (2-16 mm), and 20% sand (0.063 -2 mm). Silt and clay account for 1.6% only.

The size distribution of the sediment in the Mikeimin fan indicates that it was not a debris flow but a water flood. The silt-clay content was under 2 per cent. Mean sediment concentration for the entire flood, as reconstructed for station M a short distance upstream of the fan apex, was 10 per cent. Although this indicates

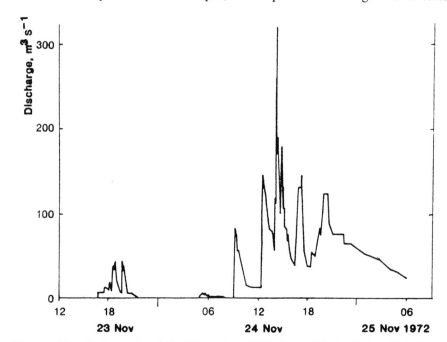

Figure 13. Hydrograph of the November 1972 flood, Wadi Watir. This flood breached the Mikeimin 1971 fan. The Mikeimin did not flow. The gauging station was located 15 km downstream of the Mikeimin junction, but the discharges reflect the flow as deduced for a gauging station 4 km west (upstream) of the junction which had been destroyed by the flood (Floods, 1972).

fairly high instantaneous concentration values at peak or during pulses, it is considerably lower than values of 60 to 90% solids reported for typical debris flows (Costa, 1986, p. 289). In addition, the cut face of the fan after the November 1972 breach revealed stratified material with considerable internal sorting (Fig. 14). The Mikeimin fan may, in fact, be similar in some respects to the "sieve" deposits (Hooke, 1967; Bull, 1977); the dry alluvium of the Watir channel during the January 1971 event could have easily abstracted the entire Mikeimin flow volume by infiltration. The sedimentary structure of the Mikeimin 1971 fan is similar to some of the Howgill Fells 1982 (Wells and Harvey, 1987) alluvial fans. In the fans primarily attributed to streamflow, several stratigraphically distinct layers which formed during this single flood event were identified.

Figure 14. Details of the breached fan scarp. The stratified water flood character of the deposit is clearly visible.

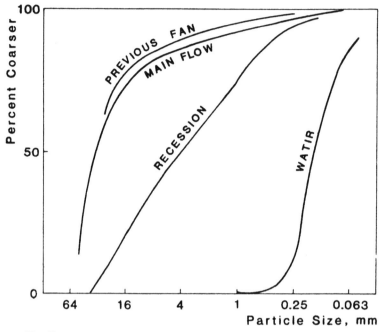

Figure 15. Representative particle size distributions, Mikeimin 1971 fan.

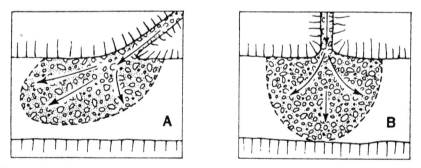

Figure 16. Junction angle and obstruction type: A - main channel narrowed; b - main channel dammed.

FLOOD GEOMORPHOLOGY
Damming by Tributaries

In addition to the evidence for repeated damming of the Watir by the Mikeimin described above, numerous other tributary junctions in the mountainous Sinai Desert were found to have relics of past obstructions. Most of these have caused only a narrowing of the main channel, though in at least two localities -- an unnamed tributary of Wadi Feiran near Wadi Nadia and Wadi Harit at Wadi Nasb -- the profile of the breached fan as extrapolated downstream coincided with a depositional remnant on the opposite side of the main channel.

The factors conducive to damming by tributaries are arid conditions, cellular rainstorms, right-angle junctions, and a narrow constricted receiving channel.

Arid conditions. The extreme ephemeral flow regime enables the geomorphic effects of an infrequent event to remain unaltered for years. Coarse sediment is available in practically unlimited quantity and detachment and entrainment are unhampered by vegetation. The result is very large sediment yields for singular, potentially damming events.

Cellular rainstorms. The localized character of cloudbursts (Sharon, 1972) provides a mechanism for the intense activation of a given tributary catchment without affecting its neighbors. The receiving main channel is likewise unaffected, due to the usually very large infiltration capacity of its alluvium.

Right angle tributary junction. Other factors equal, an event which caused damming from a right angle tributary may be short of effecting a closure if the angle of entry is less than 90 degrees (Fig. 16). Junctions of streams of dissimilar order tend to be at right angles (Abrahams, 1984, p. 177), especially if order difference is at least three. Due to considerations based on derivatives of Horton's laws of catchment morphology, this requisite also entails a substantial difference in channel slope between the tributary and the receiving channel. Such a difference contributes to the drastic loss of stream power at the point of entry and causes deposition. This localized deposition is also promoted by the lack of fines in the sediment load; if present, the fines would be winnowed out and thus lost to the obstructing fan in favor of deposition downstream.

Width of receiving channel. Many tributary fans are too small to effect a complete closure unless the main channel is locally narrow. Favorable sites are canyon-like reaches in deeply entrenched, downstream segments, a condition fulfilled by the lower Watir as well as by many other rivers such as the Colorado. The locally high relief also contributes to the high sediment yield of the tributary.

All these factors are nearly optimal at the Mikeimin-Watir confluence. Thus, the first naturally produced damming fan to be identified as a complete, untrenched landform, at least in the Negev-Sinai desert, has been found at a logical site.

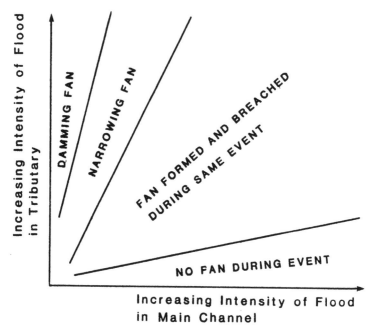

Figure 17. Schematic effect of relative flood intensity in tributary and main channel on the formation of an obstruction and its type.

Hydrologic Considerations

Hydrologic lag. The lag from a burst of intense rainfall to the hydrograph peak, for hydrographs of Horton overland flow, can be estimated for Wadi Mikeimin to be about one hour. This lag is 30% less than the lag indicated by Dunne (1978, p. 240), and takes into account the faster than average response of the Mikeimin catchment. In contrast, the lag for the entire Watir catchment, for a rainstorm covering its entire area, can be estimated at 5-15 hours.

This discrepancy in lag time leads to the conclusion that for large regional storms such as the one which covered the entire southern Sinai - southern Negev area in February 1975 (Ben Zvi and Kornic, 1976), numerous Mikeimin-type fans would be formed in the main channel. These would be breached a few hours later by the flood as it advances down the main stream. The relative intensity of the flood event in each tributary as against the main catchment is the controlling factor (Fig. 17).

Flood frequency and catchment size. Flood frequency in arid environments decreases with catchment size. For the Mikeimin-Watir situation, more floods can be expected, over a period of time, in the Mikeimin than in the Watir. If this were the only factor, our field survey of junctions in the mountains of the southern Negev and Sinai should have come up with many obstructions 'waiting' to be breached by the infrequent flood in the main channel.

In fact, relatively few modern obstructions and only one current fan dam were found. The reason is that an event of a relatively low magnitude in the main stream suffices to breach a fan dam deposited by a high magnitude event in the tributary. Most flows in the Watir are competent to breach any fan dam deposited by the Mikeimin, but only the very largest of the numerous events in the Mikeimin are powerful enough to deposit a damming fan.

A DAMMING AND BREACHING
SIMULATION MODEL

If the proportion between effective flows in the tributary and in the main stream could be established, and total, random independence of events is assumed, a series representing the fan dam history of the junction could be simulated. This history, expressed as the proportional duration of the breached state versus the dammed state, could then be checked against spatial surveys of large areas with similar conditions.

Following is an initial and simplified simulation of this type. It is assumed that, over a period of 100,000 days (274 years), a flood in the Mikeimin occurred once in two years, in the Watir once in four years, and, among these, once in six years the event was a joint one for both. For the Watir, every event is assumed to

Table 5. Hypothetical frequency of obstructing and breaching events at the Mikeimin-Watir confluence.

	Number of events per 100,000 days				
	Total	Damming	Narrowing	No effect	Breaching
Mikeimin only	91	5	18	68	
Mikeimin & Watir	46				46
Watir only	23				23
Total	160				69

be a breaching one. In the Mikeimin, only 5% of the events are assumed to be large enough to effect a complete damming of the Watir; another 20% are regarded as effecting a narrowing, and the remaining 75% (68 events, Table 5) are internal events which terminate by infiltration in the channel bed before reaching the confluence.

These assumptions entail 7-8 damming and 27 narrowing events as against 69 breaching events, some of them simultaneous. Even if it is assumed that a consecutive series of two or three narrowing events produce a damming, the total number of damming events is still much less than the potentially breaching events. Given total randomness, this means that the proportion of total times during which the confluence is breached exceeds considerably the aggregate duration of a dammed junction. In fact, this conforms to the field situation.

In one simulation run of 100,000 days with the parameters as listed in Table 5, and assuming that every third consecutive narrowing event becomes a damming event, a dam was present at the confluence for an aggregate duration of 10,414 days. This duration was composed of six periods ranging from 431 days (slightly over one year) to 3,576 days (nearly ten years.) The total 274 years also included seven periods with a breached or narrowed confluence. Four of these lasted between 19 and 25 years, one lasted 43 years, and the longest one was 102 years.

By substituting space for time, these results can be calibrated against a spatial distribution of confluence status, perhaps with the aid of modern remote sensing methods. The resulting adjustments necessary to make the model results conform to reality may help in overcoming the serious lack of information on the magnitude and frequency of floods in arid and especially hyperarid catchments.

INTRA-EVENT DAMMING AND BREACHING

A major stream like the Watir serves as a baselevel for many potentially damming tributaries. Large scale, regional storms tend to activate many of these tributaries. Therefore, the probability of the Watir encountering any tributary dam along its course, not just one -- the Mikeimin -- is much increased.

The hydrograph of the November 1972 flood in Wadi Watir, which breached the Mikeimin fan, has eight distinct and sharp peaks, six of which belong to the main, high discharge flow (Fig. 13). The gauging instrument used was a pressure device which recorded the stage every minute. Most of these sharp rises occurred over 1-3 min (Floods, 1972). The large catchment size and the characteristics of the rainstorm -- 50-70 mm total rainfall fairly unevenly distributed over 48 hours over a very large area -- lead to the conclusion that intra-event tributary damming and breaching is the main cause of these flood bores.

Some of the 'walls of water' reported by witnesses of flash floods (e.g., Hjalmarson, 1984) may have been caused by intra-event damming and breaching such as described above.

CONCLUSION

In the context of catastrophic floods in hyper-arid environments, the Mikeimin event described and analyzed in this paper is unique in several respects. It provides the first documentation of a geomorphically intact contemporary damming tributary fan in these environments. A reconstruction of the rainfall, streamflow, and sediment attributes of the event -- made possible by the total sediment yield fully preserved in the fan -- indicates that the event is on par with some of the highest magnitude flood events in small catchments in other environments. This conclusion applies to various parameters such as point rainfall, peak discharge per unit drainage area, peak flow velocity, and stream power. It is of major significance both for reappraising the role of contemporary processes, however infrequent, in the evolution of landforms in the extreme desert, as well as for engineering applications of the maximum probable flood estimate wherever needed.

A stochastic damming and breaching simulation model, conceptually developed on the basis of the Mikeimin-Watir damming-breaching relation, provides an insight into the spatial-temporal interplay of high magnitude - low frequency events in arid fluvial terrain. Its further development and application to extremely arid regions with a localized rainfall regime may provide a geomorphic substitute for, or an additive to, the very meager database on catastrophic floods in such areas.

Hydrologic and hydraulic routing procedures alone are inadequate to explain the near-instantaneous rise typical of many catastrophic floods. Intra-event damming and breaching is identified as a major cause of the 'wall of water' phenomenon in floods of arid lands. The role of the sediment, and especially of intra-event damming and breaching in catchments with above average sediment availability, is important in many desert floods, and may resolve many of the discrepancies encountered in flood analyses.

ACKNOWLEDGEMENTS

This paper is based on field work done for an M.Sc. thesis (Lekach, 1974) in a remote area and under difficult logistics. It formed a part of the Eastern Sinai Floods Research Project (1971-73) done at the Department of Geography, the Hebrew University of Jerusalem. Our associates in the project, Moshe Negev, David Sharon and Aharon Yair, spent with us many long and hot days in the field and willingly shared their expertise. Michal Kidron and 'Anat Bloch of the Cartographic Unit, Department of Geography, Hebrew University, produced the maps and the graphs. Dr. David Darom and his staff at the Photography and Scientific Drafting Unit, Faculty of Science, Hebrew University, did their best to reproduce old and mediocre photographs for this paper. To all these we express our thanks.

REFERENCES

Abrahams, A. D., 1984, Channel networks: a geomorphological perspective: Water Resources Research, v. 20, p. 161-188.

Ashbel, D., 1951, Regional Climatology of Israel (in Hebrew): Jerusalem, Hebrew University, 244 p.

Ben Zvi, A., and Kornic, D., 1976, The floods of February 1975 in Sinai and the 'Arava (in Hebrew): Hydrological Service, Water Commission, Ministry of Agriculture, Jerusalem, Report Hydro/3/1976, 8 p., 11 figures.

Benda, L. E., 1985, Delineation of channels susceptible to debris flows and debris floods, in International Symposium on Erosion, Debris Flow and Disaster Prevention, September 1985, Department of Forestry, Kyoto University, Tsukuba, Japan, 7 p.

Blair, T. C., 1987, Sedimentary processes, vertical stratification sequences, and geomorphology of the Roaring River alluvial fan, Rocky Mountain National Park, Colorado: Journal of Sedimentary Petrology, v. 57, p. 1-18.

Bull, W. B., 1977, The Alluvial fan environment: Progress in Physical Geography, v. 1, p. 222-270.

Colby, B. R., and Hubbell, 1961, Simplified methods for computing total sediment discharge with the modified Einstein procedure: U.S. Geological Survey Water-Supply Paper 1593.

Cooley, M. E., Aldridge, B.N., and Euler, R.C., 1977, Effects of the catastrophic flood of December 1977, North Rim area, eastern Grand Canyon, Arizona: U.S. Geological Survey Professional Paper 980, 43 p.

Costa, J. E., 1984, Physical geomorphology of debris flows, in Costa, J. E. and Fleisher, P. J., editors, Developments and applications of geomorphology: Berlin, Springer-Verlag, p. 268-317.

Costa, J. E., 1985a, Interpretations of the largest rainfall-runoff floods measured by indirect methods on small drainage basins in the conterminous United States, in Proceedings, China - U.S. Bilateral Symposium on the Analysis of Extraordinary Flood Events, Nanjing, China, October 1985, 42 p.

Costa, J. E., 1985b, Floods from dam failures: U.S. Geological Survey Open-File Report 85-560, 54 p.

Dietrich, W. E., and Dunne, T., 1978, Sediment budget for a small catchment in mountainous terrain: Zeitschrift fuer Geormorpologie, Supplementband 29, p. 191-206.

Dunne, T., 1968, Field studies of hillslope processes, in Kirkby, M. J., editor, Hillslope hydrology: Chichester, John Wiley & Sons, p. 227-293.

Eyal, M., Bartov, Y., Shimron, A.E., and Bentor, Y.K., 1980, Sinai - geological map: Survey of Israel, scale 1:500,000. Floods of 23-25.11.72 in relation to the Elat - Sharm-el-Sheikh road (in Hebrew): Jerusalem, Department of Geography, Hebrew University: Floods in Eastern Sinai, No. 3, December 1972, 12 p.

Gallino, G. L. and Pierson, T.C. 1984, The 1980 Polallie Creek debris flow and subsequent dam-break flood, East Fork Hood river basin, Oregon: U.S. Geological Survey Open-File Report 84-578, 37 p.

Gerson, R., 1982, The Middle East: landforms of a planetary desert through environmental changes, in T. L. Smiley, editor, The geological story of the world's deserts, Striae, v. 17, p. 52-78.

Gerson, R., and Grossman, S., 1987, Geomorphic activity on escarpments and associated fluvial systems in hot deserts as an indicator of environmental regimes and cyclic climatic changes, *in* Sanders, J. E. et al., editors, Climate: History, Periodicity, Predictability: Stroudsburg, PA, Van Nostrand Reinhold, in press.

Graf, W. H., 1971, Hydraulics of sediment transport: New York, McGraw-Hill, 531 p.

Graf, W. L., 1979, Rapids in canyon rivers: Journal of Geology, v. 87, p. 533-551.

Greenbaum, N., 1987, Point runoff in an extremely arid region: infiltration experiments in small plots in the southern 'Arava valley and their hydrological, pedological and paleomorphological implications [M.Sc. thesis, in Hebrew]: Jerusalem, Hebrew University, Department of Physical Geography, 206 p.

Harvey, A. M., 1986, Geomorphic effects of a 100 year storm in the Howgill Fells, northwest England: Zeitschrift fuer Geomorphologie, v. 30, p. 71-91.

Hjalmarson, H. W., 1984, Flash flood in Tanque Verde Creek, Tucson, Arizona: Journal of Hydraulic Engineering, v. 110, no. 12, p. 1841-1852.

Hoetzl, H., Felber, H., Maurin, V., and Zoetl, J. G., 1979, Accumulation terraces of Wadi Hanifah and Wadi Al Luhy, in Al-Sayari, S.S. and Zoetl, J.G., editors, Quaternary Period in Saudi Arabia: Wien, Springer-Verlag, v. 1, p. 202-209.

Hooke, R. LeB., 1967, Processes on arid-region alluvial fans: Journal of Geology, v. 75, p. 438-460.

Howard, A., and Dolan, R., 1981, Geomorphology of the Colorado River in the Grand Canyon: Journal of Geology, v. 89, p. 269-298.

Jahns, R. H., 1947, Geologic features of the Connecticut Valley, Mass., as related to recent floods: U.S. Geological Survey Water-Supply Paper 996, p. 68-81.

Jarrett, R. D., and Costa, J. E., 1984, Hydrology, geomorphology, and dam-break modeling of the July 15, 1982, Lawn Lake Dam and Cascade Lake Dam failures, Larimer County, Colorado: U.S. Geological Survey Open-File Report 84-612, 109 p.

Lekach, J., 1974, The January 1971 event in Nahal Mikeimin and its geomorphic significance [M.Sc. thesis, in Hebrew]: Jerusalem, Hebrew University, Department of Geography, 41 p.

Leopold, L. B., 1969, The rapids and the pools -- Grand Canyon, *in* The Colorado River region and John Wesley Powell: U.S. Geological Survey Professional Paper 669, p. 131-145.

Leopold, L. B., 1970, An improved method for size distribution of stream bed gravel: American Geophysical Union Transactions 27, p. 535-539.

Margalit, A. N., and Sharon, D., in preparation, Desert rainfall pattern associated with the Red Sea trough in Israel and southern Jordan.

Meshel Z., 1972, Agam hadash be-Midbar Sinai (A new lake in the Sinai Desert): Sal'it, v. 1, no. 5, p. 202 (in Hebrew).

Moosburner, O., 1978, Flood investigations in Nevada through 1977 water year: U.S. Geological Survey Open-File Report 78-610, 94 p.

Petts, G. E., 1984, Sedimentation within a regulated river: Earth Surface Processes and Landforms, v. 9, p. 125-134.

Schick, A. P., 1971a, A desert flood: physical characteristics, effects on Man, geomorphic significance, human adaptation -- a case study in the southern 'Arava watershed: Jerusalem Studies in Geography, v. 2, p. 91-155.

Schick, A. P., 1971b, Desert floods -- interim results of observations in the Nahal

Yael research watershed, 1965-1970: International Association of Sientific Hydrology Publication 96, p. 478-493.

Schick, A. P., in press, Floods in hyper-arid environments, *in* Baker, V. R., et al., editors, Flood Geomorphology: New York, John Wiley, chapter 12.

Schick, A. P. and Lekach, J., 1981, High bedload transport rates in relation to stream power: Catena, v. 8, p. 43-48.

Sharon, D., 1972, The spottiness of rainfall in a desert area: Journal of Hydrology, v. 17, p. 161-175.

Suwa, H., and Okuda, S., 1984, Deposition of debris flows on a fan surface, Mt. Yakedake, Japan: Zeitschrift fuer Geormophologie, Supplementband 46, p. 79-101.

Webb, R. H., 1987, Occurrence and geomorphic effects of streamflow and debris flow floods in northern Arizona and southern Utah, *in* Mayer, Larry and Nash, D. B., editors, Catastrophic Flooding: Allen and Unwin, Great Britain, p. 247-266.

Wells, S. G., and Harvey, A. M., 1987, Sedimentologic and geomorphic variations in storm-generated alluvial fans, Howgill Fells, northwest England: Geological Society of America Bulletin, v. 98, p. 182-198.

Wolman, M. G., and Miller, J. P., 1960, Magnitude and frequency in geomorphic processes: Journal of Geology, v. 86, p. 54-74.

Woolley, R. R., 1946, Cloudburst floods in Utah, 1850-1938: U.S. Geological Survey Water-Supply Paper 994, 128 p.